Essentials of Water

Water shapes the planet and all life upon it. Breaking down traditional disciplinary barriers, this accessible, holistic introduction to the role and importance of water in Earth's physical and biological environments assumes no prior knowledge. It provides the reader with a clear and coherent explanation of the unique properties of water and how these allow it to affect landscapes and underpin all life on Earth. Contemporary issues surrounding water quality – such as the rise of microplastics and climate change – are highlighted, ensuring readers understand current debates. Giving all of the necessary background and up-to-date references, and including numerous examples and illustrations to explain concepts, worked mathematical calculations, and extensive end-of-chapter questions, this is the ideal introductory textbook for students seeking to understand the inextricable links between water and the environment.

Peter D. Blanken is a professor and former chair of the Department of Geography at the University of Colorado, Boulder, where he has taught courses in climatology and biometeorology for over 25 years. His research appears in almost 150 peer-reviewed journal articles and several book chapters. He is co-author of *Microclimate and Local Climate* (2016, Cambridge University Press), and has served on the Editorial Board of the *Bulletin of the American Meteorology Society* for over 20 years.

"This book is beautifully written, meticulously researched, and a pleasure to read. It is a remarkably complete source of knowledge of water in the environment, including soils, plants, animals, water bodies, and the atmosphere. The book would be an excellent text for an undergraduate course."

Thomas A. Black, University of British Columbia

"Thales of Miletus, the ancient Greek philosopher, famously asserted that 'All is water.' This is a comprehensive, accessible, and must-have book for anyone interested in the origin and vital role of water within the intricate water cycle system, our daily life, and the broader scope of our planet."

Fei Chen, The Hong Kong University of Science and Technology

"This book is a good alternative to traditional introductory physical geography textbooks. Using water as a subject, the content is ideal for an introductory class that wants to expose a mix of students from diverse majors to learning the basics of environmental and physical principles and processes controlling the weather, climate change, landscape formation, as well as chemical and biological processes in life forms. The book will be a good resource also for high school students taking precollege courses."

Mario A. Giraldo, California State University

"Professor Blanken has compiled an excellent text spanning multiple water-related topics. While the book assumes no prior knowledge of the role of water in Earth's physical and biological environments, the technical content is in fact suitable for a wide audience. Professor Blanken's approach to interweaving water's origins, its history in sanitation and public health, and its role in global climate, hydrology, and biodiversity (among other topics) is truly unique."

Andrew Gronewold, University of Michigan

"Blanken's very informative and well-researched textbook on water uses unique presentation principles: atomic physics, chemistry, and the resistance model of electricity, Ohm's Law – cleverly applied in a variety of settings. He uses these organizing ideas in a far-ranging discussion of water, from the effects of its molecular make-up to its impact on landscape vegetation, agriculture, animals and humans, insects, amphibians, ecosystems, and society. The final chapter is devoted to climate change. The problem sets require some research and creative solutions that should help students in the real world. I enjoyed reading this book and, as an active researcher, will find it useful as a reference."

Robert L. Grossman, University of Colorado

"*Essentials of Water* is a comprehensive and systematic overview of the myriad roles of water in the environment, especially from a physical and biological perspective. This is a transdisciplinary text that will become an instant classic. Its accessible writing style, informed by decades of scholarly engagement with the topic, ensures that it can be utilized by different fields within the natural and environmental sciences. Each chapter begins with succinct learning objectives, and ends with a list of questions that serve as a review. The text is supported by ample figures, photos, and equations, with the latter being thoroughly explained in a manner that reflects an extensive teaching career. Physical and chemical concepts are clearly explained, and include linkages to broader topics, such as Earth's origins, microplastics, water pollution, and climate change. *Essentials of Water* is an appropriate text for undergraduate and graduate courses, and will be a staple reference for scholars and practitioners in fields across the natural and environmental sciences."

Paul Hudson, Leiden University

"*Essentials of Water* is just that: essential knowledge about water as a substance, its origins on Earth, and its critical functions in all life forms, especially plants – but also contains fascinating chapters on thermoregulation in animal and human physiology. This is a fine treatise on water in its many forms, fluxes, scales, and challenges. It also connects key scientific advances with water myths, history, and human concerns. It thus offers a creative, well-structured approach to the water sciences that will inspire seasoned scholars as well as university students entering the field."

James L. Wescoat Jr., Massachusetts Institute of Technology

Essentials of Water

Water in the Earth's Physical and Biological Environments

Peter D. Blanken
University of Colorado at Boulder

Shaftesbury Road, Cambridge CB2 8EA, United Kingdom

One Liberty Plaza, 20th Floor, New York, NY 10006, USA

477 Williamstown Road, Port Melbourne, VIC 3207, Australia

314–321, 3rd Floor, Plot 3, Splendor Forum, Jasola District Centre, New Delhi – 110025, India

103 Penang Road, #05–06/07, Visioncrest Commercial, Singapore 238467

Cambridge University Press is part of Cambridge University Press & Assessment, a department of the University of Cambridge.

We share the University's mission to contribute to society through the pursuit of education, learning and research at the highest international levels of excellence.

www.cambridge.org
Information on this title: www.cambridge.org/highereducation/isbn/9781108833981

DOI: 10.1017/9781108988896

First published 2024

Printed in the United Kingdom by TJ Books Limited, Padstow Cornwall

A catalogue record for this publication is available from the British Library

A Cataloging-in-Publication data record for this book is available from the Library of Congress

ISBN 978-1-108-83398-1 Hardback
ISBN 978-1-108-98437-9 Paperback

Additional resources for this publication at www.cambridge.org/blanken

Contents

Preface

Water is an element that shapes our lives and the world around us. *Essentials of Water* explores how the properties of this unique molecule have resulted in our inextricable connection to water. I am not the first to realize how important water is to life on Earth, and many books have been written on water resources, management, policy, politics, hydrology, ecology, and engineering. So why write another book about water? After 25 years of university teaching and research, I became motivated to write *Essentials of Water* for two main reasons.

First, this book fills a gap in the existing literature by examining the role and importance of water spanning both the Earth's physical and biological environments in a unique, holistic framework. Books that target one aspect of water, such as engineering hydrology or water policy and management, do so for a good reason. However, often prior coursework is required before such topics can be fully understood and appreciated. As a result, students who have an interest in water but do not have the background knowledge are excluded. This book provides the foundational requisite knowledge so these focused subtopics can be fully understood. Alternatively, this book can be viewed as a stand-alone text providing a well-rounded examination of the connection between water and life.

My second reason for writing *Essentials* is for those students for whom this book might be their first exposure to the subject. Everyone should be offered the opportunity to discover how the one molecule that we cannot live without affects nearly everything around us. This is a comprehensive book that aims to fulfill this need for an undergraduate student or anyone with an interest in the topic.

The only prerequisite to enjoy this book is an interest in water and the desire to learn how one molecule changed a planet. There is some basic chemistry and mathematics involved to help illustrate important concepts. I have included many examples, illustrations, and questions at the end of each chapter to ease any chemistry or math worries. Where possible, I have added a historical social perspective by including some relevant details behind the individuals who made key discoveries. I have also tried to include information and examples illustrating the dramatic changes to the water cycle in the changing climate.

As I assume no prior knowledge, the first section of *Essentials of Water* has chapters on topics including water's impact on life and society, the discovery of oxygen and hydrogen, and the properties of the water molecule that result in the many unique and peculiar properties it displays. Next, the role water plays in the abiotic environment is covered. Water's origin on Earth, its past and present distribution, and the water cycle are described. How water in its liquid and solid forms eroded and shaped the Earth into the many landforms we see is also illustrated. In the biotic environment, how vegetation has developed the means to lift water to tall heights and transfer liquid water from the soil to water vapor in the atmosphere is explained. The vital importance of water for thermal regulation for organisms such as insects, frogs, lizards, and humans is covered next. After a chapter on water quality, including questions of salinity, fertilizer runoff, algal blooms, and microplastics, we conclude by looking at how a changing water cycle may affect the abiotic and biotic environments in the near future. My intent in writing this book is that you can pick and choose chapters to read depending on your background and interest, and arrange chapters accordingly to supplement the typical 15- or 16-week-long semester.

I hope that you learn something about water in *Essentials of Water* that you did not know before, and that this book is the beginning of your quest to learn more and help protect this vital resource.

Acknowledgments

This book would not have been possible without the support and efforts of many. Countless individuals, inspired by their inquisitive nature, conducted experiments and communicated to us what they discovered regarding the essential aspects of water that shapes life and the landscape. Their knowledge provided the foundations of this book. The students that I have had the pleasure of teaching provided the energy and enthusiasm that keeps research alive. I will not forget a student who came to lecture dressed as a water molecule on Halloween, stating that water was her favorite molecule. My professors also transferred this enthusiasm to me, and I thank them for that and the opportunities they provided. The professional staff at Cambridge University Press provided tireless support throughout this project. In particular, the efforts of Matt Lloyd, Helen Shannon, Olivia Marsh, and Lindsay Nightingale elevated the quality of this book. Only I am at fault for remaining mistakes. Lastly, I express my thanks to my family. This book has almost become part of the family over the past few years. Their enthusiasm for learning about the natural world is truly contagious.

1 The Historical Significance of Water

Key Learning Objectives

After reading this chapter you will be able to:

1. Explain why water is such a common and recurring theme in myths and legends across cultures throughout history.
2. Describe how water has been associated with both positive and negative aspects of human activities, technology, health practices, and beliefs.
3. Understand the concept of the water balance, and how changes in water storage are connected to it.

1.1 Introduction

Water has been central to human life, thoughts, and beliefs dating back as long as records exist. Water shapes our lives and the landscape around us. Water appears to us in different forms: solid, liquid, and invisible gas; and in these forms its properties and characteristics are completely different. Water provides but can also take life. Since no other substance can lay claim to all these facts, it is no surprise that water has a long historical significance as humanity has searched for explanations for water's deep influence on life on Earth. This chapter focuses on a few of these important historical aspects of water captured in mythology, health practices, and technology prior to the nineteenth century.

Humans seek to apply logic and reasoning to explain significant events in the world around us. Before writing, oral traditions such as myths and legends served to provide such explanations, and these mythologies were often passed down through generations by references to objects or locations that symbolize these stories. The stars, visible to everyone, provided such a background for the story of water. The constellations provided the ancient Greeks with a canvas to paint and tell stories about their deities, creatures, and heroes used to explain important events and rationalize observations. Water was central to Greek mythology, and this chapter begins by explaining the connection between water and Greek mythology as represented in the constellations. As many of the characters and concepts in Greek mythology entered Roman culture and soon spread throughout the world, examples of water-related myths in other cultures, locations, and languages are also provided.

Water has been associated with both positive and negative aspects of life in Greek and other mythologies. Clean water is a requirement for good health and hygiene, but throughout history, access to reliable sources of clean water has been variable. Misunderstandings and misconceptions of water-borne diseases and contaminates were common. The rise and fall of public bathhouses, common in Roman times, can be used as an example to illustrate this point. Not only did water remove dirt from the skin's surface, it also served to purify the soul and cleanse the body of evil spirits. Such symbolic representation of water as purifying agent for body and soul remains to this day across many cultures. During the late Middle Ages, however, submersing the body was thought to spread disease and infection, so bathing and the popularity of bathhouses declined.

Exposure to contaminated water has influenced human social activity. Water sustains not only human life, but also many other forms of life, including the microbes that lived in water long before humans existed. Water, the universal solvent, also dissolves many compounds, making it difficult or impossible to detect contamination without the aid of modern scientific techniques. For thousands of years, spanning the Ancient Greeks and Egyptians to the European Colonists in North America, it was discovered largely through an accidental consequence of agriculture that drinking alcohol or tea was a safer alternative to drinking water. Although at the time it was not known why these beverages avoided the often-deadly consequences of drinking contaminated water, wine, beer, and cider fueled society and economies. One could argue that wine brought Odysseus back to Penelope, beer built the Pyramids, and cider settled North America.

Let's now look at significant examples of technologies designed to move water from one place to another to better suit human activities. Many of these inventions coincided with the development of agriculture, with origins in ancient Greek and Egyptian cultures, and no fundamental changes in the technological principles to this day. These include aqueducts artfully and carefully constructed by the Romans to funnel water from mountain source to urban bathhouse; water screws constructed to irrigate the Egyptian deserts; and terraces built by the Inca in the steep, high-altitude Andes to control water flow. During the Industrial Revolution, dams were built to provide waterpower for grist, textile, or machine mills, resulting in the manipulation of natural drainage basins and river flows. The steam engine harnessed the thermal expansion of water into steam to provide an energy source and the expansion of human activity far from sources of water.

This chapter concludes with a discussion of the water balance and the related water cycle, both of which were discussed as far back as historical records exist. The ancient Greeks, and others, appreciated the balance between light and dark, good and evil, and water inputs and outputs. The sources of water to provide river flow, the origin of water in the sky to form precipitation, and connections between these different aspects of water were questioned. In the absence of sound physical reasons, the gods or goddesses were credited with hydrologic powers. Some of these beliefs persisted as late as the seventeenth century. Connecting back to the Roman bathhouses, the water balance is described in terms of water inputs, outputs, and changes in water volumes. This is then applied to the entire globe.

1.2 Water in the Stars, Myths, and Spirits

Since water provides and takes life, shapes the landscape, and transforms between liquid, solid, and gas, it is deeply ingrained in human myth and culture. Many ancient gods are connected to water, and their stories are told by their symbolical representation in the stars. These stories and legends provided meaning, context, and explanation for events that occur in the world around us.

Perhaps the earliest known evidence of this recognition of water's significance appears in the **stellar constellations**, groups of stars that form an image when connected with imaginary lines. Inscribed clay tablets and stones dating back to 3,000 BCE from the Mesopotamian civilizations of the ancient Babylonians and Sumerians provide the first record of zodiacal "signs" dividing the Sun's apparent annual path across the sky into 12 regions recognized by the constellation that aligns with the sun's position in the zodiacal belt at the time of year. The ancient Greeks and other cultures added to the 12 zodiac constellations, with the representation of Greek deities, heroes, and creatures. The constellations not only provided a means to mark the passage of time (seasons), but also explained cultural myths and beliefs as the deities looked down upon and ruled the Earth and humankind. Of the 88 constellations currently recognized by the International Astronomical Union, 19 are directly connected to water. The largest (by area of sky) is **Hydra**, the multiheaded water snake slain by Hercules, and related is **Hydrus**, the lesser water snake constellation. The snake, or serpent, is symbolically associated with water and deities representing water across several cultures and likely explains the persistence of sea-serpent legends to this day (e.g., the symbolic Chinese Dragon or the Loch Ness Monster). This representation of water in the form of snakes and serpents could be due to their shape, resembling waves on the surface of water, as with the horizontal zig-zag symbol that represents the constellation **Aquarius** (♒), or due to the deadly power of both snakes and water.

In addition to the constellations, there are roughly 400 known water deities spanning cultures across the world, further indicating the prominence of water in human life. In Greek mythology, there are at least 33 deities with some connection to water, their names and stories still widely recognizable. Even Greek deities needed water to drink, and Aquarius was the water and wine bearer to the gods. Corvus, the crow constellation, was sent by Apollo to search for water, which he then carried to Aquarius in Crater, the cup constellation. There are several other well-known creatures associated with water in the constellations: Cancer, the crab that bit Hercules' foot; Cygnus, Zeus disguised as a swan; Delphinus, dolphin messenger of **Poseidon** (the god of water); Dorado the swordfish; and Volans the flying fish. Perhaps the best-known animal constellation associated with water is **Pisces**, represented by opposing swimming fish. To help the gods travel across the sky, the ship constellation Argo was propelled along the river Eridanus by winds filling Vela, Argo's sail.

Animals with strength and power have been associated with water in other cultures too. In Egyptian mythology, the Lioness symbolized the goddess Tefnut. One of a group of nine deities, Tefnut was goddess of moisture and rain, alongside her twin brother Shu, the god of air, and their father Ra, deity of the sun. In Hawaiian culture, shark-gods include Kamohoali'i the guardian of the Hawaiian Islands, Ka'ahupahau the protector of fishers, and Kane'apua the trickster. Native Americans also have several animal gods representing the power of water. The underwater horned serpent Misiginebig (Big Water Snake) that resides in lakes and rivers is a legend common to most North American Algonquian tribes.

Creation myths are often used to explain the origin of species. In the Arctic, Inuit recognize Sedna as the goddess of the sea and marine animals. Caught with her father in their kayak in an Arctic storm, Sedna was thrown overboard by him. As she clung in desperation to the side of the kayak, her father cut her fingers off, and each finger created the marine creatures: fish, whales, seals, and walruses. Her body sank to the ocean floor where her spirit resides in mermaid form.

Water's role in both providing and taking life is not only reflected symbolically in creatures and constellations but has also been preserved in language. This rich history can be traced through languages preserved in the written form for nearly 6,000 years. The name Poseidon may have its origins in the ancient Greek *potis*, meaning "husband", and the Doric *don*, meaning "water" or "to flow" (as in the Danube River). The Latin word ***aqua***, a common prefix today in English, may have originated from Proto-Indo-European, common ancestor to both Latin and Greek, as well as many of today's Indo-European languages. Also appearing in Proto-Indo-European is the word *wed*, the source of the Ancient Greek *ύδρο*, which likely became the prefix "hydro" in modern English. The ancient *wed* may have also been spelt *wod* and eventually become proto-Germanic words *watar* or *wasser*, then *waeter* in Old English, and subsequently "water" in modern English (Liberman, 1994).

1.3 The Two Faces of Water in Health and Hygiene

Persistent references to water that are preserved in symbols dating back thousands of years probably survived because of the importance of communicating where clean sources of water were available for consumption and cleaning. Water, however, has at times been something to avoid. Water can provide life or take it, and this dichotomy was recognized in Greek mythology. Okeanos (**Oceanus**, from which the term Ocean arises) is the god of the Earth-encircling River Okeanos, the source of all fresh water and symbolic of the eternal flow of time, and where all stars rise and set. The counterpart to the life-providing River Okeanos connecting Earth to the heavens is the River Styx, connecting Earth to the underworld. Traveling along this river would lead you to the underworld of Hades, with pain and suffering along the way. Water could be good or bad, depending on the river you chose.

Although clean water for drinking and bathing is a necessity for proper health and hygiene, this has at times been misunderstood, since bacteria and viruses were not yet discovered. Until the development of the modern disciplines of chemistry, physics, and medicine in the late nineteenth and early twentieth centuries, misconceptions about the importance of clean water abounded.

Bathing, for example, has come into and out of fashion over the course of history. Bathing has been considered an act of purification, which remains apparent today with the religious act of baptism that spans several cultures, and bathing was thought to purify and cleanse the soul of evil spirits, not just superficial dust and dirt. There are several examples throughout history of using water to cleanse more than just the external body. To become a knight of the **Order of the Bath** in England before the mid-eighteenth century, candidates took a bath for spiritual purification as part of the ceremonial. Athletes who participated in the ancient Olympic Games rubbed oil and dust on their skin to improve performance and were massaged and bathed after an event to cleanse both body and soul. The Romans adopted this Greek

Figure 1.1 The Roman Baths in the city of Bath, England, constructed around 60–70 BCE, complete with an underfloor heating system (hypocaust). Photo credit: P. D. Blanken.

bathing ritual, and would also bath after exercise to remove oil, dust, and perspiration. Since bathing took so long, it became a social activity. Dinner parties and other social gatherings were sometimes held in bathing facilities built for this purpose. Dedicated **Roman bathhouses** were constructed solely for this purpose. Being naked made you vulnerable to attack: bathing in public with many witnesses minimized this vulnerability as you exposed not only your body but also your political and social opinions. As their popularity and social importance increased, larger and more luxurious bathhouses were constructed, with features such as underfloor heating and rooms for libraries, poetry reading, and eating (Figure 1.1). For the ancient Greeks and Romans, bathing and cleanliness were integral parts of society, and regular baths were recognized as vital for good health and good politics for centuries.

Opinions of public bathing took a different turn around the fourteenth century during the late **Medieval Period**. The **Black Death**, a bubonic plague bacterial infection that spread rapidly to humans primarily through infected flea bites, killed one-third (25 million) of the European population (Glatter and Finkelman, 2021) (Figure 1.2). Public bathhouses closed as it was believed that bathing in water opened pores that allowed the disease to enter the body, just as it spread through the open wounds of the infected. Public bathing fell out of practice in the West until the nineteenth century when baths and bathhouses appeared once again. Few people had access to clean, warm water, a vessel large enough to put it in, and the time and staff required for the lengthy procedure of taking a bath, so regular bathing was usually restricted to the wealthy. Even among the wealthy, infrequent bathing allowed the perfume industry to develop products to mask what must have been very unpleasant odors. Throughout this time

Figure 1.2 The Spreuer Chapel covered bridge in Lucerne, Switzerland, contains 67 paintings themed "Dance of Death", painted by Casper Meglinger between 1625 and 1635. The paintings show that no one, not even a king, was safe from the bubonic plague and death itself. Photo credit: P. D. Blanken.

the washing of hands and food was recognized as important for health, but washing in rivers or other bodies of natural water was the best many could do. Then, as now, rivers served both as a source of water and as the means to dispose of waste including raw sewage, so some knowledge of hydrology was required to locate a safe source of water.

Concurrent with the avoidance of full body bathing during the later Medieval Period, the consumption of water was also avoided because of the correlation between being ill (or dying) following the consumption of polluted water. This practice of avoiding water from natural sources has a history that far pre-dates the Black Death, and the consumption of fermented drinks with weak alcohol content such as ciders, ales, or wine has long been preferred over water. Even small quantities of alcohol dissolved in water act as an antiseptic by entering bacterial cells, breaking down protective membranes, and then dissolving, or denaturing, the proteins within, thus making the water safer to drink.

The consumption of wines, ales, and spirits appears in many stories written long before proper water treatment. The ancient Greek epic poem *The Odyssey*, attributed to Homer in the eighth century BCE, mentions the process of "mixing wine", where water was added to dilute the wine for consumption at the many feasts described. In Egypt, likely as an accidental consequence of storing grain in large porous ceramic vessels, the ancient Egyptians discovered beer as the stored grains produced alcohol upon fermentation. The brewing and consumption of **hecht** (beer) are represented in the hieroglyphics, including the communal sharing of hecht by those who built the Pyramids some 5,000 years ago. Beer was a source of nutrition, was consumed by all, including children, and was often used as a form of compensation. In

Figure 1.3 Tea leaf harvesting, Doi Ang Khang, Chiangmai, Thailand. Black, white, green, and Oolong tea varieties all come from the same species *Camellia sinensis*, with the age of the leaves picked and the processing determining the tea variety.
Photo credit: Jung Getty/Getty Images Creative.

North America thousands of years later, many drank cider instead of water. As the population increased in the eighteenth and nineteenth centuries, water became increasingly polluted and unsafe, and low-alcohol ciders made from fermented apples that grew well in the region's climate and soil were a safer option. Cider provided a tasty, nutritious drink that stored well and often served as currency, as did the beer in Egypt.

Nonfermented beverages also have a long and rich history of providing alternatives to water consumption. Drinking tea was (and still is) extremely popular, and tea remains the most widely consumed beverage in the world, with more than three billion cups consumed daily worldwide (Hicks, 2009). Originally recognized and developed for its medicinal qualities, tea became a popular drink in China between the seventh and tenth centuries, and across Europe in the late sixteenth century, as it provided a safe and healthy alternative to questionable water sources. The boiling of water for tea kills pathogens, and the natural compounds found in green tea (made from the mature unfermented leaves of *Camellia sinensis*; Figure 1.3) have antimicrobial, antioxidant, and anti-inflammatory properties (Hamilton-Miller, 1995; Reygaert, 2014).

1.4 Altering Water's Path

Not only has the significance of clean water been long recognized as a requirement for proper health and sanitation, the potential to harness water's energy as it flows from one location to another has also had a significant impact on agriculture and industry. Witnessing the force of water flowing over a waterfall, the erosion of the landscape by flash floods, or the destructive impacts of large waves on ships and shores inspired many to consider the possibility of harnessing such energy. Often this required altering the natural course of water to better meet agricultural or industrial needs.

Figure 1.4 Sketch of a water screw, or Archimedes screw. Credit: ibusca/Getty Images Creative.

Throughout history, there have been attempts (some successful, others not) to move water to where someone preferred it to go. One of the earliest references to a mechanical device designed to relocate water for irrigation purposes is the Egyptian **water screw** (Figure 1.4). Appearing in ancient Egypt around the third century BCE, the device consists of a long helical-shaped screw fastened to a shaft housed within a cylinder (an auger). Rotation of the shaft results in the upward movement of water from the bottom to the top of the cylinder. When placed at a shallow angle, rotating the shaft could move water to higher elevations for irrigation purposes. The ancient Greeks adopted this invention, and it was described by **Archimedes** who later was assumed to have invented the device himself. The simple but effective technology of the water screw and later the **water wheel** remain virtually unchanged to this day and are used in wind and water turbines (hydroelectric power), pumps, and other machinery.

Related to agriculture, the ancient Greeks have records of rotating a heavy stone connected by a gear to a wheel that was rotated by the force of flowing water to grind grain into flour. The earliest descriptions of these **grist mills** appear around the third century BCE in works of Philo of Byzantium. Their ability to produce large quantities of flour with little human effort must have had a significant impact on food availability and quality of life.

The Romans were famous for their construction of extensive **aqueduct** (water channel) systems that transported water from the source to city centers, often to provide water for their popular bathhouses. Several of these aqueducts remain in good condition today after nearly 2,000 years of operation (Figure 1.5). The South American Inca were also experts in designing hydraulic systems to strategically move water from one location to another (Figure 1.6). The Inca designed and constructed terraces in the steep high-elevation Andes that permitted both adequate water supply and drainage for crops. They also used their knowledge of hydrology to provide clean fresh water supplies using a complex system of diversion and infiltration canals, some of which pre-dates the Incan culture. Water diverted from mountain headwater streams onto mountain slopes during the wet season is naturally filtered as it recharges downslope aquifers, extending the supply of water when needed during the dry season (Ochoa-tocachi et al., 2019).

Regions in northeastern North America during the late nineteenth century witnessed a proliferation of settlement, in part made possible by the numerous rivers capable of providing sufficient year-long water flow to power various industries. The waterpower provided by these

Figure 1.5 The Roman Aqueduct of Segovia, Spain, standing 28.5 m high and 728 m long, with 167 arches. Built during the first century BCE without mortar, the aqueduct carried water from mountain springs 17 km away to Segovia's fountains, public baths, and private homes, until 1973. Photo credit: P. D. Blanken.

Figure 1.6 A 500–600-year-old Incan fountain in Peru that is still used to provide fresh water from mountain streams to local populations. Photo credit: P. D. Blanken.

rivers was not only used to grind grain but also to fuel the booming textile industry in the New England region. An example of how water can transform a region, the large mills that processed plant fiber into fabric, and the manufacturing of machine tools required for such processing, changed the social and physical fabric of the eastern United States. To provide a consistent supply of water, thousands of **dams** were constructed that flooded upstream regions and decreased flow downstream. In New England, the highest density of dams in the country (0.015 dams per square kilometer) was installed, decreasing watershed area to an average of only 44 km^2 (Graf, 1999). The importance of the removal of these dams to restore the riparian environments (e.g., fish populations) here and in other regions is now being recognized (Graber, Chipman and Fox, 2016). Textile mills also had societal and economic impacts. Although some made fortunes, the majority, especially women, were employed with minimal pay, long hours,

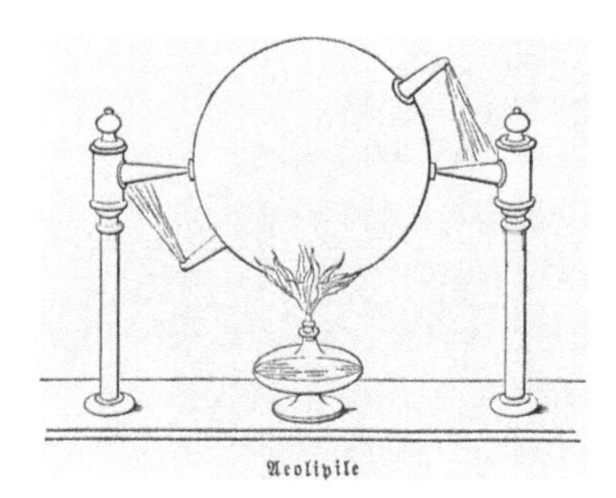

Figure 1.7 1874 wood engraving of an aeolipile or Hero engine. Credit: ZU_09/ Getty Images Creative.

and dangerous conditions. Young women, so-called "Mill Girls", accounted for 75 percent of the workforce during the textile boom (Montrie, 2004).

The power provided by flowing water for these grist and textile mills required that the mills be co-located by reliable sources of fast-flowing rivers, such as those found in eastern North America. The invention of the **steam engine**, however, allowed for the expansion of industry to locations without such rivers. Instead of harnessing the water flow, the harnessing of the energy associated with a change of state in water from liquid to gas soon changed both the atmosphere and the world. As water is heated to boiling point, the separation between water molecules as they leave the liquid state and expand to the vapor state creates an increase in pressure when the vapor (steam) is contained within a tight chamber. The controlled release of the steam past a gear or blade attached to a shaft can cause the shaft to spin to power a saw, pump, wheel, or a turbine to generate electricity.

The history of the steam engine is nicely summarized by Riznic (2017). The first known steam engine was the **Aeolipile**, invented by the Greek inventor Heron (Hero) of Alexandria in 62 BCE (Figure 1.7). Water was heated in a kettle above a fire. Steam exiting the kettle entered a sphere from a small tube that ran through its center and also acted as an axis, allowing the sphere to spin as steam was released from two opposing nozzles. It was hundreds of years before steam was used for practical purposes. In 1606, Jeronimo de Ayanz y Beaumont developed a steam engine to pump water from flooded mines. Later (1698), Thomas Savery's improved steam engine design relied on a vacuum created when steam condensed to pump water. Later versions took advantage of both the expansion of air when heated and the contraction of air when cooled by having the process confined within a cylinder above a piston connected to a drive shaft. This original design was made by Thomas Newcomen in 1705 and improved in 1765 by **James Watt**.

It was harnessing of the thermal properties of water, with the original intention of pumping water, that led to the development of the steam engine and the start of the **First Industrial Revolution** (1760–1840). People no longer needed to locate industry near sources of water for power, since the steam engine allowed for a mobile source of energy free from cascading water flows.

1.5 The Water Balance and Cycle

The numerous cultural associations between water and myths, deities, and creatures have connections to the concepts of a balance between water inputs and outputs, and water cycling between land, air, and ocean. Without today's understanding of the chemical and physical properties of water, it was difficult to explain (for example) how water can fall out of the sky as precipitation with no apparent origin; or how water can emerge under pressure from the ground. Water can vanish from lakes and puddles, but where does it go? Water can suddenly condense on a cool surface, appear as a dense fog, or quickly become solid enough to walk across. What are the sources of rivers, and where do they go? Reasons for the movement of water between locations and forms must have seemed mysterious. Without a physical explanation, myths and legends provided some rationale and logic.

It is no surprise that the ancient Greeks thought about the water balance and cycle (see review by Duffy, 2017). As mentioned, the River Okeanos represented a continuous Earth-encircling surface river where the river's output at one end was the input at the other. Other rivers, such as the subterranean River Styx leading to the underworld, had no apparent origin. Rivers were believed to receive water from subterranean connections to the oceans. Rain was thought to supply enough water for watering plants, but not to fill rivers. At this time, there was no understanding of water flow through roots and vascular plant tissue, so rain had to be the only source of water for plants. This idea of two separate water cycles, one below ground and one above, was prevalent for centuries, but lacked a plausible explanation for river and spring formation in mountain regions. Rivers were thought to originate from seawater moving underground and being raised against the force of gravity to emerge as springs in mountain regions. This **reverse hydrologic cycle** struggled to find a plausible explanation for how water would flow backwards from the sea to the underground, then upwards against the force of gravity. As late as the seventeenth century, the Roman god **Neptune** appears in sketches, moving water from the depths to the surface to form rivers and springs (Duffy, 2017).

The French scientist **Pierre Perrault** (ca. 1611–1680) proposed the concept of a **drainage basin** to determine whether river discharge could be explained by precipitation alone. Subsequently, cartographers' maps of river networks and drainage basins began to form the notion of a global water cycle where the horizontal flow of water between land and oceans was connected by vertical exchanges between the atmosphere and groundwater. The existence of such a water cycle required an explanation of changes of the state of water between liquid, solid, and gas, together with the conservation of energy and mass. Experiments and evidence were also required, and much in vogue during the late seventeenth century, when the **Age of Enlightenment** (1685–1815) began.

The concept of a balance of water between inputs and outputs across a defined spatial and temporal scale is a foundational concept of the global water cycle. This is especially relevant today: climate change, coupled with Earth's changing landscape, means the water cycle is undergoing rapid alteration. The **water balance** compares the flux (mass or volume per time period) of water inputs to outputs in a volume or system over a period of time. Imagine a bathtub with water entering from a faucet above and exiting from a drain below. As long as the rate of water input is exactly equal to the rate of water output, the volume of water in the bathtub will remain constant. The volume can be indicated by the depth of water, which could be zero (no water) if initially no one plugs the drain, and the water input and output rates are equal. There are two options to achieve the desired goal of filling the tub. One is to decrease (or stop) the water output by plugging the drain; the other is to increase the input of water to a rate greater than the output rate. With both options the result is the same; the volume of water in the bathtub increases. When the bathtub is full, a constant water level can be achieved when the input and output rates are equal. To drain the tub, the water output rate must exceed the water input rate.

In nature this concept is no different. Bathtubs are represented by lakes and oceans, glaciers and ice sheets when frozen, or water vapor in the atmosphere. These are reservoirs, or volumes of water, stored in each location, time, and state (liquid, solid, or gas). The water entering the bathtub and the water leaving it represent the fluxes, or exchanges of water, between locations and reservoirs. These exchanges are volumes of water over time (a rate of exchange) and storage of water depends not only on the dimensions of the tub but also the relative rates of water input(s) to output(s). A change in water volume (hence water level) will occur whenever the rates of inputs and outputs are different. When the volume of water in a reservoir is constant, the **residence time** (or **flushing time**), the length of time the water stays in the storage volume, can be estimated by dividing the volume of water stored by the volume flux of water out of the reservoir (equal to the inflow volume flux when the volume is constant). In North America, water in Lake Ontario has a residence of 6 years (1,640 km^3 divided by 273 km^3 per year) compared with 191 years for Lake Superior (12,000 km^3/63 km^3 per year). (Resident time can be calculated using mass instead of volume if assuming a constant density.)

To understand the water balance requires that the region of interest and time period be defined as well as a knowledge of all inputs and outputs and volume of the water being stored. This is difficult. Over the entire Earth, the water balance is always balanced since inputs equal outputs with no change in the overall volume of water. As we focus on smaller regions, such as river drainage basins, it is reasonable to question how large rivers can continue to flow during extended periods without precipitation. Such spatial variation becomes apparent when looking at differences between precipitation and evaporation across latitudinal zones (Figure 1.8). Now that we know that horizontal and vertical water exchange are connected, such that water stored in groundwater and soil can maintain river base flow during dry periods, measuring all the water inputs, outputs, and change in water storage remains elusive; it is very hard to "close" the water balance to this day. Measuring precipitation over a continent is much harder than over a small drainage basin owing to the variation in precipitation that occurs across such a large scale. In many regions, precipitation is not measured. Water stored in groundwater aquifers and surface soil moisture has few measurements and large spatial variability. Transpiration, evaporation that occurs from within leaves of vascular plants such as trees, has only recently been measured. The majority of the world's rivers are ungauged, and many gauging networks in remote regions have been discontinued for economic reasons.

To complicate matters further, the volume of water stored in a reservoir can change even when the inputs and outputs of water are equal. This is due to the unique properties of the water molecule. Consider ice floating in a full glass of water, and think of the analogous situation with melting ice shelves in the polar regions. The ice floats because it has a lower density than liquid water, and the **Archimedes Principle** states that the upward buoyancy force keeping the ice afloat is equal to the downward force of the ice, which is equal to the weight of the water displaced by the ice (weight equals mass multiplied by gravitational acceleration). The

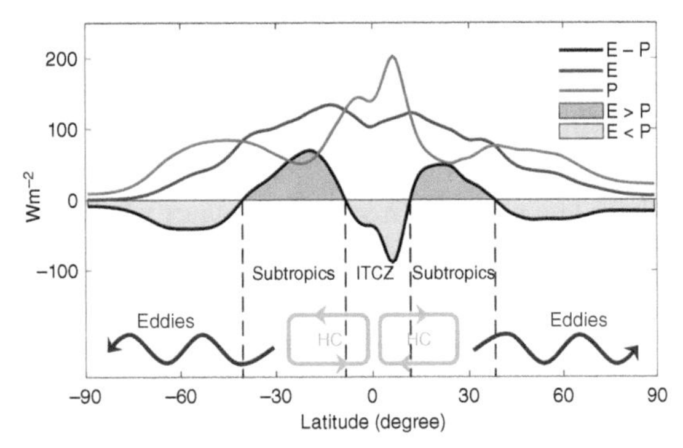

Figure 1.8 The global water balance expressed as the vertical input of water from precipitation (P) and outputs from evaporation (E) in units of watts per square meter of energy exchange. To maintain the global water balance, surplus water ($E < P$) moves to locations of water deficits ($E > P$) by horizontal transport through the atmosphere (eddies, Hadley cells [HC] near the Intertropical Convergence Zone [ITCZ]) and surface flows (streams and rivers). From Siler, Roe and Armour (2018). © American Meteorological Society.

rise in water level that occurs when you first add the ice to the glass, however, results from the volume of water displaced by the ice, not the weight of displaced water. For the same weight of water, the volume is larger when the water is frozen than when liquid (the ice has a lower density). Therefore, once the ice melts, the volume of water displaced does not increase (i.e., the glass does not overflow) since the volume of water displaced by the ice is more than the volume of liquid water created upon melting. Even the extra 10 percent of ice volume above the surface when melted will not result in water overflowing the glass. Why, then, are sea levels rising with climate change? Liquid water expands (the density decreases) as it warms, and this thermal expansion is a significant factor in sea level rise, in addition to the increase in sea level resulting from the additional water provided by meltwater inputs from land-based melting ice sheets and glaciers.

1.6 SUMMARY

The importance of water in everyday life and its influence on the world around us is reflected in cultures across time and throughout the world. Taking a place in the heavens above are the myths and legends told by the ancient Greeks, as represented by the constellations, many of which are related to water. Myths provide a means to explain and rationalize aspects of life that otherwise have no explanation. The creation of powerful gods and creatures with capacities equal to the powers that water possesses reflects the ability of water to give or take life, and this can be seen in ancient cultures across the world.

Water has been mythologized as both a positive and a negative force. Rainfall is necessary, but too much rain causes flooding. We need water to drink, but too much can drown us. Clean water is needed to support life, but contaminated water will take life. Illness and death associated with the drinking of water contaminated with microbes invisible to the human eye resulted in a movement to consume teas and alcohol-containing wines, beers, and ciders. The boiling of water and the natural compounds contained in tea, and the alcohol produced by fermentation of grapes (wine), grain (beer), or apples (cider), all served as antiseptics.

With the development of agriculture and civilization came the need to move water from one location to another. The ancient Greeks used water wheels that were spun by passing water to grind grain into flour. The Romans constructed aqueducts to relocate water downslope from source regions to urban centers. The Egyptians designed water screws, or augers, to move water upslope for irrigation purposes. The Inca channeled water high in the Andes to provide fresh water for irrigation in remote locations. In North America, rivers were dammed to provide hydraulic power for numerous textile, grain, and industrial mills. The development of the steam engine and its use in train locomotives, tractors, pumps, mills and so on meant that industry no longer had to be located by sources of water, but the power of water would be available anywhere.

The water balance concept and the cycling of water across the landscape can also be traced back to ancient Greece. At a given location and time, the volume, or storage, of water will increase if the rate of water input exceeds the output; decrease if outputs exceed inputs; or remain constant if inputs equal outputs. This fundamental concept is easy to understand yet difficult to quantify in nature. The many inputs and outputs of water have large spatial variation and are often difficult to measure (if they are measured at all). The volume of water stored also varies with spatial and temporal scales and with changes in water density. In the context of climate change, understanding how the water balance and water cycle are changing is more important now than at any time in history.

This chapter provided a brief summary of some of the significant aspects related to water through human history up to the early nineteenth century. The next chapter continues exploring water, starting at the **Age of Enlightenment** (1685–1815). Following the **Scientific Revolution** (1543–1687), and overlapping the **Industrial Revolution** (1760–1840), this was a prolific time for advances in the understanding of all of the natural and social sciences, including the properties of water that helped to explain many of its mysterious characteristics.

1.7 QUESTIONS

1.1 Why are themes related to aspects of water so prevalent in mythology, legends, and deities throughout the course of history for so many cultures?

1.2 Myths and legends related to some aspect of water appear in nearly all cultures through time. Where you are originally from, or where you currently reside, find and describe a local myth (not a misconception or misunderstanding) or legend that is clearly related to some aspect of water.

1.3 Myths involving fog are plentiful, and many are associated with darkness, ghosts, and the unknown in general. Describe why such negative and mysterious aspects are so commonly associated with the formation of fog.

1.4 Public bathing has at times been something to seek and at other times something to avoid. What were some causes leading to the rise and demise of the once highly popular social activity of the Roman public bathhouses?

1.5 Throughout history and across the globe, it was once common for everyone, including children, to consume drinks containing alcohol. Why was this such a common and acceptable practice?

1.6 By controlling the flow of water, we can enable large ships to avoid rapids and waterfalls by being raised and lowered through one or more of a series of steps called "locks". These locks can be found on the Great Lakes in America and around the world (e.g., the Welland Canal in Ontario, Canada, connecting Lakes Erie and Ontario). Describe how these systems operate without the use of external power.

1.7 Calculate the residence time (in days) of water in the atmosphere, using current estimates that the volume of water stored in the atmosphere is 12.6×10^3 km^3, the volume of precipitation falling over the oceans is 403.5×10^3 km^3 yr^{-1}, and the volume of precipitation falling over the land is 116.5×10^3 km^3 yr^{-1} (values are taken from van der Ent and Tuinenburg, 2017).

1.8 The water levels (therefore the volume of water stored) on the North American Great Lakes are usually highest in the summer (July) and lowest in the winter (February). Use the concept of the water balance to explain how this annual water level fluctuation can be explained in terms of seasonal changes in water inputs and outputs from these lakes.

1.9 In either the location where you are originally from or where you currently reside, find a map showing your local drainage basin (watershed). Your task is to determine the water balance for your watershed over the course of one year. What are the major sources of water inputs and outputs? Are they being measured now? What additional measurements are required to adequately measure the complete water balance, including any storage terms?

1.10 A full glass of water with ice cubes will not overflow once the ice melts. A full glass of salty water, however, will overflow once the ice melts. Why? Note that the ice contains only fresh water.

REFERENCES

Duffy, C. J. (2017) 'The terrestrial hydrologic cycle: an historical sense of balance', *WIREs Water*, 4, pp. 1–21. doi: 10.1002/wat2.1216.

van der Ent, R. J. and Tuinenburg, O. A. (2017) 'The residence time of water in the atmosphere revisited', *Hydrology and Earth System Sciences*, 21, pp. 779–790. doi: 10.5194/hess-21-779-2017.

Glatter, K. A. and Finkelman, P. (2021) 'History of the plague: An ancient pandemic for the age of COVID-19', *The American Journal of Medicine*, 134, pp. 176–181. doi: 10.1016/j.amjmed.2020.08.019.

Graber, F. J. M. B. E., Chipman, K. H. N. J. W. and Fox, C. S. S. C. A. (2016) 'River restoration by dam removal: Enhancing connectivity at watershed scales', *Elementa: Science of the Anthropocene*, 4, pp. 1–14. doi: 10.12952/journal.elementa.000108.

Graf, W. L. (1999) 'Dam nation: A geographic census of American dams and their large-scale hydrologic impacts', *Water Resources Research*, 35, pp. 1305–1311.

Hamilton-Miller, J. M. T. (1995) 'Antimicrobial properties of tea (*Camellia sinensis* L.)', *Antimicrobial Agents and Chemotherapy*, 39, pp. 2375–2377.

Hicks, A. (2009) 'Current status and future development of global tea production and tea products', *Au Journal of Technology*, 12, pp. 251–264.

Liberman, A. (1994) *An Analytic Dictionary of English Etymology*. Dictionary Society of North America. doi: 10.1353/dic.1994.008.

Montrie, C. (2004) ‘I think less of the factory’, *Environmental History*, 9, pp. 275–295.

Ochoa-tocachi, B. F. et al. (2019) ‘Potential contributions of pre-Inca infiltration infrastructure to Andean water security’, *Nature Sustainability*, 2. doi: 10.1038/s41893-019-0307-1.

Reygaert, W. C. (2014) ‘The antimicrobial possibilities of green tea’, *Frontiers in Microbiology*, 5, pp. 1–8. doi: 10.3389/fmicb.2014.00434.

Riznic, J. (2017) ‘Introduction to steam generators – from Heron of Alexandria to nuclear power plants: Brief history and literature survey’, in Riznic, J. (ed.) *Steam Generators for Nuclear Power Plants*. Elsevier, pp. 3–33. doi: 10.1016/B978-0-08-100894-2.00001-7.

Siler, N., Roe, G. H. and Armour, K. C. (2018) ‘Insights into the zonal-mean response of the hydrologic cycle to global warming from a diffusive energy balance model’, *Journal of Climate*, 31, pp. 7481–7493. doi: 10.1175/JCLI-D-18-0081.1.

2 The Composition of Water

Key Learning Objectives

After reading this chapter, you will be able to:

1. Summarize the contributions of Scheele, Lavoisier, Priestley, and Cavendish to the discovery of the composition of water.
2. Describe the structure of the atom and the contributions of Rutherford, Geiger, Marsden, and Bohr.
3. Compare and contrast metallic, ionic, and covalent chemical bonds.
4. Discuss how hydrogen and oxygen bond to form water, and the resulting unique characteristics of water.

2.1 Introduction

Water has long been recognized as a substance required for life and for shaping the Earth's surface. The chemical composition and structure of water, however, is a recent discovery. During the Age of Enlightenment (1685–1815), the popular philosophy was that the "evidence of the senses" was the primary source of knowledge. Laboratory experiments were popular. These experiments resulted in the discovery of oxygen and hydrogen and a description of their characteristics. Only in the late nineteenth and early twentieth centuries was our current understanding of the atomic structure developed, leading to our understanding of how atoms bond to form molecules like water.

This chapter describes the key individuals whose experiments resulted in the discovery of hydrogen and oxygen, and thus the composition of water. The discovery of oxygen and hydrogen as unique elements with distinct properties, coupled with a model of the structure of the atom, provides the foundational knowledge required to understand how atoms bond. Understanding these concepts is a prerequisite for understanding the unique properties of water.

2.2 The Discovery of Oxygen and Hydrogen

It is well known that **oxygen** and **hydrogen** atoms combine to form water molecules. In the eighteenth century, few of the elements that appear on today's **Periodic Table of the Elements** had yet been discovered. This, however, was an active time for both social and scientific activity

Table 2.1 Major contributors to the discovery of oxygen and hydrogen

Contributor	Birth–death	Contribution
Carl Scheele	1742–1786	Performed experiments that identified "fire air" as the gas that supported combustion.
Antoine-Laurent Lavoisier	1743–1794	Performed experiments showing the mass of some elements increased and produced an acid after combustion. Named the substance in the air that caused this "oxygen" (acid maker), and later named "hydrogen" (water generator) as the gas that produced water when burned with oxygen.
Joseph Priestley	1733–1804	Discovered that "dephlogisticated air" (oxygen) was produced by burning mercuric oxide, and that this air supported both combustion and respiration. Also showed that plants produced oxygen through the process of photosynthesis.
Henry Cavendish	1731–1810	Discovered the chemical composition of water by burning hydrogen gas in the presence of oxygen.
Elizabeth Fulhame	Unknown	Discovered the role of water oxidation reactions, catalytic reactions, and photoreduction.

across Europe, with the Age of Enlightenment (1685–1815) and Industrial Revolution (1760–1840) closely following the **Scientific Revolution** (1543–1687). Empirical evidence and reasoning underlay the pursuit of knowledge. Several individuals performed experiments involving air and the combustion of various materials, often metals, resulting in the discovery of oxygen and hydrogen (Table 2.1). In addition to identifying the composition of the materials and discovering new elements contained in them, these experiments also revealed that combustion ceased when the air supply was exhausted.

Carl Scheele (1742–1786), like most chemists at the time, was convinced that air consisted of two types, one that supported combustion and one that did not. When Scheele heated oxides of mercury, silver, and gold, in his laboratory in Uppsala, Sweden, he discovered that the gas that formed supported combustion and respiration better than common air (West, 2014a). Scheele referred to this gas as "fire air" and estimated that the ambient air contained 25 percent of this gas.

Scheele also performed experiments on iron rusting, resulting in our understanding of the role that dissolved oxygen plays in forming liquid water. He observed that when iron gets wet, additional water is produced, furthering the rusting (oxidation) of the iron. Today, we know that the dissolved oxygen in the water, $O_2(aq)$, initially in contact with the solid iron Fe(s), creates a solution of dissolved iron Fe(aq), releasing **electrons** (e^-) in the process:

$$\mathrm{Fe(s)} \rightarrow \mathrm{Fe^{2+}(aq)} + 2\mathrm{e^-}.$$

These electrons are then available to combine with dissolved hydrogen H(aq) and the dissolved oxygen in the water to produce additional liquid water $H_2O(l)$:

$$4\mathrm{e^-} + 4\mathrm{H^+(aq)} + \mathrm{O_2(aq)} \rightarrow 2\mathrm{H_2O(l)}.$$

This observation showed the importance of electrons in chemical reactions long before the concept of electrons first appeared in the model of atomic structure.

Scheele's prolific research laid the foundation for the discovery of oxygen and its role in chemical reactions including the formation of water. Likely as a result of regular exposure to such chemicals as mercury, lead, and arsenic, Scheele died at the early age of 43.

Antoine-Laurent Lavoisier (1743–1794) and **Joseph Priestley** (1733–1804) were performing similar experiments in France and England, respectively, at roughly the same time. In 1772, Lavoisier observed that burning phosphorus produced phosphoric acid, and the mass of the phosphorus increased after burning. This increase was puzzling. Popular at this time was the **Phlogiston Theory**, which stated that materials burned because they contained a combustion-supporting material called phlogiston. According to the theory, a substance's mass should always decrease after combustion since the phlogiston was released into the air and combustion ceased when it was all consumed. Lavoisier concluded that the increase in mass of the phosphorus was due to the equal addition of mass from something in the air, thus laying the foundation for the "conservation of mass" concept in chemical reactions. Lavoisier named the gas that was removed from the air during combustion **oxygen**, meaning "acid generator", in reference to the phosphoric acid produced (see review in Vera, Rivera and Núñez, 2011).

Joseph Priestley also experimented with the combustion of materials. In 1774, Priestley focused sunlight on solid mercuric oxide placed in a pool of mercury inside a sealed glass dome with a candle. Vapor produced by burning the mercuric oxide caused a candle flame to burn much more intensely than one burning in regular air. Priestley observed that a mouse placed in the sealed chamber where the combustion occurred survived four times longer than a mouse placed in a sealed chamber containing only fresh air. A mouse sealed in a chamber containing fresh air was revitalized after a plant was placed inside the chamber, a key discovery in our understanding that photosynthesis produces oxygen.

As in Lavoisier's experiments, Priestley observed that contrary to the Phlogiston Theory, for some materials the mass increased, and that the volume of air within a sealed chamber decreased after combustion. Priestley named the oxygen-rich air produced by burning mercuric oxide **dephlogisticated air**, reasoning that the air supported combustion so well since this phlogiston-depleted air could readily absorb the phlogiston produced by the burning material.

In 1793, Priestley met with **Elizabeth Fulhame** who was conducting experiments on the chemical reduction of metals to stain cloth. Impressed with her research, Priestley encouraged her to publish her results that rejected the Phlogistic Theory and demonstrated the importance of water in chemical reactions. Fulhame also showed that some metallic salts could be reduced to pure metals in aqueous solutions at room temperature using light, instead of the high temperatures required for processing metals by smelting, thus discovering the concepts of catalysis and photoreduction. She showed that water was required for and regenerated through oxidation reactions, an important result for understanding the composition of water.

There was considerable debate and uncertainty regarding the role each of Scheele, Lavoisier, and Priestley played in the discovery of oxygen (Cassebaum and Schufle, 1974). Scheele's experiments preceded those of Lavoisier and Priestley, but a letter he wrote to Lavoisier about them never arrived. Scheele's results sat in a printer's office for two years, delaying publication until 1777: after Priestley's and Lavoisier's results had been independently published in 1775 and 1776, respectively. Adding to the confusion was a meeting between Priestley and Lavoisier in Paris in 1774. At this meeting, Priestley replicated his experiments for Lavoisier, thus showing Lavoisier how to make oxygen (Lucibella, 2010). Regardless of the timing and circumstances of these events, the three could be considered equal contributors to oxygen's discovery: Scheele performed the first experiments to isolate the gas; Lavoisier named the gas oxygen; and Priestley first formally published the results.

Water contains hydrogen as well as oxygen, and this gas was discovered by Henry Cavendish (1731–1810). The word **hydrogen** literally means "water producer", from the Greek "hydor" and the French "-gen" (generator/producer). Whereas Scheele's, Lavoisier's, and Priestley's experiments produced inflammable oxygen, Cavendish produced a highly inflammable gas he referred to as "inflammable air". **Robert Boyle** (1627–1691) first produced hydrogen gas, but he did not recognize its properties, and it was Cavendish who first correctly identified hydrogen as a unique element. Cavendish discovered that when hydrogen gas burned in oxygen, water condensed from the air. In other words, adding oxygen (O) to hydrogen gas (H_2) produced water, H_2O:

$$H + H + O \rightarrow H_2O$$

with the ratio of two parts hydrogen to one part oxygen, correctly hypothesized by Cavendish. Cavendish was the first to analyze the chemical composition of atmospheric air, finding that it is mostly nitrogen and oxygen gas. He also devised a method for measuring the density of gases, in addition to several other significant discoveries and inventions related to the characteristics and properties of gases (see West, 2014b).

2.3 The Atomic Structure of Oxygen and Hydrogen

By the end of the eighteenth century, oxygen and hydrogen had been identified as unique elements and their basic properties were known. It was also known that water was comprised of oxygen and hydrogen. Atomic structure was not yet understood. In the early twentieth century, **Ernest Rutherford** (1871–1937), **Hans Geiger** (1882–1945), and **Ernest Marsden** (1889–1970) measured how alpha radiation scattered after passing through a thin sheet of gold (Table 2.2). They discovered that instead of passing straight through, some of the beams of radiation were deflected back towards their source. This series of experiments, known as the **Rutherford gold foil** or **Geiger–Marsden experiments**, resulted in a fundamental understanding of the atomic structure. Rutherford proposed his **Rutherford model** that atoms contained a dense, positively charged **nucleus**.

Niels Bohr studied mathematics, physics, and astronomy, and was interested in numerous topics including why water has such a high surface tension. Given his interest in astronomy, Bohr proposed that negatively charged electron particles with negligible mass stabilize the atom and circle the nucleus in discrete orbits called **electron shells**. Bohr showed that the number of electrons in the outer (or **valence**) orbit influenced the atom's chemical properties. The modification of Rutherford's model of the atom to include electrons resulted in the **Rutherford–Bohr model**, published in 1913. Subsequent research revealed that electrons most often occur in pairs, each with a "spin" opposite to the other to counteract the similar negative charge. With their low mass, electrons contribute almost nothing to the atomic mass and are attracted to the nucleus by electrostatic forces. In the 1920s, it was realized that an electron could not

Table 2.2 **Major contributors in the discovery of the atom's structure**

Contributor	Birth–death	Contribution
Ernest Rutherford	1871–1937	Directed the Geiger–Marsden gold foil experiment in 1909. From the results of that experiment, he developed the **Rutherford model** where the atom contains a dense nucleus with a positive charge.
Hans Geiger	1882–1945	Together with Marsden under Rutherford's direction, conducted the Geiger–Marsden gold foil experiment providing the evidence used in Rutherford's model of the atom.
Ernest Marsden	1889–1970	Together with Geiger under Rutherford's direction, conducted the Geiger–Marsden gold foil experiment providing the evidence used in Rutherford's model of the atom.
Niels Bohr	1885–1962	Added electrons in stable orbits around the nucleus to the structure of the atom, resulting in the **Rutherford–Bohr model** of the atom.

be treated as a particle with position and velocity, since the electrically charged electron would lose energy as it orbits the nucleus, and therefore eventually spiral into the nucleus and collapse the atom. Instead of treating electrons as orbiting particles, **Quantum Theory** was developed, describing the probability of finding an electron's location away from the nucleus. This atomic orbital shell or electron cloud model represents our current understanding of the structure of atoms, with mathematic equations used to describe the probability of finding an electron at any location around the nucleus.

The cumulative research of Rutherford, Geiger, Marsden, and Bohr resulted in a conceptual model of the atom that is now generally accepted. An atom consists of a nucleus, which contains **protons** and **neutrons**, and electrons positioned beyond the nucleus. Protons have a positive electrical charge, and neutrons are electrically neutral particles with a slightly greater mass than a proton. Electrons have a negative electrical charge and a mass much less than a proton or neutron. The number of protons in the atom is known as the **atomic number**, and atoms of different elements have different atomic numbers. The number of electrons in an electrically neutral atom is equal to the number of protons, and the **atomic mass number** is the sum of the number of protons and neutrons. For example, hydrogen has one proton, so it has the atomic number 1. Oxygen has eight protons, so it has the atomic number 8. Atoms of the same element may have different numbers of neutrons but the same number of protons and electrons, and these variations are called **isotopes**. Therefore, the atomic mass number will change depending on the number of neutrons present but the atomic number will remain the same. Note that the atomic mass is often given either as an integer or as the average value of all isotopes of the element based on the abundance of each isotope.

Hydrogen is the simplest and most abundant element in the universe and has therefore been used in many atomic structure studies. It has three naturally occurring isotopes. The most common, with an abundance of 99.98%, is ^{1}H, or **protium**, without any neutrons (Figure 2.1). The atomic mass, the sum of the number of protons and neutrons, can be used to easily identify isotopes of the same element. Hydrogen with one neutron is ^{2}H, or **deuterium**, while with two

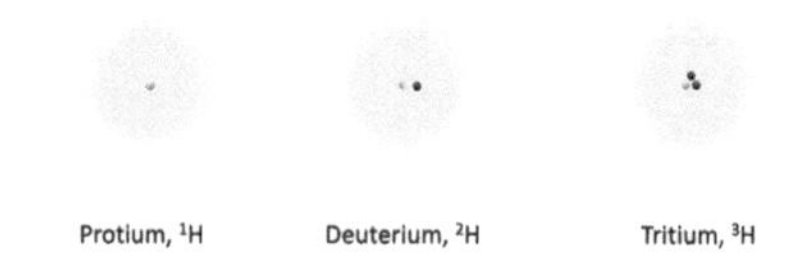

Figure 2.1 Atomic nucleus (yellow) structure of hydrogen and its common isotopes. The white spheres represent protons, the black spheres neutrons.

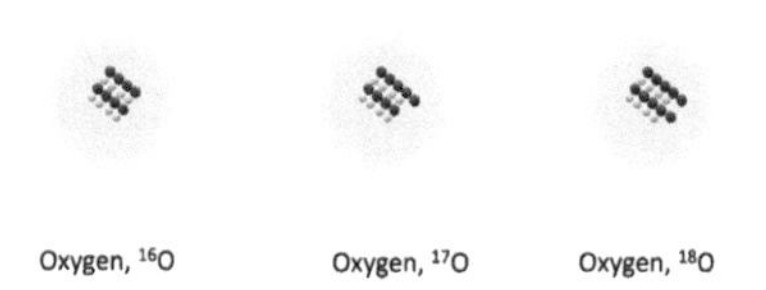

Figure 2.2 Atomic nucleus (yellow) structure of oxygen and its common isotopes. The white spheres represent protons, the black spheres neutrons.

neutrons it is ^3H, or **tritium**. In each hydrogen isotope there is one proton and one electron. **Oxygen** in its most abundant (99.76%), stable form with eight protons and eight neutrons is ^{16}O; add an extra neutron, ^{17}O; add two extra neutrons, ^{18}O (Figure 2.2). No special names are given to these isotopes of oxygen. The atomic mass of a water molecule changes depending on which isotopes of oxygen and hydrogen are combined. Processes such as evaporation, transpiration, and condensation may vary depending on the water molecule's mass. Therefore, measurements of oxygen and hydrogen isotopes can help discern aspects of the water cycle.

2.4 Valence Electrons

Electrons play an important role in several properties of the atom, especially in the bonding between atoms to form **molecules**. Electrically neutral atoms have an equal number of protons (positive charge) and electrons (negative charge), yet the electrons are often not equally distributed around the nucleus. The number of electrons located furthest from the nucleus influences how atoms can combine to form molecules. Since electrons tend to occur in pairs, when an atom gains or loses an electron, the net electrical charge will change. This exchange or sharing of electrons occurs in the region furthest from the nucleus, since those electrons share less of an attraction to the positively charged protons in the nucleus. This outermost region is called the **valence** (or **orbital**) **shell**, and electrons located in the valence shell are called **valence electrons**. The number of electrons in this outermost region is indicated by the **valence number**. The Periodic Table of the Elements is arranged with all of the elements in a row (called **Periods**) having an equal number of electron shells, and all elements in a column (called **Groups**) having the same chemical and physical characteristics primarily due to the number of electrons in the

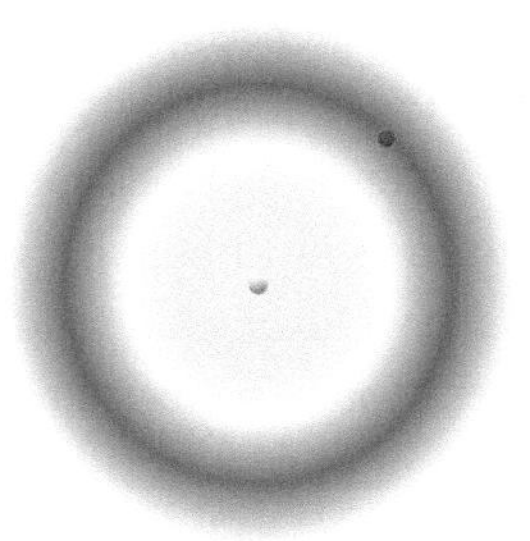

Figure 2.3 The atomic structure of hydrogen including the positively charged proton (white sphere) in the nucleus (yellow). One electron (red sphere) is in the one orbital shell (blue region).

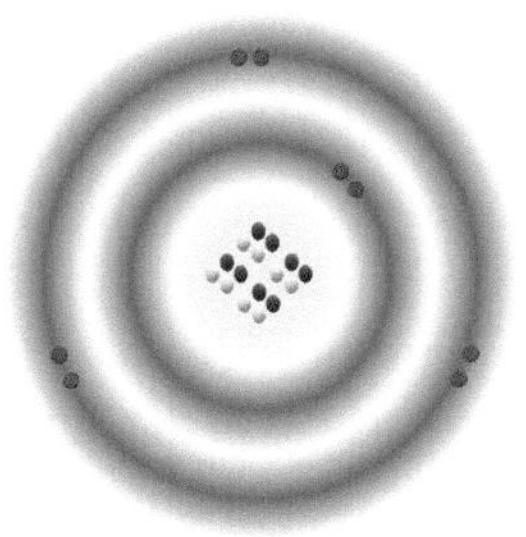

Figure 2.4 Atomic structure of oxygen including eight positively charged protons (white spheres) and eight neutrons (black spheres) in the nucleus (yellow). Two electrons (red spheres) are located in the inner orbital shell (inner blue region) and six are located in the outermost (valence) orbital shell (outer blue region).

valence shell. Except for the **transition metals** (Groups 3 through 12), the valence number for an element can be determined on the Periodic Table by reading the last digit of the group number (vertical columns). For example, hydrogen (Group 1) has one valence electron; oxygen (Group 16) has six valence electrons. The transition metals have a variable number of valence electrons so the number cannot be read directly from the Periodic Table.

The number of valence electrons for hydrogen and oxygen is key to understanding how the atoms bond to create water. Hydrogen has only one electron, so it can only have a valence number of one (Figure 2.3). Oxygen, however, has eight electrons, but it is listed under Group 16 on the Periodic Table, so it has six valence electrons (Figure 2.4). This tells us that all the electrons are not in the same region surrounding the nucleus. Six are in the outermost region, and two are closer to the nucleus, so oxygen has two electron orbital shells since the inner shell can contain a maximum of two electrons.

Each electron shell corresponds to a **principal quantum number** describing the electron's energy state, and each shell can contain a finite maximum number of electrons (e_{max}). For example, an atom with two electron shells has a maximum of eight electrons as given by the following general equation (Eq. 2.1):

$$e_{max} = 2(n)^2 \tag{2.1}$$

where n is the integer electron shell number. Note that in Eq. (2.1), e_{max} will always be in multiples of two since electrons tend to occur in pairs. In the case of oxygen $n = 2$, $e_{max} = 8$, but there are only six electrons in the valence shell. This means that the valence shell is "open" and available to share or exchange two electrons with another atom(s). In contrast, an atom with a "closed" electron shell has its valence shell occupied with the maximum number of electrons, so it is stable and chemical bonding is not likely. In general, elements common in biogeochemical cycles such as carbon, nitrogen, and oxygen tend to prefer a maximum of eight valence electrons and will bond with other atoms to ensure that eight valence electrons are always present; this is known as the **Octet Rule**.

Knowing the number of electrons an element has and how many are in the valence shell tells us how chemically reactive the element is. For example, the gas neon (atomic number 10) has ten protons and electrons. Neon is listed under Group 18, so it has eight valence electrons and two orbital shells. Using Eq. (2.1), $e_{max} = 2(2)^2 = 8$, the maximum number of valence electrons. Since eight valence electrons are already present, neon has a closed valence shell. Neon is an example of the noble or inert gas that is stable and chemically inert because its valence shell is occupied with the maximum number of electrons. Hydrogen, on the other hand, has a single electron shell with space for two electrons ($e_{max} = 2(1)^2 = 2$). Because it has only one, it is chemically reactive and "open" to share an electron and form a bond with another atom or molecule. Oxygen has two vacant electron spaces in the outer ring, making it also chemically reactive and likely to bond with another atom or molecule.

2.5 Types of Chemical Bonds

The Ancient Greek philosopher **Aristotle** described fire, air, earth, and water as the four elements from which all substances were derived. As of today, 118 elements are described in the Periodic Table, and from the bonding of these elements, through the formation of chemical bonds, all substances are derived. These chemical reactions can join atoms to create new molecules, such as when oxygen and hydrogen combine to create water. There are three general types of bonds between atoms resulting in formation of different species or compounds: metallic, ionic, and covalent.

2.5.1 Metallic Bonding

Around 95 of the 118 known elements (about 80 percent) are broadly classified as metals. Metals have certain characteristic properties: high density, high melting and boiling temperatures, and being excellent conductors of heat and electricity. Many of these properties are the result of the bonding between metal elements known as **metallic bonding**. When several metal atoms combine, they share or "pool" their valence electrons away from the nucleus in a delocalized manner (Figure 2.5). The **electron sea model** conceptualizes these electrons moving freely, being shared amongst the many atoms, thus forming several bonds and explaining why metals are such good electrical conductors. This metallic bonding through a shared pool of valence electrons also explains how different metal atoms can be readily combined to form **alloys** (a metal comprising two or more elements), such as bronze (copper and tin), or brass (copper and zinc).

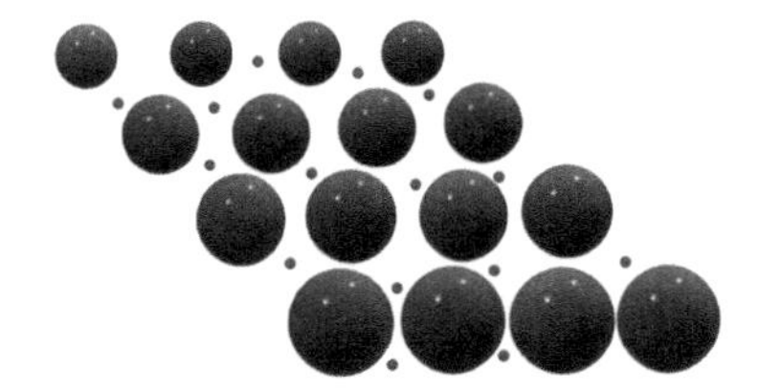

Figure 2.5 Metal atoms (dark spheres) share electrons (red spheres) in a metallic bond.

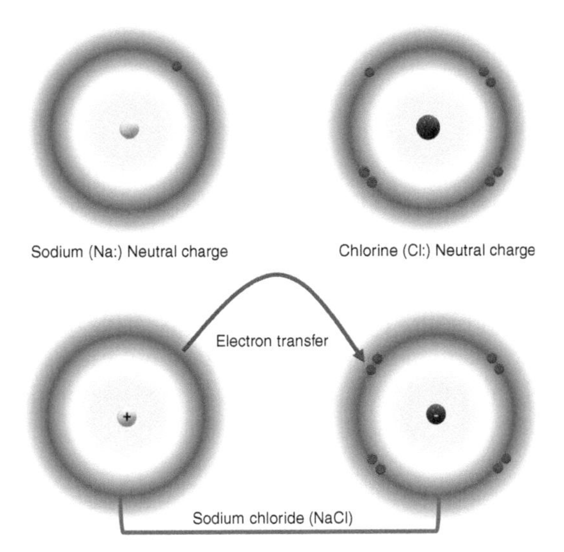

Figure 2.6 Ionic bonds involve a transfer of a valence electron from a metal to a nonmetal atom. In this example, the transfer of an electron from sodium to chlorine results in the formation of two ions and the common table salt compound, sodium chloride. For simplicity, only the valence electrons are shown.

2.5.2 Ionic Bonding

There are 17 nonmetal elements in the Periodic Table, including oxygen and hydrogen. These nonmetal atoms can form **ionic bonds** with metal atoms through the transfer of valence electrons between atoms. An **ion** is an atom or molecule with a net electrical charge due to the gain (an **anion** has a net negative charge) or loss (a **cation** has a net positive charge). In contrast to metals, nonmetals have a low density, low melting and boiling temperatures, and are poor conductors of heat and electricity. Other than for the inert noble gases in the nonmetal group, these atoms have openings in their valence shell to exchange or share valence electrons with other atoms. Examples of compounds formed through an ionic bond of a metal to a nonmetal include sodium chloride (table salt; NaCl), potassium chloride (KCl), and magnesium oxide (MgO).

In the formation of an ionic bond, a valence electron is transferred from the metal to the nonmetal atom. For example, a sodium atom (atomic number 11) has only one electron in its valence shell. This unpaired valence electron is readily donated, and when donated to a chlorine atom (for example), the sodium atom now has a positive charge since there are still 11 protons but now only 10 electrons. The sodium atom becomes a positively charged ion. The chlorine atom, by receiving an electron from the sodium atom, becomes a negatively charged ion, since it now has

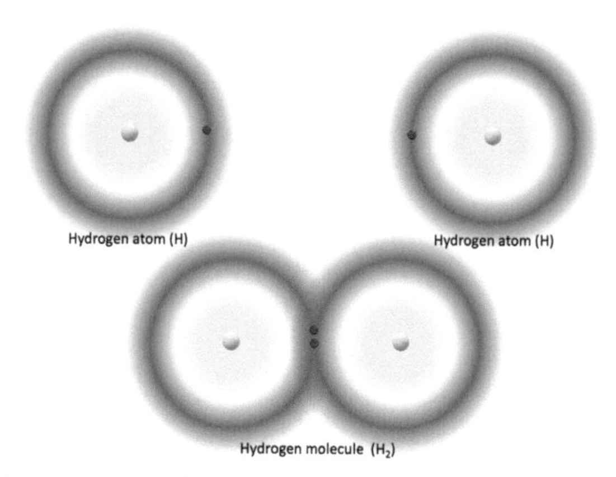

Figure 2.7 Covalent bond between two hydrogen (H) atoms to form molecular hydrogen (H_2) is achieved through the sharing of one valence electron (red sphere) from each atom (white sphere is the proton). This is a single covalent bond.

18 electrons and 17 protons. The resulting NaCl is stable because the combined total number of protons is equal to the total number of electrons, but each atom is now an ion, hence the term ionic bonding (Figure 2.6). Most compounds formed with ionic bonding easily dissolve in water.

2.5.3 Covalent Bonding

Since neither oxygen nor hydrogen are metals, metallic or ionic bonding cannot explain how they form water. Instead of bonding through the sharing of electrons displaced from their nucleus (metallic bonding) or through the donation of electrons from one atom to another (ionic bonding), **covalent bonding** involves a sharing of valence electrons. Covalent bonds are common in the nonmetal atoms when the number of valence electrons is fewer than the maximum possible in the valence shell, e_{max}. For example, hydrogen has one valence electron in its one electron shell, but it has the capacity for two. Bonding to another atom with a similar deficit in valence electrons by the sharing of a valence electron satisfies the valence electron deficiency for both atoms. In such a bonding arrangement, the molecule is stable since its valence electron capacity is filled.

Conceptually, pairs of valence electrons are shared between the two atoms, so when bonded, both have a filled valence shell. The two hydrogen atoms (H) covalently bond to form the hydrogen molecule H_2 (molecular hydrogen or hydrogen gas) in what is referred to as a **single covalent bond** since one electron pair is shared (Figure 2.7). Similar for oxygen; an oxygen atom lacks two valence electrons. The sharing of two electron pairs results in the **double covalent bond** (four electrons in total) creating molecular oxygen, O_2.

2.5.4 Polarity and Covalent Bonds

Although valence electrons in a covalent bond are shared between two atoms to fill the valence shells, the probable location of those electrons may not be distributed equally around the nucleus. When two of the same atoms bond, such as H_2 or O_2, a **nonpolar covalent bond** forms. Since the

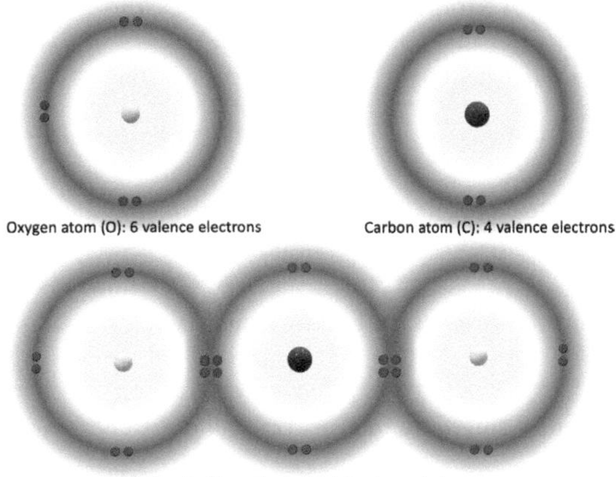

Figure 2.8 Covalent bonding between one carbon (C) atom and two oxygen (O) atoms to form carbon dioxide (CO_2). Since two pairs of electrons are shared in each carbon–oxygen bond, there are two double covalent bond. Since the arrangement of the carbon molecules bonded to oxygen is linear (symmetrical), no polarity develops; this is a nonpolar bond. In this arrangement, each atom now has eight valence electrons, satisfying the Octet Rule. For simplicity, only valence electrons are shown.

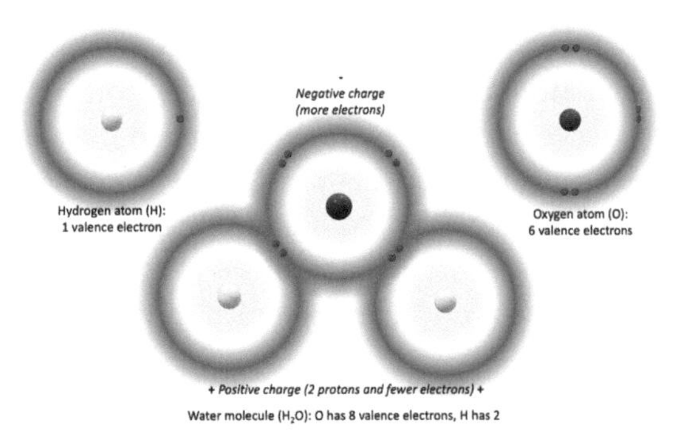

Figure 2.9 Two single covalent bonds between hydrogen atoms (H) and one oxygen atom (O) to form water (H_2O). Since the arrangement of the hydrogen molecules bonded to oxygen is nonlinear (bent), polarity develops. In this arrangement, each atom has a full (closed) valence shell, so the molecule is stable. For simplicity, only the valence electrons are shown.

valence electrons are shared equally between each atom, there is no net electrical charge, or polarity, across the molecule. A linear arrangement of the atoms across the molecule forms. This is also the case when two oxygen atoms join carbon to form carbon dioxide (Figure 2.8).

When there is a difference in the number of valence electrons in atoms that covalently bond, a **polar covalent bond** forms. In this case, the net electrical molecular charge is zero, but polarity develops across the molecule. The side of the molecule having a larger share of the valence electrons has a slight negative charge, leaving the protons on the other side to provide a slight positive charge. The electrons are shared, but not equally around the nucleus. Such is the case for the water molecule (Figure 2.9), and it has several implications for the properties of water (see Chapter 3).

A summary of the three types of chemical bonds is given in Table 2.3.

Table 2.3 **Summary of atomic bond types**

Bond type	Atoms (elements)	Valence electrons	Example
Metallic	Metal-to-metal	Shared amongst atoms	Copper (Cu) + tin (Sn) = bronze (CuSn)
Ionic	Metal-to-nonmetal	Transferred from the metal to the nonmetal atom resulting in ion formation	Sodium (Na) + chlorine (Cl) = sodium chloride (table salt) ($Na^{+}Cl^{-}$)
Covalent*	Nonmetal-to-nonmetal	Shared between two or more atoms	Hydrogen (H) + oxygen (O) = water (H_2O)

*Covalent bonds can be single, double, or triple, and with nonpolar or polar electrical charge.

2.6 Bonding between Oxygen and Hydrogen: Water

Now that we know hydrogen's and oxygen's atomic structure and the nature of chemical bonding between them, the water molecule can be described. The water molecule is the result of two single covalent bonds between an oxygen atom and two hydrogen atoms. Each hydrogen atom shares one electron with the oxygen atom, so the valence electron deficit for each hydrogen atom is fulfilled. The oxygen atom also has a filled valence shell since it now has eight valence electrons, with one supplied from each hydrogen atom (Figure 2.10). Named after the chemist **Gilbert Lewis** (1873–1946), a **Lewis diagram** is used to illustrate the bonding and geometry between atoms, as shown in Figure 2.10. In a Lewis diagram, each dot around the chemical symbol represents a valence electron, and each solid line represents a pair of valence electrons covalently bonded to the other element (each line represents one covalent bond).

In water, the hydrogen atoms are not located at opposite sides of the oxygen atom, as one might expect. The oxygen atom has four pairs of electrons (three valence pairs) and since electrons all have a negative charge, they repel each other. In three dimensions, a tetrahedron arrangement (with an angle between bonds of 109.5°) provides the maximum separation of electron pairs. When the oxygen atom bonds with two hydrogen atoms, the bond angle between the hydrogen–oxygen–hydrogen atoms decreases to roughly 104.5° to provide maximum electron separation as a result of additional negative charges imposed by the hydrogens' electrons. Oxygen, with its greater share of electrons, has a much larger electronegative charge than either hydrogen atom, so a negative charge develops near the oxygen atom, and a positive charge near the hydrogen atoms. This bent arrangement results in the development of polarity across the water molecule, in contrast to the linear arrangement in carbon dioxide, which has no polarity (see Figures 2.8–2.10).

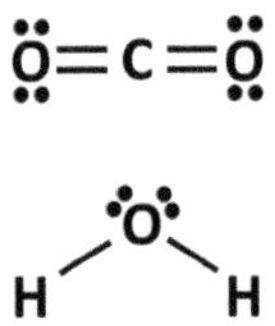

Figure 2.10 Lewis diagrams for carbon dioxide (CO_2; top) and water (H_2O; bottom). Each dot represents one valence electron. Each solid line represents a pair of shared electrons (a covalent bond), and the angle between the bonds is shown by the angle of the solid lines.

2.7 SUMMARY

Experiments conducted during the eighteenth century involving the combustion of metals resulted in the discovery of oxygen ("fire air") and hydrogen ("water generator"). Key contributors to these discoveries were Carl Scheele, Antoine-Laurent Lavoisier, Joseph Priestley, Henry Cavendish, and Elizabeth Fulhame. During the late nineteenth and twentieth centuries, the basic structure of the atom was proposed by Ernest Rutherford, Hans Geiger, and Ernest Marsden. Their experiments revealed that the bulk of an atom's mass resides in the nucleus that consists of positively charged protons and neutrons without an electrical charge but with a mass equal to protons. Subsequent research by Niels Bohr revealed that negatively charged electrons are found around the nucleus in a cloud-shaped pattern. The number of protons defines the element, with the number of electrons equal to the number of protons (so the net electrical charge is neutral), but the number of neutrons can vary resulting in isotopes of the same element.

The number of electrons in the outer or valence shell of an atom plays an important role in chemical bonding between atoms. Metallic bonding occurs between atoms in the metal group of the Periodic Table, where many electrons are shared between atoms. Ionic bonding occurs between atoms in the metal and nonmetal groups, where valence electrons are transferred from one atom to another. Many compounds formed by an ionic bond such as sodium chloride easily dissolve in water. Covalent bonds form between atoms that share valence electrons is common in atoms that have an equal deficit in the number of valence electrons.

The bonding of two hydrogen atoms to one oxygen atom to form the water molecule is in the form of two single covalent bonds. Although this is a stable arrangement, the position of the valence electrons is not symmetrical. There are more electrons clustered around the oxygen atom, resulting in a slight negative charge at the oxygen side and a slight positive charge from the protons at the hydrogen side of the molecule. This asymmetric arrangement results in polarity across the molecule, referred to as a bipolar charge.

The individual properties of hydrogen and oxygen, and the additional properties that result from the ability of these three atoms to bond together to form water, are fundamental in understanding the unique properties of water that are described in Chapter 3.

2.8 QUESTIONS

2.1 Describe how Lavoisier's results from his phosphorus combustion experiments that conflicted with the predictions based on Phlogiston Theory laid the foundation for the discovery of oxygen.

2.2 What were the contributions of Elizabeth Fulhame to our understanding of the properties and composition of water? Research and summarize the contribution(s) of another individual from an underrepresented group that provided important research leading to our current understanding of the properties of water.

2.3 Scheele, Lavoisier, and Priestley all contributed to the discovery of oxygen. If the discovery should be credited to only one person, which of these three do you think it should be, and why?

2.4 Imagine you could travel back in time to the laboratories of Scheele, Lavoisier, Priestley, and Fulhame. Which modern instrument or technique would you bring with you to demonstrate to them other elements and their properties that were not yet discovered?

2.5 Describe how Niels Bohr modified Rutherford's model of the atom.

2.6 Discuss why valence electrons play such an important role in chemical properties of an element, including whether the element is stable or unstable.

2.7 Sodium (Na) appears under Group (column) 1 and Period (row) 3 in the Periodic Table. How many electron shells and valence electrons does sodium have? What is the maximum number of electrons sodium can have?

2.8 Describe why a covalent bond between hydrogen and oxygen forms water, instead of a metallic or ionic bond.

2.9 Although impossible, imagine that the bond between hydrogen and oxygen to form water was not a covalent bond, but instead a metallic or ionic bond. Describe how the characteristics of the water molecule would change with these different types of bonds.

2.10 If water has an equal number of protons and electrons, why does one side of the molecule have a weak positive electrical charge and one side a weak negative electrical charge?

REFERENCES

Cassebaum, H. and Schufle, J. A. (1974) 'Scheele's priority for the discovery of oxygen', *Journal of Chemical Education*, 52, p. 442.

Lucibella, M. (2010) 'August 1774: Priestley isolates a new "air", leading to discovery of oxygen', *APS News*, 19(August–September), p. 2.

Vera, F., Rivera, R. and Núñez, C. (2011) 'Burning a candle in a vessel, a simple experiment with a long history', *Science and Education*, 20, pp. 881–893. doi: 10.1007/s11191-011-9337-4.

West, J. B. (2014a) 'Carl Wilhelm Scheele, the discoverer of oxygen, and a very productive chemist', *American Journal of Physiology – Lung Cellular and Molecular Physiology*, 307, pp. L811–L816. doi: 10.1152/ajplung.00223.2014.

West, J. B. (2014b) 'Henry Cavendish (1731–1810): Hydrogen, carbon dioxide, water, and weighing the world', *American Journal of Physiology – Lung Cellular and Molecular Physiology*, 307, pp. 1–6. doi: 10.1152/ajplung.00067.2014.

3 The Properties of Water

Key Learning Objectives

After reading this chapter, you will be able to:

1. Summarize how the attraction between water molecules through the hydrogen bond results in the cohesive and adhesive properties of water.
2. Define surface tension and explain why it develops at the air–liquid water interface and how it influences aspects of water cycling.
3. State which compounds best dissolve in water, which best dissociate in water, and why water is referred to as the universal solvent.
4. Explain why liquid water reaches its maximum density at roughly 4 °C and provide examples of the effects of this property.
5. Discuss thermal properties such as specific and volumetric heat capacity, and how the large values for liquid water affect heating and cooling rates.
6. Describe how the state of water in liquid, solid, or vapor forms varies with temperature, pressure, and the solute concentration.

3.1 Introduction

The individual properties of hydrogen and oxygen atoms, and the properties when they combine to form water and interact with other molecules, affect Earth's **abiotic** and **biotic** environments profoundly. Properties such as density, heat capacity, and the release or requirement of energy to change between liquid, solid, and gas mean that water molecules have many unique characteristics. These include a unique change in liquid water's density with temperature, cohesion between water molecules and adhesion to other molecules, and the fact that many compounds dissolve and dissociate in liquid water. The chemical properties at the molecular level provide water with the ability to continuously cycle between liquid, solid, and vapor forms and therefore influence both the landscape and life on Earth.

An understanding of the properties of water underpins topics from how water formed on Earth to weathering and erosion, to how sweating keeps us cool. This chapter reviews hydrogen attraction, or hydrogen bonding, between water molecules. Hydrogen bonding, with the dipole electrostatic charge on water molecules, is used to explain why liquid water clings to itself and other surfaces. Why so many compounds are isolated or broken down once dissolved in water, and why the density of water does not simply decrease with temperature, is explained.

Finally, the thermal properties of water and the temperature and pressures associated with water states are reviewed.

3.2 Hydrogen Attraction between Water Molecules

The attraction between water molecules comes from the difference in electrostatic charge across the water molecule resulting from the covalent bond between hydrogen and oxygen. The hydrogen atom is electrically neutral. Since a pair of valence electrons is more stable than just one, hydrogen readily gives up its single electron to become a net electrically positive atom (the cation H^+) or takes another electron to become a net electrically negative atom (the anion H^-).

When the hydrogen atom in a molecule has a net positive charge, it readily electrostatically bonds to another molecule that is more electronegative, thus forming a **hydrogen bond**. Although referred to as a bond, this is not equivalent to covalent, ionic, or metallic bonds (Chapter 2) between atoms where electrons are exchanged or shared. Hydrogen bonds result from electrostatic attraction between molecules, thus are weaker than the covalent bonds between atoms that form the water molecule. The oxygen side of one water molecule is more electronegative than the hydrogen side of another. Therefore, the oxygen atom of the water molecule can bond to the hydrogen atom of another, and each hydrogen atom can bond to the oxygen atom of a neighboring water molecule. Each water molecule can form hydrogen bonds to three other water molecules, creating a group of four (Figure 3.1). In liquid water, these weak hydrogen bonds break and reform often and easily. The hydrogen bonding of two water molecules $(H_2O)_2$ is known as a **water dimer** (dimer refers to the combination of two molecules), and this simple structure has been used in many studies to examine the properties of the hydrogen bond (e.g., Buckingham, 1991).

As mentioned, the hydrogen attraction (bond) is much weaker than covalent, metallic, or ionic bonds. The covalent bond within the water molecule (**intramolecular**) between hydrogen and oxygen is strong: 492 kJ per mol. In comparison, the hydrogen bond between liquid water molecules (**intermolecular**) is much weaker: only 23.3 kJ per mol (Chaplin, 2010), roughly 5 percent of the strength of the intramolecular covalent bond (Silberberg and Amateis, 1996).

The weak intermolecular hydrogen bonds are flexible to bend and deform to form liquid water yet have sufficient rigidity and strength to allow water molecules to bond together in a solid structure: ice. Or, with sufficient energy to break the hydrogen bonds, water may

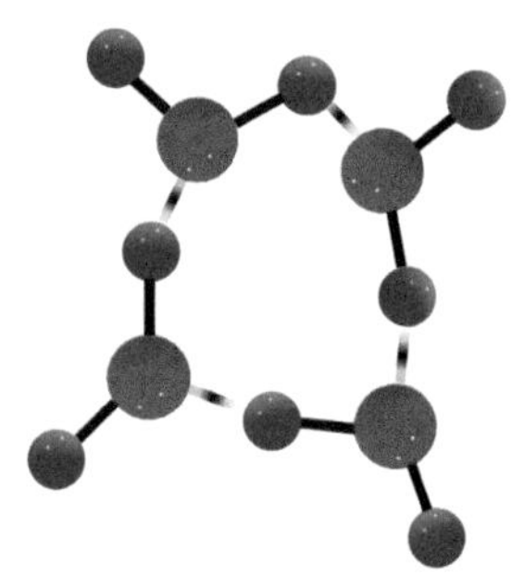

Figure 3.1 Hydrogen bonds (faded lines) connecting four water molecules to create liquid water. Oxygen atoms (red) are covalently bonded (solid lines) to hydrogen atoms (blue).

evaporate or sublimate, liberating individual water molecules to the gaseous form from their liquid (**evaporation**) or solid forms (**sublimation**), thus allowing for Earth's water cycling. In addition, the combination of bonding within and between water molecules results in unique physical and chemical properties that affect Earth's abiotic and biotic environments.

3.3 Cohesion and Adhesion

As a result of the water molecule's dipole electrostatic charge and hydrogen bonding between molecules, liquid water molecules are attracted to other water molecules (**cohesion**) or other molecules (**adhesion**). A common example illustrating water's cohesion and adhesion properties is the curved top of the upper water surface in a narrow clear-glass cylinder, known as the **meniscus** (Greek meaning crescent) (Figure 3.2). The water molecules adhere to the glass, which contains the dipole silicon dioxide (or silica, SiO_2), silicon (Si) covalently bonded to two oxygen atoms. The positive hydrogen atoms in the water molecule are electrostatically attracted to negative oxygen atoms in the silicon dioxide, thus creating the hydrogen bond. This results in water clinging to the sides of the glass and curving upwards. The smaller the diameter of the glass cylinder, the more curved the surface becomes. Owing to the dipole nature of water, nearly any dipole surface in contact with liquid water will allow a meniscus to form.

When the adhesive force attracting water to the sides of the vessel exceeds the cohesive force within the liquid water, water is pulled upwards against the sides of the glass in a concave-up shape. The height to which water clings to the sides is determined by a balance between the cohesive force between molecules and the force of gravity pulling them down. In contrast,

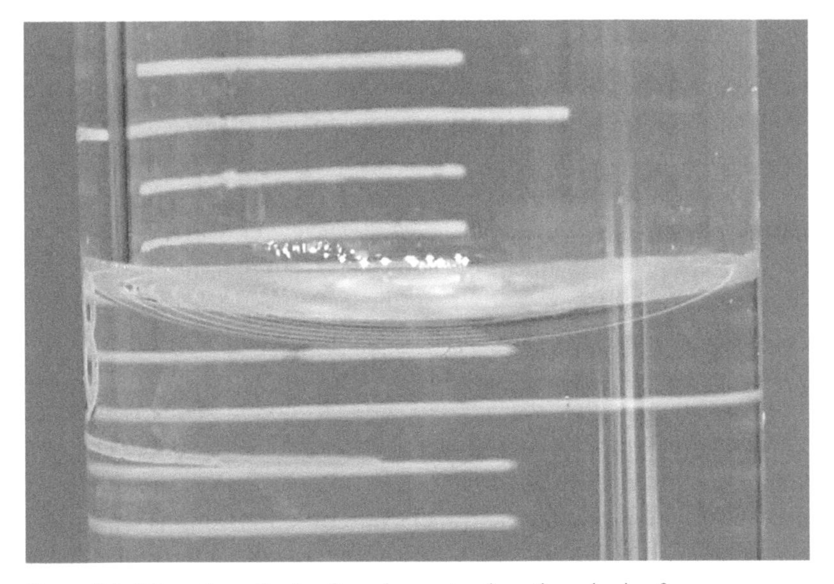

Figure 3.2 When the adhesive force is greater than the cohesive force, a "concave-up" shape results. Photo credit: P. D. Blanken.

Figure 3.3 When the cohesive force is greater than the adhesive force, a "concave-down" shape results. Photo credit: P. D. Blanken.

when the cohesive force exceeds the adhesive force, the fluid is pulled downwards against the sides in a concave-down shape. Such is the case for mercury in a glass cylinder or a drop of water on a flat surface such as a leaf (Figure 3.3).

At the surface where the liquid water encounters air, the abrupt transition from the internal cohesive force to its absence gives water its **surface tension** property. Within the liquid water, there are many hydrogen bonds since each water molecule is surrounded by others, cumulatively providing a strong cohesive force. At the water's surface, there are fewer adjacent molecules, so the attractive cohesive force is directed sideways and downwards, resulting in a drop of water's spherical concave shape (Figure 3.3). The water molecules at the surface that are pulled sideways and downwards must be replaced by other molecules. This requires the breaking of cohesive hydrogen bonds, and the energy required to do so determines the surface tension in units of joules per square meter of water surface area (J m^{-2}). For a fluid to be stable in terms of energy, the surface area tends to be as small as possible. For water, surface tension at 20 °C is 7.28×10^{-2} J m^{-2} (surface tension decreases with temperature) which is high compared with most fluids, owing to the multiple hydrogen bonds between molecules (Silberberg and Amateis, 1996). Liquid mercury has a much higher surface tension (48.6×10^{-2} J m^{-2}) due to its strong metallic bonds.

There are practical examples of the importance of adhesion, cohesion, and the related surface tension properties of water in the water cycle. One example relates to the ascent of water in tall, vascular plants such as trees (Chapter 11). Trees do not have pumps, like the human heart, to move water from the roots to the leaves, and the force exerted by the weight of the atmosphere at sea level can only raise a column of water to roughly 10 m. How then can water ascend to the tops of trees that are taller than this? This is detailed in Chapter 11, but Dixon and Joly's (1894) "On the ascent of sap" gives a good introduction which formed the basis for the **Cohesion–Tension** Theory. The theory states that within a leaf, evaporation within stomatal cavities results in formation of a meniscus and surface tension at the air–water interface. This increase in surface tension at the site of evaporation within the leaf (transpiration) results in a "chain reaction" through the entire water column within the xylem (vascular tissue that conducts water; Chapter 11). Owing to cohesion, surface tension is transferred through the continuous water column within the xylem, all the way down the roots. If the tension within the water column exceeds the cohesive force, the water molecules separate, and an air pocket forms. The xylem tissue would thereafter be incapable of moving water. According to this theory, the adhesion of water to the xylem tissue coupled with hydrogen bonding acting through cohesion explains how water ascends in tall trees. This theory has provoked debate (e.g., Zimmermann et al., 2004) but support remains widespread (e.g., Brooks, 2004). Water flow through vegetation is discussed in Chapter 11 and transpiration in Chapter 12.

Another example of the consequences of hydrogen attraction between water molecules is precipitation formation. Chapter 7 discusses how very small droplets of water can form in the atmosphere while retaining their spherical shape, allowing the formation of precipitation. Surface tension plays a key role as it influences the droplet's size between competing rates of evaporation and condensation. The droplet's surface tension increases as its diameter decreases, and this has a direct effect on evaporation rate, relative humidity, and whether the droplet will become large enough to fall as precipitation. If it were not for hydrogen's ability to covalently bond to oxygen to form the water molecule, and to attract adjacent water molecules to form liquid water with a high surface tension, the Earth's water cycle (and associated abiotic and biotic systems) would be completely different. Trees could not be tall, and rain could not form.

3.4 The Universal Solvent

Liquid water can isolate and break apart some molecules, so it is often known as the "universal solvent". The terms solute, solvent, and solution are commonly used, and are demonstrated when we mix sugar into tea or add salt to boiling water. The **solute** is the substance being dissolved: it is often a solid (e.g., sugar or salt), but can be solid, liquid, or gas. The solute occurs in much smaller quantities than the **solvent** (the substance the solute dissolves into), which can also be solid, liquid, or gas. When the solute is dissolved in the solvent, a **solution** is produced, in which the solute's molecules or atoms are uniformly distributed, surrounded by the solvent.

What determines whether something dissolves in something else? Clearly table sugar or sucrose (a polar molecule consisting of a molecule of glucose with a strong covalent bond to a molecule of fructose) dissolves in liquid water, as does table salt (a compound consisting of a sodium ion ionically bonded to a chloride ion). When either sugar or salt is stirred into liquid water in small quantities, the solids apparently disappear. **Compounds** (atoms from more than one element bonded together) such as sugar or salt that readily dissolve in water are **hydrophilic** (water loving), whereas compounds that do not, such as oils and fats, are **hydrophobic** (water fearing). Even though both sugar and salt are hydrophilic, a difference between them develops as they dissolve.

Sugar consists of **dipole molecules**, meaning that one side of the molecule has a slight positive electrostatic charge and one side a slight negative electrostatic charge. The overall charge is zero. Dipole molecules are soluble in solvents that are also dipole molecules, such as water (Chapter 2). Given the opportunity to be electrostatically attracted to either a positive or negative charge, water molecules surround the individual molecules, forming a **hydration shell** around them keeping them isolated and dispersed in water, thus forming a solution. When sugar dissolves in water, the molecule does not change. The glucose is still covalently bonded

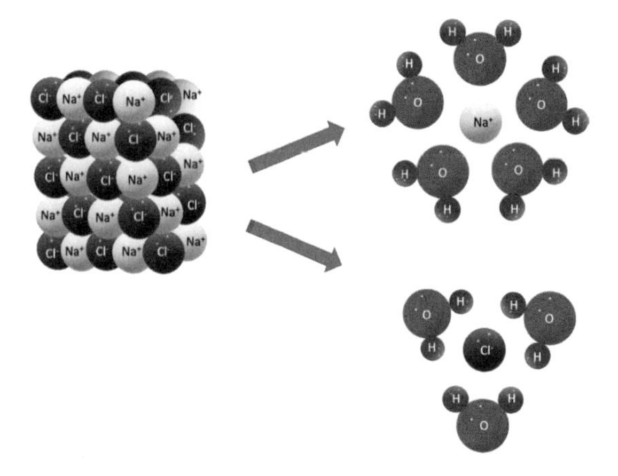

Figure 3.4 NaCl crystal (left) dissolving and dissociating when liquid water is added. The sodium (Na^+) and chloride (Cl^-) ions are surrounded by water molecules due to the polarity of the molecules and the charge on the ions.

to fructose, with each sucrose molecule dispersed and separated from each other. When solute molecules separate in this way, they **dissolve** in the process of **solvation** (**hydration** if water is the solvent).

When salt dissolves in water, however, it undergoes an additional change. Sodium chloride (NaCl) forms through an ionic bond, not a covalent bond as in the case of sucrose. When placed in liquid water, the ionic bond between the sodium and chloride is broken and the molecules split into sodium and chloride atoms, each with a net charge (ions). The sodium **cation** (positively charged) and the chloride **anion** (negatively charged) are both surrounded and isolated by water molecules in a hydrogen sphere (Figure 3.4). The salt not only dissolves but also **dissociates** into atoms. When these atoms are ions, the solution is a good electrical conductor known as an **electrolyte**. If the compound formed through ionic bonds, dissociation is also likely. Dissociation is reversible, meaning that the dissociated atoms can reform into their original structure. For example, salt can be recovered after liquid water evaporates.

Not all compounds dissolve or dissociate equally in water, and some do not dissolve at all (**insoluble**). The **solubility** expresses how the properties of the solute and solvent determine how well a compound dissolves, expressed in grams (or moles) of solute per mass (or per volume) of solvent. (A mole is a unit based on Avogadro's number, the number of particles found in 12 grams of carbon-12, used to convert between the number of atoms or molecules and grams.) The maximum solubility defines when saturation is reached. As described above, dipolar solute molecules dissolve well in other dipolar molecules (water), so the aphorism "like dissolves like" is commonly used. To dissolve, the electrostatic attraction of the solvent must be greater than the intermolecular (to dissolve) or intramolecular bonds (to dissociate) of the dipolar molecule. A compound formed through strong metallic bonds, in addition to being mostly nonpolar, far exceeds that attraction to water molecules, so metals do not dissolve in water, yet water often participates in chemical reactions (such as rusting) to form new compounds. A compound formed through dipolar covalent bonds may dissolve but the internal bonds are usually too strong for the molecule to dissociate. The weaker ionic-dipole compounds may be weak enough for water to break them apart into individual atoms. Several other properties also determine solubility, such as temperature and pH, but the key point to remember is that most dipoles will easily dissolve in water since it too is a dipole.

3.5 Density

The **density** (ρ, defined as mass (m; kg) divided by volume (V; m^3); $\rho = m/V$, with units of kg m^{-3}) of liquid water does not always increase with temperature. Mass is simply the amount of the substance present, and this does not change with temperature unless chemical reactions occur that transfer mass through the creation of new molecules. The volume of space that the substance occupies, however, can change with changes in temperature, especially for a gas but also for liquids or solids. Most objects increase their volume as temperature increases, therefore decreasing the object's density. The thermal expansion of steel that may result in the bending and buckling of railroads on very hot days shows that even high-density solids are not immune to temperature-induced density changes. Logically, the greater the molecular motion as indicated by the object's temperature, the larger the volume of space they occupy, hence the decrease in density as temperature increases. Liquid water, however, does not follow this temperature–density relationship.

The density of liquid water increases with temperature, but only to a certain point. Once this temperature is reached, density decreases. This odd behavior is related to both the intermolecular hydrogen bonds and the shape of the molecule. The covalent bond between hydrogen and oxygen, and the resulting dipole charge, give liquid water the general properties shown in Table 3.1. The density of liquid water between roughly 0 and 10 °C is easy to remember: 1 gram per cubic centimeter (1 g cm^{-3}), equivalent to 1 gram per milliliter, or 1,000 kg m^{-3}. One liter (1000 cm^3) of water therefore weighs 1 kg. The actual value is 998.2071 kg m^{-3} for pure water, at the **normal temperature and pressure** (NTP) of 20 °C and 1 atmosphere (sea level, 101.325 kPa). In its liquid state, water's density decreases as temperature increases as in most other compounds, but only when the temperature exceeds 3.98 °C. If the temperature is below 3.98 °C, density again decreases (Figure 3.5).

Table 3.1 Various thermal, physical, and chemical properties of pure liquid water

Property	Value
Maximum density	999.975 kg m^{-3} at 3.98 °C
Boiling-point temperature	99.974 °C at 101.325 kPa
Melting or freezing-point temperature	0 °C at 101.325 kPa
Molar mass	18.01527 g mol^{-1}
pH	6.9976 at 25 °C
Specific heat – liquid	4.187 kJ kg^{-1} K^{-1} at 15 °C
Specific heat – solid	2.108 kJ kg^{-1} K^{-1}
Specific heat – vapor	1.996 kJ kg^{-1} K^{-1}
Triple-point temperature	0.01 °C
Triple point pressure	611.657 Pa
Critical-point temperature	373.946 °C
Critical point pressure	22.06 MPa
Latent heat of vaporization	2256 kJ kg^{-1} at 100 °C
Latent heat of melting	334 kJ kg^{-1} at 0 °C

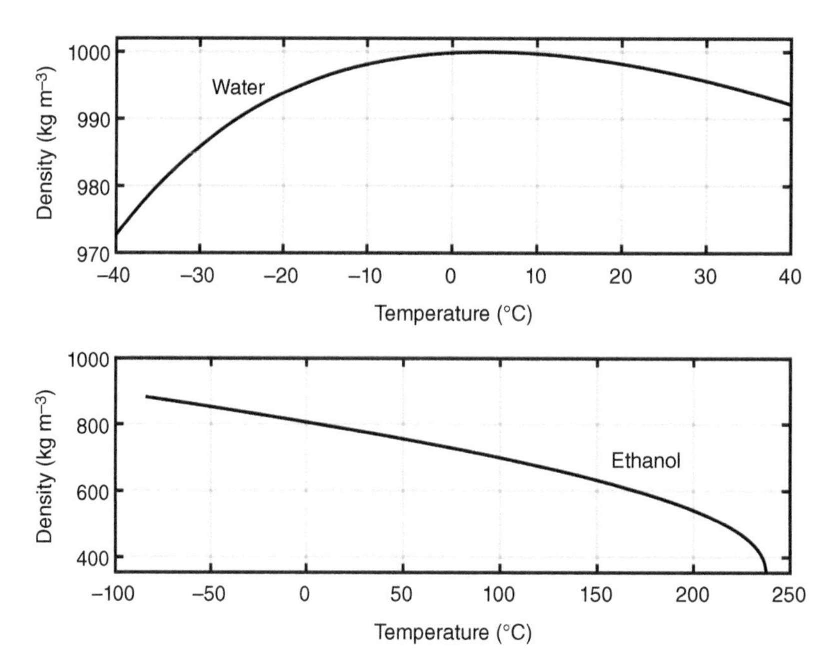

Figure 3.5 Density of pure liquid water and ethanol as a function of temperature. Although both have polar molecules with hydrogen-oxygen bonds, only water shows the unique density variation with temperature.

Hydrogen bonding between water molecules can help explain this peculiar behavior. If it were hydrogen bonding alone, then other hydrogen-bonded liquids such as ethanol (C_2H_6O), or any molecule with a hydrogen atom bonded to an oxygen or nitrogen atom, should show a similar temperature–density relationship, but they do not (Figure 3.5). Ethanol's density decreases as temperature increases, despite its hydrogen bonds. As a result of the orientation of the hydrogen bonds, water has its maximum density at 3.98 °C. Above this temperature, increased molecular motion separates molecules, thus decreasing density. Below this temperature, the clusters of molecules begin to form with greater separation as the temperature lowers, thus decreasing density.

Water freezes when the temperature is at or slightly below 0 °C (depending on the solute concentration). When water changes from a liquid to a solid, a major change in density and its response to temperature occurs. Changes in density as water changes state between solid, liquid, and gas are a result of changes in spacing, hence volume, between water molecules (Figure 3.6). Above 100 °C at sea level, pure water as a gas (water vapor) has no hydrogen bonds between molecules, thus the molecules are spaced widely apart, are free, and the density is low (assume mass is constant). At temperatures below 100 °C, kinetic energy decreases, allowing water molecules to be close enough for intramolecular hydrogen bonding to occur. When liquid, every water molecule has at least one hydrogen bond to another water molecule, with the number of hydrogen bonds increasing as the temperature decreases (e.g., up to a cluster of four molecules; Rastogi, Ghosh and Suresh, 2011). Hydrogen bonding between

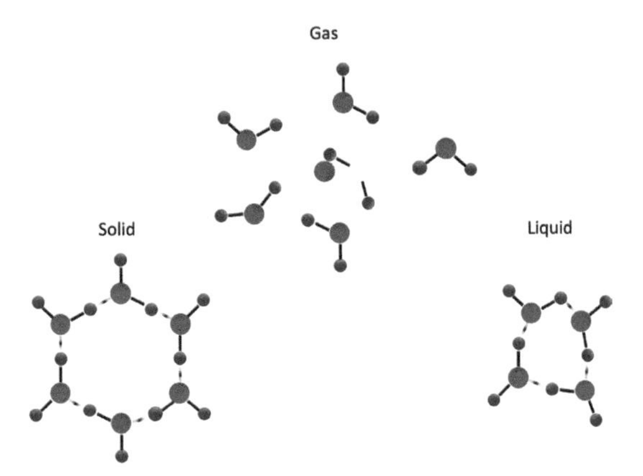

Figure 3.6 Arrangement between water molecules in vapor, liquid, and solid forms. Dark lines are covalent bonds between oxygen (red) and hydrogen (blue) atoms. Gray lines are weaker hydrogen bonds between water molecules.

molecules continues as temperature decreases to near 4 °C, decreasing the volume of water, hence increasing density. As temperature decreases from 4 to 0 °C, the volume increases (thus density decreases; Figure 3.7) as the hydrogen bonds begin to arrange water molecules in a unique lattice structure (Figure 3.8). When pure water (at sea level) freezes at 0 °C, a sharp decrease in density occurs. Further decreases in temperature tighten and shrink this lattice structure resulting in a slight increase in density.

As a result of water's unique temperature-related density changes, two properties emerge with critical implications for the role of water in abiotic and biotic environments. First, water's maximum density is achieved nearly 4 °C above freezing point (0 °C). This has profound implications for the **thermal stratification** of water bodies, which in turn has profound biological and ecological implications for aquatic systems. If liquid water had its maximum density close to 0 °C, then any vertical mixing of liquid water due to density differences would not occur. Lakes would always have warmer lower-density water atop cooler high-density water. Water rich in dissolved oxygen, nutrients, phytoplankton, and zooplankton would not mix to lower levels based on vertical water temperature gradients (Chapter 5).

Second, when water freezes, its density changes drastically (see Figures 3.7 and 3.8), decreasing 9 percent from roughly 999.8 kg m^{-3} to 917.0 kg m^{-3}. The result of this density change is that ice floats on liquid water with only 9 percent of the ice above the water surface. The biotic and abiotic implications of these density changes are significant. Imagine if water behaved like other substances and did not form a lattice structure with a lower density in its solid compared with its liquid form (only five known elements have a lower density in solid than liquid form: arsenic, bismuth, gallium, germanium, and silicon). If ice had a higher density than liquid water, ice forming on the surface of a lake would immediately sink to the bottom to the detriment of any aquatic organism below. If water freezes within plant or animal cells, its expansion is almost always very harmful (think of frostbite). At the cellular level, a suite of complex biochemical processes has evolved in plants routinely exposed to freeze–thaw cycles. Examples of freezing

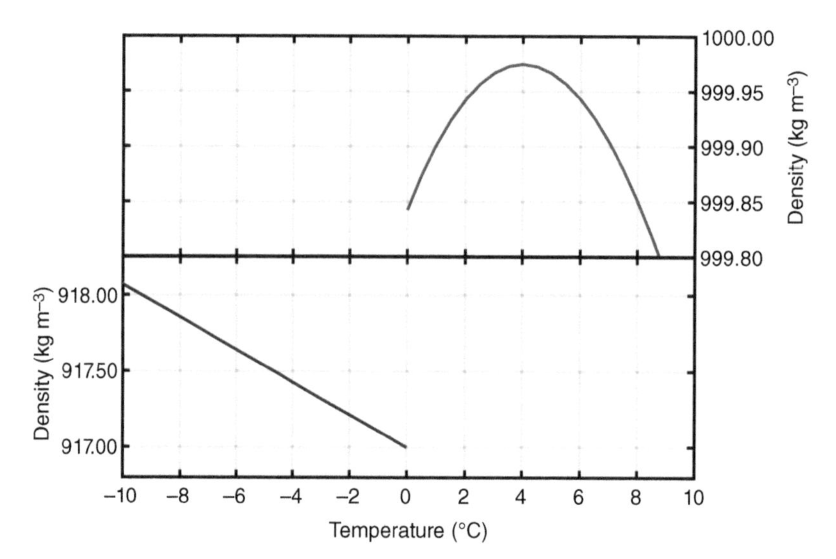

Figure 3.7 The density of water decreases abruptly as water changes state from liquid (red line) to solid (blue line).

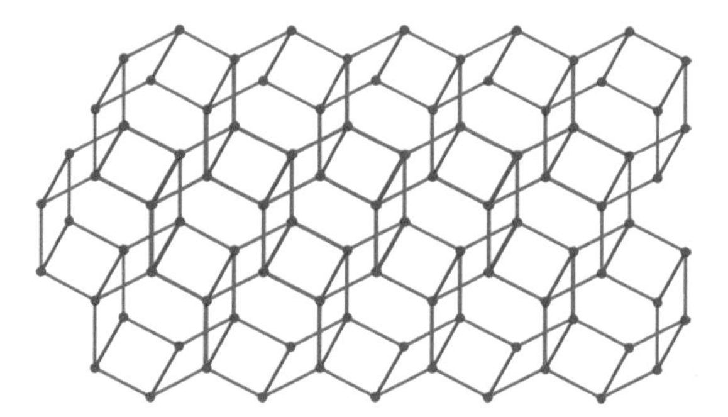

Figure 3.8 Lattice structure of ice showing the widely spaced water molecules (red circles) connected by hydrogen bonds (blue lines).

tolerance mechanisms include cells' solute concentration increasing to lower the freezing point, water being removed from inside cells, and proteins being produced to prevent the formation of ice crystals.

Weathering, erosion, and landforms are also affected by the difference in density between liquid and solid water. In fact, it was once believed that the reason Earth is not in a constant state of glaciation was because ice floats: "*The predominant fact is the floating power of ice. Hereby the water is screened from further attacks of the cold air, and dispersal is provided in the puzzling conditions of ground or anchor ice*" (Croft, 1913). The heat released when water freezes (discussed below) was also recognized as "*...an effective safeguard against sudden and extensive solidifying*" (Croft, 1913). The changes in density and volume that occur as water freezes also play a role in soil formation. Without this, the breaking down of parent material, the first step in the formation of soils, would be limited. The expansion of water when it freezes within small cracks and fissures provides sufficient force to shatter even the hardest rocks. Freeze–thaw activity can result in the mixing of soils, a process known as **cryoturbation**. In fact, entire landscapes known as **periglacial environments** contain landforms created by freeze–thaw activity (Chapter 10).

3.6 Specific and Volumetric Heat Capacity

As a consequence of water's covalent bonds between hydrogen and oxygen, water has a high **specific** and **volumetric heat capacity**. The amount of heat (energy, in units of work, joules, J) required to change the temperature by one kelvin (K) is known as the **heat** or **thermal capacity** C (J K^{-1}). Often, the heat capacity of a material is measured at a starting temperature of 25 °C and at sea-level pressure (1 atm; 101.3 kPa). The heat capacity depends on the amount of material, so thermal capacity is often expressed on a mass basis: the specific heat capacity, C_p (J g^{-1} K^{-1}). When the heat capacity is expressed on a volume basis, it is known as the volumetric heat capacity, C_v (J m^{-3} K^{-1}). Conversion from specific to volumetric heat capacity is easy when the density of the material is known.

Liquid water has one of the highest specific heat capacities of any natural substance. One gram of liquid water requires 4.187 joules of energy to raise the temperature by one kelvin (same as one degree Celsius), 4.187 J g^{-1} K^{-1} (Table 3.1). A decrease of one kelvin releases 4.187 joules per gram of water. Since liquid water has such a high density (roughly 1,000 kg m^{-3}), the volumetric heat capacity is very high, 4,186,000 J m^{-3} K^{-1}. This high specific heat and density explains why water is such a good temperature regulator. It takes a lot of energy to warm water, and a lot of energy is released when water cools. Therefore objects, living or not, that have a large volume of liquid water are buffered against fluctuating temperatures.

3.7 The Three States of Water

Water changing between liquid, gas, and solid helps to shape Earth's landscapes and influences many aspects of life. Liquid water's high density together with its high specific heat capacity tells us that a large amount of energy is required when the temperature is raised. Again, this is a consequence of hydrogen bonding between molecules. For water to change state from liquid to gas (evaporation), from solid to liquid (melting), or from solid directly to gas (sublimation) also requires energy to break the hydrogen bonds between water molecules. Such energy-requiring changes from higher organized state to lower disorganized state are known as **endothermic processes**. As the **First Law of Thermodynamics** (the conservation of energy) requires, the same amount of energy is released when hydrogen bonds form or break as water changes state from gas to liquid (**condensation**), liquid to solid (**freezing**), or gas directly to solid (**deposition**). Such energy-releasing changes of state from a lower disorganized state to a higher organized state are known as **exothermic processes**. The energy required (or released) to transition between states without changes to the molecule itself is known as **latent heat**, with "latent" meaning

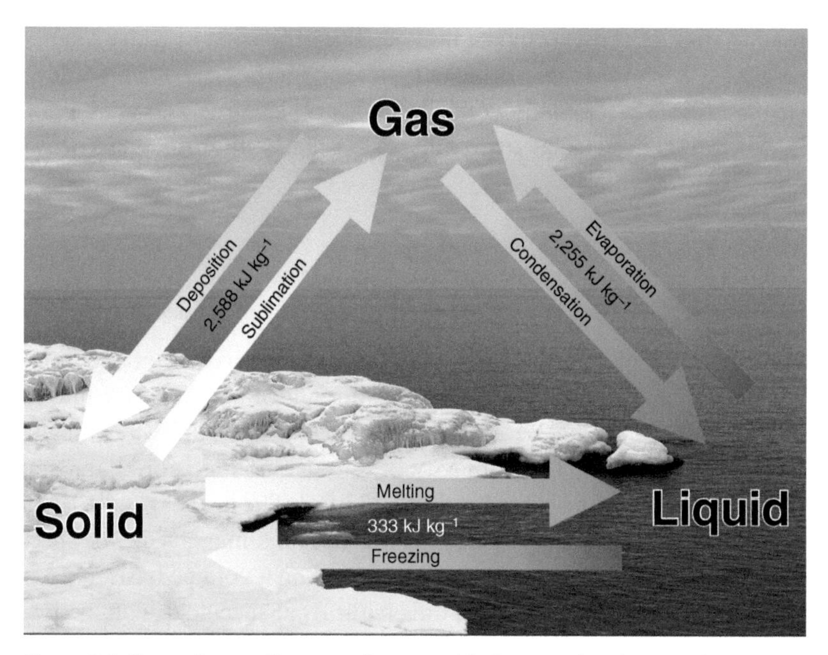

Figure 3.9 State-change diagram of water with the associated approximate latent heat required or released as water cycles between solid, liquid, and gas (vapor). Photo credit: P. D. Blanken.

"hidden" (Figure 3.9). Chemical reactions resulting in new molecules forming also require or release energy and are also referred to as endothermic or exothermic reactions, respectively (Chapter 4).

The latent heat required (or released) for water (or any molecule) to change states varies with temperature and pressure. The melting point for pure water at sea level is 0 °C, but the melting temperature decreases with pressure. Similarly, the boiling point for pure water is 100 °C at sea level. As pressure decreases, so does the boiling point. This is why it takes longer for food to cook at a high elevation. In Denver, Colorado, at 1,600 m above sea level, the reduced atmospheric pressure means that water can form bubbles with less energy input than at sea level. Therefore, the boiling point of pure water in Denver is about 95 °C. Many people think that since the boiling point is lower, food will cook more quickly, but the opposite is true. Food will take longer to cook since the water will not get above 95 °C. The energy from the stove or fire is being used to break hydrogen bonds and not to raise the water temperature.

Water is the only molecule that exists naturally on Earth in solid, liquid, and gaseous forms. When looking at a glass of water containing ice on a warm day, you can see two states of water (liquid and solid), but not the water vapor which is invisible to the human eye (Figure 3.10). Evidence of water vapor exists on the sides of the glass where small water droplets indicate **condensation**, the change of state from a gas to liquid releasing latent heat in the process. Condensation occurs when the **dew-point temperature** is reached after moist air contacts the cooler surface of the glass (Chapter 7). The ice floats in the water owing to the lower density of

Figure 3.10 Evidence of all three states of water is apparent in a glass of ice water on a warm day. Photo credit: P. D. Blanken.

water in its solid form, and only a small (9 percent) increase in the volume of water in the glass will occur when the ice melts (most of the liquid water volume is already displaced by the ice cubes; this has implications for sea level rise from melting sea ice). What might not be apparent is that the melting ice in the glass cools the liquid water not by releasing cold ice water into the liquid; the liquid temperature decreases since energy is required to break the hydrogen bonds in the lattice ice structure. This energy is supplied by the liquid water, hence the liquid cools.

3.8 State Diagram

The **state diagram** (also known as a **phase** or **triple point diagram**) shows how the state of a water (or any) molecule varies with temperature (x-axis) and pressure (y-axis) (Figure 3.11). As discussed, hydrogen bonding plays a key role in whether water exists in the solid, liquid, or gaseous states. To alter the arrangement between water molecules requires or releases energy, and this depends on two variables, temperature and pressure. The **triple point** refers to the temperature and pressure where the molecule exists in thermodynamic equilibrium simultaneously in all three states. For pure water, the triple point occurs at a temperature of 273.16 K (0.0075 °C) and a pressure of 0.611 kPa (0.006 atm). The **critical point** is the temperature and pressure where liquid and vapor states can coexist. For pure water, this occurs at roughly 647 K (374 °C) and 22,089 kPa (218 atm).

Water as the universal solvent rarely exists in its pure form. Creating an aqueous solution changes the melting and boiling points and shifts the boundaries between the states shown on the phase diagram (Figure 3.12). Dissolving and dissociating NaCl in water results in a decreased melting-point temperature and an increased boiling-point temperature. This fact is used in our daily lives. Salt (road salt crystals or calcium chloride ($CaCl_2$) de-icing or anti-icing solutions) is applied to melt ice or prevent ice from forming by interfering with the formation of the lattice structure. For cooking, table salt (NaCl) is commonly added to boiling water, raising the boiling point so the stove's energy is used to raise the water temperature instead of breaking hydrogen bonds between liquid water molecules.

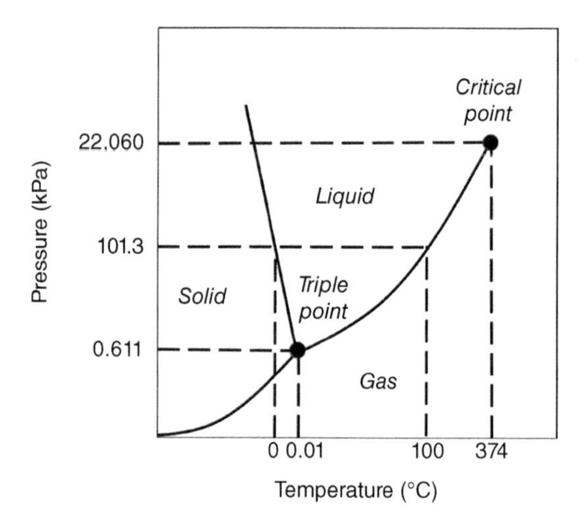

Figure 3.11 State diagram for pure water. The range for different states of water (liquid, solid, gas) corresponding to the vapor pressure and temperature are shown, as well as the triple point (all three states coexist) and critical point (distinction between gas and liquid vanishes).

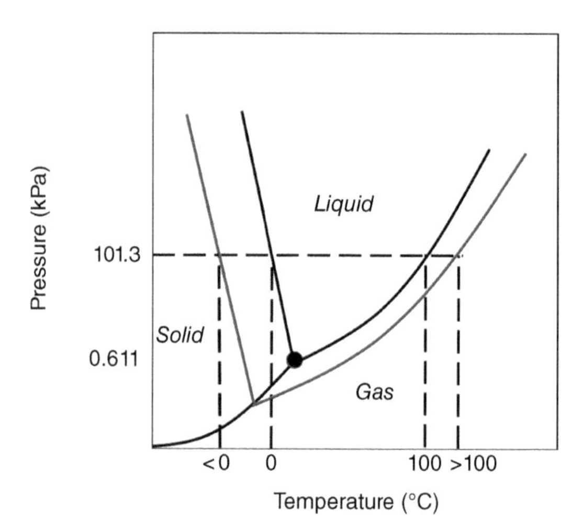

Figure 3.12 State diagram for water showing the extension of the liquid phase for a solution such as seawater (red lines) compared with pure, fresh water (black lines). For the same pressure, the freezing-point temperature decreases below 0 °C and the boiling-point temperature increases above 100 °C.

The extension of the range of temperatures that liquid water exists through the creation of a solution has significant implications. Because of salts dissolved in the oceans, their freezing point is roughly −2 °C. The lower freezing point means that water in polar regions remain ice-free longer and later in the winter season, which has cultural as well as physical significance (see Section 3.9). The **freezing-point depression**, the change in the temperature at which an aqueous solution becomes a solid relative to the freezing point for the pure liquid, can be calculated using Eq. (3.1):

$$\Delta T_{\mathrm{f}} = iK_{\mathrm{f}}m \tag{3.1}$$

where ΔT_{f} is the freezing-point depression (°C or K), i is the **van't Hoff factor**, K_{f} is the freezing-point constant for water (1.86 °C kg mol^{-1}), and m is the molarity of the solution (moles of the solute per kilogram of the solvent). For seawater, average salinity is about 3.5 percent, equivalent to roughly 0.6 moles of NaCl per liter (~1 kg) of water (not all salt in the oceans is NaCl), so $m \sim 0.6$ mol kg^{-1}. For compounds that dissolve well in water, the van't Hoff factor is generally equal to the number of ions formed in solution: since NaCl dissociates into Na^+ and Cl^- when dissolved in water, two ions are formed so $i = 2$. Therefore, for seawater the freezing-point depression is:

$$\Delta T_f = iK_f m = (2)(1.86\ °\text{C kg mol}^{-1})(0.6\ \text{mol kg}^{-1}) = 2.23\ °\text{C}.$$

Since the freezing point of pure water is 0 °C, and our calculation gave ΔT_f = 2.23 °C, and 0 – 2.23 °C = −2.23 °C, this means that seawater with an average salinity equivalent to 0.6 molar NaCl will freeze at −2.23 °C, not 0 °C.

3.9 A Northwest Passage Example

The lowering of the freezing point in the polar oceans results from water's ability to easily dissolve and dissociate salts such as NaCl. In addition to affecting the albedo (the reflected solar radiation), heat exchange, and ecology of these regions, there are also cultural, political, and economic effects that can be illustrated by an examination of the Arctic's **Northwest Passage**.

The search for a sea route between Europe and Asia shorter than traversing around the southern tip of Africa (Cape of Good Hope) or South America (Cape Horn) extends back to the fifteenth century. Conflict between the English, Dutch, French, and Spanish over lucrative trading in the Asian and South Pacific regions meant that anyone finding a shorter route across the northern polar region would have a trading advantage. Several unsuccessful expeditions led to great suffering and loss of life, but also resulted in today's political shaping of North America. In 1497, Italian explorer **John Cabot** (born **Giovanni Caboto**; ca. 1450–1500) was commissioned by England's **King Henry VII** (1457–1509) to search for a shorter route to Asia by sailing west, not east, from Europe. Cabot mistook the eastern coast of North America for Asia and claimed the land which eventually became Canada for England. Under commission from the Dutch East India Company, **Henry Hudson**'s (ca. 1565–1611) search for the Northwest Passage in 1606 led him through today's Hudson River in today's New York State. The Dutch claimed these regions and established New Amsterdam, later renamed New York when claimed by England in 1664. Later expeditions by Hudson commissioned by England took him on a more northern route, resulting in the exploration of, and English claim to, regions including Hudson Strait, Hudson Bay, and James Bay, where he was set adrift by a mutinous crew never to be seen again.

A major motivation for the Northwest Passage search was the commonly held belief that the Arctic Seas were ice-free, and that the **Open Polar Sea** existed in the high northern latitudes (Wright, 1953; Luedtke, 2015). At the time it was believed that ice could only form near landmasses, and therefore ice could not exist near the open ocean of the North Pole. Only fresh water could freeze, so it was thought, so seawater away from the coastal influence of freshwater runoff must remain ice-free. In addition, it was thought at that time that 24 hours of summer daylight would be sufficient to melt any ice. Ironically, the sea ice extent in the northern polar region during much of the seventeenth and eighteenth centuries was exceptionally high, owing to the extended period of cooling referred to as the **Little Ice Age**, from roughly 1300 through 1850 (Fauria et al., 2010). The warming that began towards the end of this period further fueled the notion that a vast, open polar sea existed. It is again ironic that with current climate change, the Arctic Sea could be ice-free in summer by 2050 (Thackeray and Hall, 2019) permitting the true opening of travel, commerce, and associated environmental impacts in the Northwest Passage.

3.10 SUMMARY

The covalent bonding between hydrogen and oxygen to form water, and the attraction of the hydrogen atom to other water molecules or to different molecules, result in water's unique chemical and physical properties. An understanding of these properties helps our understanding of how water shapes Earth's landscape and why water plays such a critical biological role.

Since the hydrogen atom has a net positive charge, it is electrostatically attracted to any other atom or molecule that has negative electrical charge. The dipole electrostatic charge on the water molecule developed through the covalent bonding between hydrogen and oxygen facilitates the attraction of one water molecule to another. The hydrogen bond between water molecules is much weaker than the covalent bond within the water molecule, so liquid water can bend and form tightly curved surfaces as expressed by the large surface tension across the liquid–air interface. This hydrogen bond, coupled with the dipole electrostatic charge of the water molecule, explains how water molecules bond to each other (cohesion) and other molecules (adhesion). Cohesion, adhesion, and surface tension are all properties that are central to the Cohesion–Tension Theory of transpiration.

The strong covalent bond within the water molecule and the resulting asymmetric distribution of valence electrons results in a dipole molecule that very effectively dissolves or dissociates other covalently bonded dipole compounds. The term "universal solvent" is used to describe water's ability to dissolve so many compounds. The high solubility of compounds such as salts and sugars in water has shaped Earth through several means, including the weathering of compounds in rock and soils (making them available for vegetation), the transport and dissolving of salts in the oceans (affecting the water's density and freezing point), and the movement of water within plant and animal cells (Chapter 11).

The arrangement and spacing between water molecules affect both density and the energy required or released as water changes states. Unlike most compounds, the density of solid water is much less than that of liquid water, owing to the lattice structure that forms when water freezes. When liquid, water contracts until a temperature of approximately 4 °C is reached and expands with subsequent warming. These density changes affect processes such as thermal stratification and vertical mixing and the floating of ice on liquid water. Thermal stratification and the formation of surface ice, in turn, affect radiation and heat exchange between surface and atmosphere.

Water is the only substance that naturally exists in liquid, solid, and gas state on Earth. The state (or phase or triple-point) diagram graphically shows the state of water as a function of temperature and pressure. The amount of energy required to raise the temperature of liquid water, the specific heat capacity, is very high, owing to the hydrogen bonds between molecules. Energy is required or released when water changes state, with the most energy required when water sublimates from solid to vapor forms. Water's ability to dissolve many compounds disrupts the formation of hydrogen bonds between water molecules, thus decreasing the freezing-point temperature and increasing the boiling-point temperature of liquid water.

With this understanding of these properties of water, and an explanation as to why they occur, an appreciation of why and how water is the most important molecule that shaped the Earth and life on it is within reach.

3.11 QUESTIONS

3.1 What is a hydrogen bond between water molecules, and how does it differ from the covalent bond within a water molecule?

3.2 Explain the roles of surface tension, cohesion, and adhesion in the meniscus that forms in a narrow glass graduated cylinder, in a drop of water on a waxy leaf surface, and within the xylem tissue of vascular plants.

3.3 What is the difference between dissolving sugar and salt in water, and why do these different changes occur?

3.4 Imagine a hypothetical situation where water was not a dipole molecule. How would this change the properties of water discussed in this chapter? What would be the significance of this to the abiotic and biotic environments?

3.5 Imagine a hypothetical situation in which the density of liquid water reached its maximum at 0 °C and monotonically decreased as the temperature increased. How would this change in the density of water with temperature affect the abiotic and biotic environments?

3.6 If the density of ice was greater than the density of liquid water, how and why would Earth's climate change? Would there be any ecological impacts? Why or why not?

3.7 How many joules of energy are required to raise the temperature of a liter of pure liquid water from 20 to 25 °C? How many joules of energy are released if the water cools from 25 back to 20 °C?

3.8 Looking at the state diagram for pure water, at a constant pressure, what changes in temperature are required to change state from solid to liquid to gas? At a constant temperature, what changes in pressure are required to change state from gas to liquid to solid? How could the liquid state of water be extended (e.g., a lower freezing-point temperature and high boiling-point temperature)?

3.9 Great Salt Lake in the western United States has a salinity of 4.6 moles of (mostly) NaCl per liter of water in some locations. Compared with fresh water, calculate the temperature at which this water would freeze.

3.10 The Caspian Sea freezes at a temperature of roughly 0.7 °C below the freezing point of pure water. Based on this freezing-point depression, estimate the salinity in moles per kilogram, assuming the majority of the salt is NaCl.

REFERENCES

Brooks, J. L. (2004) 'The cohesion-tension theory', *New Phytologist*, 163, pp. 451–452. doi: 10.1111/j.1469-8137.2004.01160.x.

Buckingham, A. D. (1991) 'The hydrogen bond, and the structure and properties of H_20 and $(H_20)_2$', *Journal of Molecular Structure*, 250, pp. 111–118. doi: 10.1016/0022-2860(91)85023-V.

Chaplin, M. F. (2010) 'Water's hydrogen bond strength', in Lynden-Bell, R. et al. (eds.) *Water and Life: The Unique Properties of H_2O*. CRC Press, pp. 69–86. doi: 10.1201/EBK1439803561.

Croft, W. B. (1913) 'The maximum density of water', *Nature*, 91, p. 505.

Dixon, H. H. and Joly, J. (1894) 'On the ascent of sap', *Philosophical Transactions of the Royal Society of London Series B*, 186, pp. 563–576.

Fauria, M. M. et al. (2010) 'Unprecedented low twentieth century winter sea ice extent in the Western Nordic Seas since A.D. 1200', *Climate Dynamics*, 34, pp. 781–795. doi: 10.1007/s00382-009-0610-z.

Luedtke, B. (2015) 'An ice-free Arctic Ocean: History, science, and scepticism', *Polar Record*, 51, pp. 130–139. doi: 10.1017/S0032247413000636.

Rastogi, A., Ghosh, A. K. and Suresh, S. J. (2011) 'Hydrogen bond interactions between water molecules in bulk liquid, near electrode surfaces and around ions', in Moreno-Pirajan, J. C. (ed.) *Thermodynamics: Physical Chemistry of Aqueous Systems*. InTech, pp. 351–364. Available at: www.intechopen.com/books/927.

Silberberg, M. S. and Amateis, P. (1996) *Chemistry: The Molecular Nature of Matter and Change*. Mosby.

Thackeray, C. W. and Hall, A. (2019) 'An emergent constraint on future Arctic sea-ice albedo feedback', *Nature Climate Change*, 9, pp. 972–978. doi: 10.1038/s41558-019-0619-1.

Wright, J. K. (1953) 'The open Polar Sea', *Geographical Review*, 43, pp. 338–365.

Zimmermann, U., Schneider, H., Wegner, L. H. and Haase, A. (2004) 'Water ascent in tall trees: Does evolution of land plants rely on a highly metastable state?', *New Phytologist*, 162, pp. 575–615. doi: 10.1111/j.1469-8137.2004.01083.x.

4 The Origin of Water

Key Learning Objectives

After reading this chapter, you will be able to:

1. Understand the conditions when Earth formed that resulted in the formation of water.
2. Describe how extraterrestrial sources contributed to the water on Earth.
3. Discuss how the composition of gases in the atmosphere changed over time as the Earth's atmosphere developed.
4. Identify the evidence used to indicate the earliest time that liquid water appeared on Earth.

4.1 Introduction

Where did water come from? This perplexing question is complex since it requires knowledge of the origins of hydrogen and oxygen and the conditions that allowed them to bond and form water and ultimately the water cycle. Knowing when liquid water first formed also provides clues as to the origins of life on Earth. Evidence for the existence of liquid water, past or present, on other planets is a major clue for the possibility of extraterrestrial life. Finding water means finding life, and past and recent missions to Mars speak to the seriousness of this question, since finding evidence of water provides strong evidence that life may have existed there and elsewhere.

The search for the origin of water requires looking billions of years into Earth's past. This chapter begins by examining the origins of hydrogen and oxygen gas as Earth formed into a planet. The chemical reactions and physical processes involved at the time of Earth's formation to form water are discussed. The controls and processes of chemical reactions such as endothermic and exothermic reactions, temperature, and the proportions and concentrations of reactants are summarized in the context of the conditions needed to form liquid water. Lastly, geological evidence indicating when liquid water first appeared is described. Throughout this chapter, the conditions that allowed water to exist on Earth, setting the stage for the beginning of life and the subsequent dramatic alteration of the newly forming planet, are illustrated.

4.2 The New Earth

The new Earth not only contained sufficient quantities of hydrogen and oxygen, but also had the necessary conditions required for the chemical bonding to form water. After the **Big Bang** occurred roughly 13.8 billion years ago, the simple low-mass elements such as **hydrogen** and **helium** formed in the expanding Universe once sufficient cooling allowed fusion between neutrons and protons (forming hydrogen) and subsequent hydrogen fusion (forming helium). When stars formed, **oxygen** formed from the nuclear fusion within the interior of stars was distributed through solar systems as stars exploded during their supernova stage. Together these three elements are the first (hydrogen), second (helium), and third (oxygen) most abundant elements in the Universe, but hydrogen and oxygen had to be available in sufficient quantities under the right conditions on the new Earth to form water.

A star such as our **Sun**, because of its large mass, plays a large role in the formation of the planets that subsequently orbit it. Earth's Sun formed about 4.6 billion years ago. Like most stars, the Sun is about 70% hydrogen and 28% helium (by mass). During formation, a large **protoplanetary disk** of gas and small amounts of dust orbited the Sun, attracted by its increasing mass. Over time, gravitational forces in the denser regions of the protoplanetary disk along with electrostatic attraction may have caused orbiting gases and dust to aggregate into the precursor of planets, **planetesimals**. Gravity continued to condense material into these planetesimals, and the more massive they became, the greater the gravitational force that further attracted dust, gas, and perhaps even water that formed extraterrestrially. In addition to the formation of water from chemical reactions within Earth as it was forming and cooling, some water was likely imported from space from the many collisions between planetesimals, comets, and asteroids swirling within the protoplanetary disk (Feaga et al., 2007). Although **asteroids** and comets do contain water and have impacted Earth, the frequency of impacts is low, with strong comet showers only occurring roughly once every 100 million years (Heisler, Tremaine and Alcock, 1987). Asteroid and comet impact during the formation and at end of the formation of Earth was much greater, however, and the frequent collisions with small planetesimals rich in water could have significantly increased the water content above what was contained within the primordial Earth at that time (Morbidelli et al., 2000).

The formation of a **solar system** with either solid or gas planets represents a competition between opposing forces as described by the **nebular hypothesis**. Counter to the gravitational force that attracts matter are the **solar winds** that remove matter. If the gravitational force continues to attract matter to the point where the planetesimal is large enough to become round from its own gravity and is distant from other planetesimals, it becomes a **terrestrial planet**. Planets known as **gas giants** are sometimes referred to as failed stars since, like a star, they consist largely of hydrogen and helium gas. The Universe, stars, and four of the nine planets in our solar system (the gas giants Jupiter, Saturn, Uranus, and Neptune) are all comprised primarily of hydrogen gas. On the surface of the new Earth, however, hydrogen gas was largely stripped away by the solar winds, and within the hot planetary core, hydrogen and oxygen existed but not in their gaseous form. As the surface cooled, any hydrogen or helium gas released would have escaped to space. The Earth would need time to cool before some early semblance of an atmosphere with surface liquid water could appear.

4.3 The Origin of Hydrogen Gas on Earth

Earth's hydrogen along with other elements such as carbon, nitrogen, and some oxygen would have initially been released as a gas into the newly forming atmosphere by volcanic outgassing.

As Earth's cooling formed the **lithosphere**, large quantities of gas collected during Earth's formation were trapped and released through violent volcanic eruptions, thus forming the early **atmosphere**. Today, volcanic eruptions release gases including carbon dioxide (CO_2), sulfur dioxide (SO_2), hydrogen sulfide (H_2S), and large quantities of water vapor (H_2O), but volcanic activity and outgassing would have been much more prevalent in the young hot Earth than today.

Hydrogen was released into the early atmosphere in molecular form bonded to other atoms (e.g., oxygen to form H_2O; sulfur to form H_2S), while hydrogen was in the liquid phase under the high temperatures closer to Earth's core. The hydrogen-based molecules that formed were much heavier than the lighter hydrogen gas which escaped Earth's gravity and was lost to space. Although the current concentration of hydrogen gas in the atmosphere is very small (~1 ppm by volume, or 0.0001%), it could have been much larger, greater than 30%, in Earth's early primordial atmosphere (Tian et al., 2005). Since H_2 has such a low molecular weight, the relatively high initial H_2 concentration in the atmosphere would have steadily decreased as the lightweight molecule escaped Earth's gravitational pull. Currently, approximately 3 kg of H_2 escape the Earth's atmosphere to space every second (Catling and Zahnle, 2009).

4.4 The Origin of Oxygen Gas on Earth

Oxygen in its molecular gaseous form (O_2) has only recently (relatively speaking) been in the atmosphere at its current 21 percent concentration (21 percent of all the air molecules in the lower atmosphere by volume, the **homosphere**, are O_2). To maintain O_2 concentrations at this value requires a balance between sources of O_2 to the atmosphere (e.g., photosynthesis) and sinks (or losses) of O_2 from the atmosphere (e.g., respiration, organic matter decay, fossil fuel combustion). Such balance between O_2 production and loss has not always been the case in Earth's history. Oxygen gas first appeared in Earth's atmosphere about 2.4 billion years ago during the **Paleoproterozoic Era** when the continents began to stabilize. At that time, **anaerobic bacteria** (bacteria that do not require molecular oxygen) in the oceans that were protected from harmful UV radiation evolved the capacity for photosynthesis. First **prokaryotes** (single-cell organisms without organelle membranes), then later **eukaryotes** (multicell organisms with organelle membranes) produced O_2, a toxin to anaerobic organisms, through photosynthesis. During the **Precambrian Era**, communities of **cyanobacteria** (aquatic prokaryotic bacteria commonly referred to as **blue-green algae**, but not actually algae), living in shallow water where they created **stromatolite** structures of layered cemented sand grains, gradually increased the atmospheric O_2 concentration (Figure 4.1). Although large quantities of water existed as the Earth's atmosphere was forming, it was all water vapor that originated from the constant volcanic eruptions (the Steam atmosphere; Figure 4.1). There was no surface liquid or ice water present, since the Earth's surface temperature at that time was far above 100 °C (Sleep et al., 2001). Therefore, there was no water cycle or energy exchange as water changed states.

Much of the O_2 produced by cyanobacterial photosynthesis would have been absorbed in the oceans through the oxidation of dissolved iron Fe(aq):

$$4\text{Fe}(\text{aq}) + \text{O}_2(\text{aq}) \rightarrow 2\text{Fe}_2\text{O}. \tag{4.1}$$

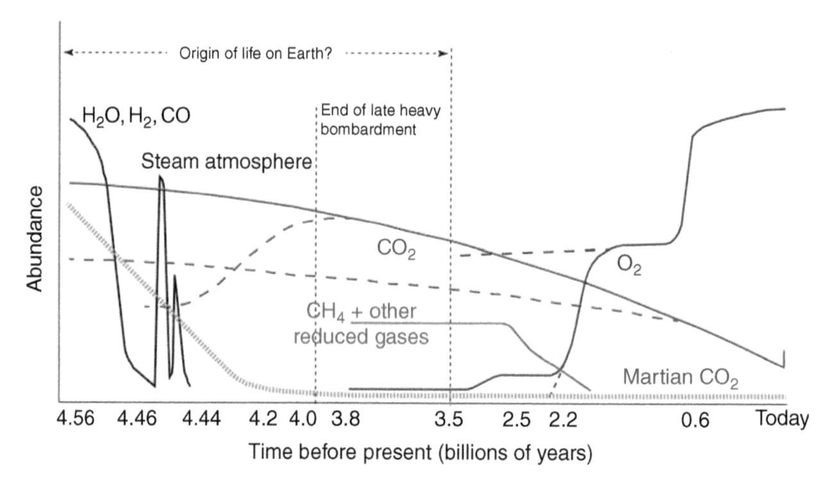

Figure 4.1 Relative abundance of oxygen and other gases in Earth's atmosphere. The CO_2 concentration on Mars is shown for comparison. Reproduced with permission from ten Kate and Reuver (2016).

Figure 4.2 The Hamersley Banded Iron Formation in Australia's Karijini National Park. Photo credit: Lon Abbott and Terri Cook.

The oxygen produced in the oceans by the photosynthetic cyanobacteria reacted with the dissolved iron to form insoluble iron oxide minerals (Eq. 4.1) that precipitated out of solution, preserved as iron-rich grey layers ($2Fe_2O$) that alternate with iron-depleted red chert layers (SiO_2) in banded iron formations (Figure 4.2). The banding is due to fluctuations in the cyanobacteria population. The O_2 they produced was toxic, resulting in self-poisoning and population declines. When the oxygen levels declined with the decrease in O_2 production and the removal of O_2 by the formation of iron oxides, the cyanobacteria population increased until the O_2 they produced once again reached toxic concentrations. This cycle continued until **aerobic** cyanobacteria evolved with the capability to tolerate oxygen (Cloud, 1973).

With the adaptation of cyanobacteria to the oxygen it produced, eventually O_2 began to accumulate in the atmosphere. The increase in atmospheric O_2 provided a means for life to propagate and diversify from single-cell aquatic bacteria to multicellular aquatic phytoplankton, all producing O_2. This significant event when O_2 production resulted in the sudden increase in atmospheric O_2 concentration roughly 2.4 billion years is known as the **Great Oxidation Event** (Lyons et al., 2014) with the proliferation of multicell eukaryote life and the mass extinction of many anaerobic bacteria.

Concurrent with the increase in the atmosphere's O_2 concentration from photosynthesis, abiotic processes contributed to liquid water production. **Methane** (CH_4), present in the atmosphere from volcanic eruptions and lithospheric degassing, chemically reacted with the steadily increasing O_2 to produce CO_2 gas and liquid water (Eq. 4.2):

$$CH_4(g) + 2O_2(g) \rightarrow CO_2(g) + 2H_2O(l). \tag{4.2}$$

This conversion of methane to carbon dioxide favored cooling the atmosphere since CO_2 is a less effective greenhouse gas than CH_4. This additional cooling made possible by the extra O_2 released by billions of single-cell bacteria could have been sufficient to induce Earth's first global Ice Age, the **Huronian Glaciation Event**, completely covering Earth with ice for the first time, which coincided with the Great Oxidation Event (Tang and Chen, 2013). Ironically, once again the photosynthetic production of O_2 resulted in cooling and ice cover that severely limited photosynthesis. However, some bacteria survived, and about 1,500 million years after the ice receded, terrestrial vegetation first appeared on Earth (Morris et al., 2018). With the presence of terrestrial vegetation, photosynthesis flourished, once again adding oxygen back into the atmosphere, ultimately reaching today's concentration of roughly 21 percent.

4.5 Hydrogen and Oxygen Bonding to Form Water

To understand how hydrogen and oxygen gas initially combined to form water vapor in the new Earth's atmosphere, we first need to examine some basics about chemical reactions and what controls the rate of the reaction.

Energy is required to break existing bonds between molecules to allow new molecules to form. The **Collision Theory** in chemistry states that to form new molecules (**products**) from existing atoms or molecules (**reactants**), the reactants must collide with sufficient energy to break any existing bonds. For example, consider molecular hydrogen gas (H_2) and molecular oxygen gas (O_2) as the reactants. To form water as the product of the reaction between H_2 and O_2, the Collision Theory states that the reactants first must collide. A high concentration of the reactants increases the probability of a collision, and as the density of Earth's forming atmosphere increased, so did the probability of H_2 and O_2 colliding. Collisions required sufficient energy to break the existing hydrogen-to-hydrogen and oxygen-to-oxygen bonds before they could bond to form different molecules such as H_2O.

The energy required to break the covalent bond, called the **bond energy** or **bond dissociation energy**, between two H atoms is 436 kJ for each mole of hydrogen gas. The bond dissociation energy between two O atoms in the double covalent bond is 498 kJ for each mole (following the conservation of energy, 436 and 498 kJ per mol are released when hydrogen or oxygen gas form, respectively). These fairly large energies required to break the bonds in hydrogen and oxygen gas and start the reaction to form water would be (and still are) present in the Earth's core where temperature and pressure are high, thus explaining how water formed early in Earth's history and is still released in large quantities from today's volcanic eruptions. In today's atmosphere, however, the temperature and concentration of H_2 and O_2 are both too low to provide the required bond dissociation energy – thankfully, since explosive conditions would otherwise exist (see below).

4.6 Endothermic and Exothermic Reactions

The **activation energy** (E_a; J mol^{-1}) is the minimum energy required to start a chemical reaction. This is equal to the bond dissociation energy when a chemical reaction begins after the breaking of the reactants' chemical bonds. Usually, the activation energy of the **reactants** is not equal to the energy produced when the bonds in the new molecule(s) (the **products**) are formed (Figure 4.3). The difference between the stored chemical energy in the products and the reactants is known as the change (Δ) in **enthalpy** (H; $\Delta H = H_{products} - H_{reactants}$), where enthalpy is defined as the total energy of the system (at constant pressure). Changes in energy due to chemical reactions should not be confused with changes in energy due to changes of state where no chemical reaction occurs.

When the reactants are at a lower energy level than the products (ΔH is positive), energy is required for the chemical reaction to occur, and the reaction is **endothermic**. In an endothermic reaction, energy often in the form of heat from the surrounding environment is required to keep the reaction going, and therefore the temperature in the surrounding environment decreases. An example of an important endothermic reaction involving water is photosynthesis (Eq. 4.3):

$$6CO_2 + 6H_2O \overset{\text{energy}}{\rightarrow} C_6H_{12}O_6 + 6O_2. \tag{4.3}$$

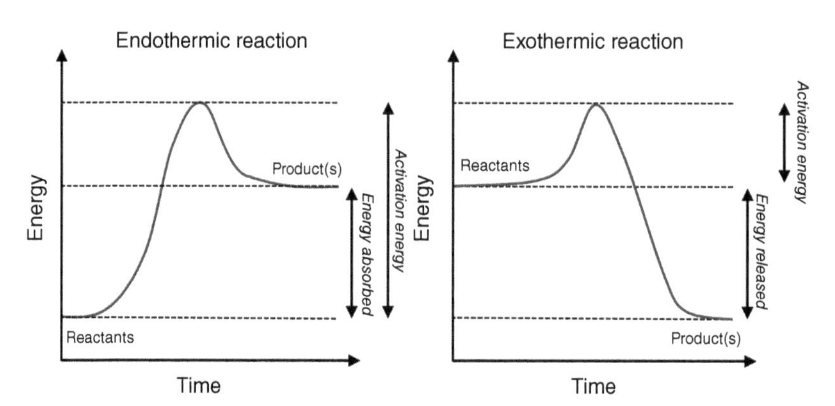

Figure 4.3 Endothermic and ectothermic reactions showing the activation energy required and the energy absorbed (endothermic) or released (exothermic) as reactants are chemically converted to product(s).

In photosynthesis, the products of sugars and oxygen gas are at a much higher energy state than the reactants, with approximately 15 million joules of energy required to produce 1 kg of glucose (sugar). This energy is supplied not from the surrounding environment but the Sun in the form of **photosynthetic active radiation** (wavelengths 400–700 nm).

When the reactants are at a higher energy level than the products (ΔH is negative), energy is released when the chemical reaction occurs, and the reaction is **exothermic**. In an exothermic reaction, energy often in the form of heat is released to the surrounding environment and therefore the temperature increases. Energy is constantly being released as the reaction occurs and products form, and this energy, in turn, can serve as the activation energy, thus promoting a chain reaction. Examples of exothermic chemical reactions include any combustion reaction (defined as requiring oxygen), or the reverse process of photosynthesis, respiration. In respiration, energy (heat), water, and CO_2 are released as glucose reacts with oxygen. Thus, respiration (Eq. 4.4) is technically a form of combustion:

$$C_6H_{12}O_6 + 6O_2 \rightarrow 6CO_2 + 6H_2O + \text{energy}. \tag{4.4}$$

Another example of an exothermic reaction is the formation of water from the reaction between hydrogen and oxygen gas (again, this is technically combustion) (Eq. 4.5):

$$2H_2(g) + O_2(g) \rightarrow 2H_2O(l) + \text{energy} \tag{4.5}$$

where approximately 16 million joules of energy is released for each kilogram of liquid water formed; an explosive release of energy. Think of the energy released and the volume of water produced when the Hindenburg Airship, filled with roughly 2×10^5 m^3 of hydrogen gas, suddenly exploded. Or think of hydrogen fuel cells that produce energy with only water as the chemical product, a huge benefit in dry regions or for potential long-distance space travel.

Exothermic reactions that produce water and release energy are common. Recall from Chapter 2 that hydrogen literally means "water generator", and the discovery of oxygen was a result of combustion experiments that produced water as a product of the reaction. As Earth was forming, the chemical reactions between hydrogen, oxygen, and many other elements would have resulted in the formation of water as they continue to do so today. Examples of such reactions appear in the combustion of fuels such as natural gas (methane; CH_4) (Eq. 4.6), propane (C_3H_8) (Eq. 4.7), or gasoline (mostly C_8H_{18}) (Eq. 4.8) with each combining forms of hydrogen and carbon with oxygen, releasing energy and water as products of the reaction:

$$CH_4 + 2O_2 \rightarrow CO_2 + 2H_2O + \text{energy} \tag{4.6}$$

$$C_3H_8 + 5O_2 \rightarrow 3CO_2 + 4H_2O + \text{energy} \tag{4.7}$$

$$2C_8H_{18} + 25O_2 \rightarrow 16CO_2 + 18H_2O + \text{energy}. \tag{4.8}$$

Thankfully, the spontaneous combustion of hydrogen gas does not happen under the current conditions in the atmosphere. For reactions to occur, the Collision Theory states that the reactants must collide with sufficient activation energy to break the existing bonds, or else the reactants will rebound off each other with no reaction as H_2 and O_2 do in the atmosphere today.

4.7 Chemical Reaction Rates

When chemical reactions occur, the rate at which new products are created depends on the reactants' temperature, the concentration and relative proportions, the surface area, and if present, a catalyst. **Temperature** is essentially a measure of the average kinetic energy of a substance ($K_E = \frac{1}{2}mv^2$ where m is mass and v is velocity). Therefore, an increase in the temperature of a gas indicates that the atoms (or molecules) are traveling at a higher velocity, assuming constant mass. As a result, atoms (or molecules) have a better chance of attaining sufficient activation energy to start a reaction. One of the earliest formulations of the relationship between the **chemical reaction rate** (k; L mol^{-1} s^{-1}) and temperature was published in 1889 by the Swedish physicist and chemist **Svante Arrhenius** (1859–1927), who also described the connection between atmospheric CO_2 and temperature (Arrhenius, 1869). The equation that Arrhenius developed to predict the rate of a chemical reaction based on temperature is known as the **Arrhenius equation** (Eq. 4.9):

$$k = Ae^{-\frac{E_a}{RT}} \tag{4.9}$$

where A (L mol^{-1} s^{-1}) is a constant experimentally determined for each chemical reaction (also known as the pre-exponential or frequency factor), E_a is the activation energy (J mol^{-1}), R is the universal gas constant (8.314 J mol^{-1} K^{-1}), and T is temperature (K). The Arrhenius equation shows that the rate constant, and therefore the reaction rate, increases with temperature. In fact, at temperatures around 300 K, k roughly doubles for every 10 K (10 °C) increase in T, a popular rule-of-thumb used by many chemists and biologists today. For example, the average rate of a chemical reaction estimated from Eq. (4.9) with $A = 4 \times 10^5$ L mol^{-1} s^{-1}, $E_a = 5 \times 10^4$ J mol^{-1}, and $T = 20$ °C (293 K) is 0.49 mL per mole per second (Eq. 4.10):

$$k = Ae^{-\frac{E_a}{RT}} = 4 \times 10^5 \frac{\text{L}}{\text{mol s}} \times e^{-\left(\frac{5 \times 10^4 \frac{\text{J}}{\text{mol}}}{8.314 \frac{\text{J}}{\text{mol K}} \times 293\,\text{K}}\right)} = 0.00049 \frac{\text{L}}{\text{mol s}}. \tag{4.10}$$

Repeating the calculation at a 10 °C higher temperature (30 °C) results in $k = 3.3$ mL mol^{-1} s^{-1}, nearly doubling the chemical reaction rate (3.3/1.8 = 1.83).

The concentration of the reactants also affects the reaction rate. The greater the concentration of a gas, the greater is the probability that molecules will collide with sufficient activation energy. Therefore, as the concentration of gases increased as Earth was forming, so would the chemical reaction rates. Doubling the concentration of hydrogen (or any other molecule) doubled the likelihood of hydrogen colliding with another element and therefore the likelihood of a chemical reaction. Increasing the concentration of oxygen gas with the Great Oxidation Event increased reaction rates involving combustion. Higher concentrations of hydrogen, then later oxygen, increased the rate and number of chemical reactions including those resulting in the formation of water.

It is not only the amount or concentration but also the relative proportion of the reactants that controls the chemical reaction rate. **Stoichiometry** refers to the study of the relationship between the quantities of the reactants and products before, during, and after a chemical reaction occurs. For the formation of liquid water, Eq. (4.5) shows that two molecules of H_2 gas combined with one molecule of O_2 gas produce two molecules of liquid water. The number of molecules required to write a balanced equation, hence conserve mass, is known as the **stoichiometric number**. For the production of liquid water these numbers are 2:1:2 (H_2:O_2:H_2O). In this proportion of hydrogen to oxygen gas, all the oxygen is used to produce water. What is preventing the hydrogen and oxygen gas in the atmosphere from reacting to form liquid water? As discussed, temperature plays an important role in chemical reactions, and an exceedingly high temperature of over 500 °C is required for H_2 and O_2 to spontaneously react (autoignition) to produce water, which would be in vapor form at such a high temperature.

Earth's surface temperatures were likely well above 1,200 °C as the planet was cooling during the Precambrian Era, so the spontaneous H_2:O_2 reaction could have contributed to the production of water vapor during this "Steam atmosphere" period (Figure 4.1), but complex molecules were also required for life to exist. After sufficient surface cooling and the appearance of liquid water and the formation of the water cycle, lightning may have supplied the energy sufficient for reactions between inorganic molecules to produce the amino acids necessary for early life. The **spark of life**, or, more formally, the **Miller–Urey** experiments performed in 1952 consisted of heating a small flask of liquid water to simulate evaporation from the primordial ocean (Figure 4.4). A mixture of methane, ammonia, and hydrogen was contained in another flask to simulate the conditions in the early atmosphere. A continuous spark between a pair of electrodes simulated lightning, and cooling of the evaporated water allowed the gases to condense and return to the flask containing water. After a week, the water in the flask was analyzed and found to contain the building blocks for life: amino acids, sugars, and lipids (Miller, 1953). Unknown to Miller and Urey in 1952, it was later found that the glass flask contained silica and other minerals required for life that dissolved into the water solution (i.e., the minerals in the glass flask represented the rock) (Criado-Reyes et al., 2021).

The surface area of the reactants affects the reaction rate, with the rate proportional to the surface area of the reactants. Think of the explosive nature of coal, sugar, or even flour dust. With the relatively small surface area relative to volume for a lump of coal, a grain of sugar, or a kernel of wheat, each will burn but not explode when exposed to a flame or spark. If finely ground into large quantities of dust when mined or when processed, however, the large surface area relative to volume can result in deadly and catastrophic explosions. There have been numerous examples of such explosions in coal mines, sugar mills, and food processing facilities.

The last control on chemical reaction rate is a **catalyst**, defined as a substance that increases the reaction rate without being altered by the reaction. A spark or flame is often mistaken as a catalyst. Often the creation of the spark or flame involves a catalyst – such as when striking a match – but the flame itself can increase the reaction rate by increasing temperature as described by the Arrhenius equation and as demonstrated in the Miller–Urey experiment. A catalyst increases the reaction rate by providing an alternative reaction pathway with a lower activation energy. For example, common hydrogen peroxide (H_2O_2) left over long periods of time will produce liquid water and oxygen gas with an activation energy of approximately 75 kJ mol^{-1} (Eq. 4.11):

$$2H_2O_2(l) \xrightarrow{75\ \text{kJ mol}^{-1}} 2H_2O(l) + O_2(g). \tag{4.11}$$

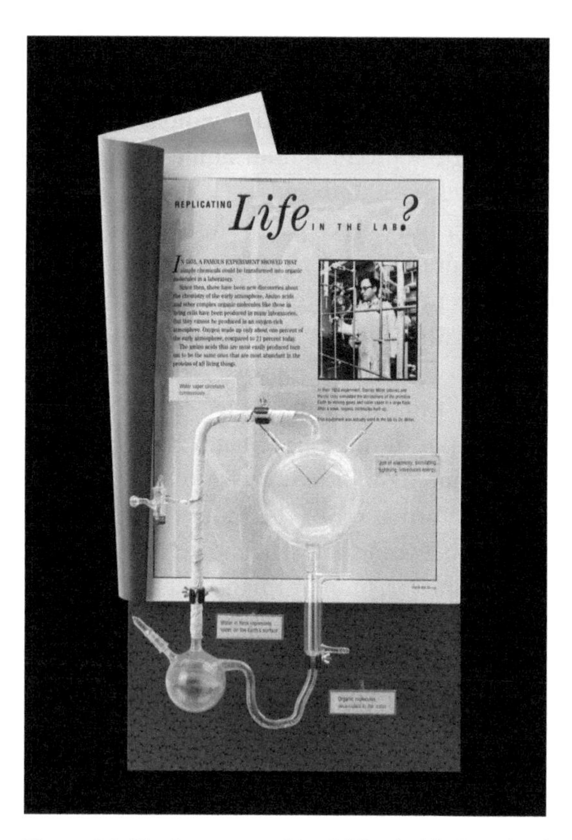

Figure 4.4 Equipment used by Miller in his "spark of life" experiment. Photo credit ESI.1_002 © Denver Museum of Nature & Science.

Adding the catalyst manganese dioxide (MnO_2) increases the reaction rate by reducing the activation energy to approximately 58 kJ mol^{-1} (Eq. 4.12):

$$2H_2O_2(l) + MnO_2(s) \xrightarrow{58\,\text{kJ mol}^{-1}} 2H_2O(l) + O_2(g) + MnO_2(s). \tag{4.12}$$

Using the Arrhenius equation (Eq. 4.9) with the same value for A for both the catalyzed and uncatalyzed reactions at a constant temperature of 20 °C (for example), the ratio of k with the catalyst (k_{catalyst}) to k without the catalyst ($k_{\text{no catalyst}}$) is 1,070, meaning the reaction is 1,070 times faster with the MnO_2 catalyst added (Eq. 4.13):

$$\frac{k_{\text{catalyst}}}{k_{\text{no catalyst}}} = \frac{Ae^{-\frac{E_a}{RT}}}{Ae^{-\frac{E_a}{RT}}} = \frac{e^{-\left(\frac{58{,}000\ \text{J mol}^{-1}}{\left(8.314\ \text{J mol}^{-1}\text{K}^{-1}\cdot 293.15\ \text{K}\right)}\right)}}{e^{-\left(\frac{75{,}000\ \text{J mol}^{-1}}{\left(8.314\ \text{J mol}^{-1}\text{K}^{-1}\cdot 293.15\ \text{K}\right)}\right)}} = 1{,}070. \tag{4.13}$$

4.8 Earliest Evidence of Liquid Water

For liquid water to be present on Earth's surface, sufficient cooling must have occurred. What happened to cause water vapor to condense, precipitation to form, surface liquid water to gather, and the water cycle to develop? Hydrogen and later oxygen were present in Earth's history, but were there sufficient quantities to account for the volume of liquid water currently on Earth?

To answer the first question, geological evidence tells us that the formation of the **lithosphere** permitted sufficient surface cooling for condensation to occur not long after the formation of Earth. The continents formed much later, and the Earth's surface has always been covered more by water than land (Figure 4.5). This remains true today, with 70 percent of Earth's surface area covered by oceans.

To answer the second question, it was mentioned at the beginning of this chapter that asteroid and comet impacts with Earth as it was forming could have increased the water content above what was contained within the primordial Earth. The hydrogen and oxygen already present during Earth's formation, however, were likely more than adequate to produce the current volume of water with only a small (5 percent) contribution of water from asteroids or comets (Piani et al., 2020).

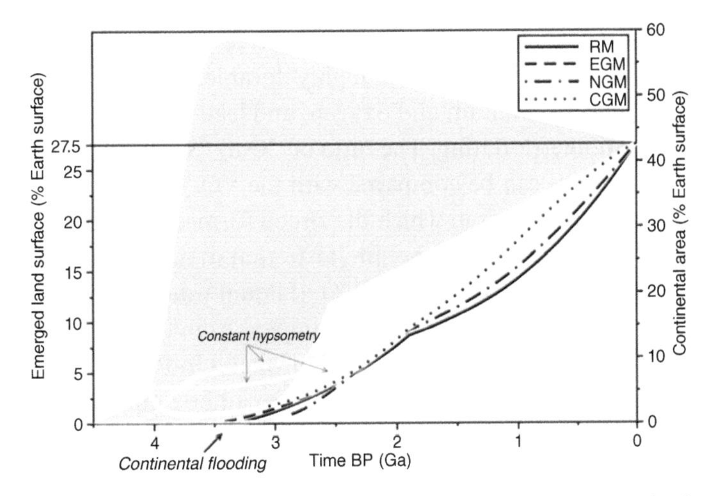

Figure 4.5 Throughout history, Earth's surface has been dominated by water. The fraction of land emerged depends on changes in both sea level and the continental area. The figure shows estimates of the percentage of Earth's surface emerged above sea level (lines; left axis) and the range in continental area estimates (shaded area; right axis) in billions of years (Ga) before the present (BP). Different continental growth models (lines; different models indicated by their abbreviations) generally agree as to the rate and time when the continents first appeared roughly 3.5 Ga BP. At this time the land was flooded since sea levels were at their maximum. Reproduced with permission from Flament et al. (2013).

The responses to both fundamental questions on the origins of water require information about physical conditions on Earth billions of years ago. **Isotopes**, elements with the same number of protons but different number of neutrons (Chapter 2) found preserved in rock provide much of the key information required to answer this question. These can be **stable isotopes**, meaning that they remain indefinitely in their form, or they can be **unstable isotopes**, decaying at a precise rate into other elements by changing the number of protons. An unstable isotope releases protons and/or neutrons to return to a more stable configuration, and by doing so releases energy in the form of radioactivity – **gamma radiation**. The rate at which an unstable isotope decays to a stable configuration is known as the **half-life**, the time it takes for half the material to return to a stable configuration. These decay rates are known for several elements and are constant over time. Hydrogen with one proton (^{1}H; protium) or one proton and one neutron (^{2}H: deuterium) and common isotopes of oxygen, ^{16}O, ^{17}O, and ^{18}O, are stable. Hydrogen with one proton and two neutrons (^{3}H; tritium) is unstable with a half-life of 12.32 years. How can isotopes of hydrogen and oxygen be used to understand the age and physical conditions pertinent to the origin of water?

A crystal known as **zircon**, together with the stable isotopes of oxygen, provides the key to look far into the past to understand when and how liquid water first appeared. Zircon is a nesosilicate mineral, zirconium silicate ($ZrSiO_4$), a combination of zirconium, silicon, and oxygen. Using radiometric dating of the conversion of uranium to lead (U–Pb) zircon crystals (not complete rocks, only crystals, since no known rocks have been preserved during the first 0.5 billion years of Earth's history) have been dated to be 4.36 billion years old, the oldest known material on Earth. Zircon crystals are highly durable, resistant to heat and corrosion, and contain small amounts of uranium and oxygen, and lead as a product of uranium decay, allowing for the U–Pb radiometric dating. The ratio of ^{18}O to ^{16}O that is incorporated and preserved into the crystal as it grows can be compared with the ^{18}O:^{16}O ratios in magma and liquid water to understand the environment in which the zircon formed. If the zircon formed with liquid water present, the ^{18}O:^{16}O ratio would be similar to that in liquid water where evaporation removes the lighter ^{16}O over the slightly heavier ^{18}O. If liquid water was not present, the zircon's ^{18}O:^{16}O ratio would match that found in magma sampled from today's volcanic eruptions.

By knowing the age of the zircon crystal through Ur–Pb radiometric dating, and the ^{18}O:^{16}O ratio, the date when liquid water was on Earth can be estimated. The ratio of ^{18}O:^{16}O is compared to a known, standard reference known as the $\delta^{18}O$ **oxygen isotope ratio** (Eq. 4.14). In zircon crystals, the $\delta^{18}O$ ratio at the time of formation is preserved and therefore can be correlated with the age of uranium–lead (U–Pb) to estimate the crystal's age (Valley, 2002).

$$\delta^{18}O(\text{‰, or parts per thousand}) = \left(\frac{\left(\frac{^{18}O}{^{16}O}\right)_{\text{sample}}}{\left(\frac{^{18}O}{^{16}O}\right)_{\text{standard}}} \right) \times 1{,}000. \tag{4.14}$$

The relative abundance of ^{18}O to ^{16}O in a sample compared with a standard reference value (often the **Vienna Standard Mean Ocean Water**, VSMOW, is used) can be useful to

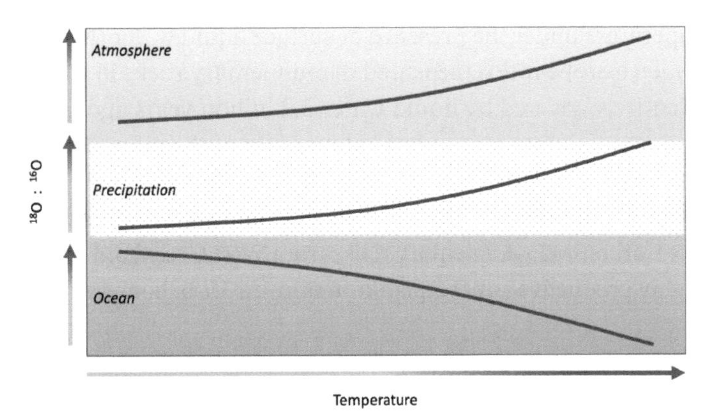

Figure 4.6 Oxygen isotopes in water can be used to help understand the water cycle and temperature. Lines show the relative changes in the ^{18}O:^{16}O ratio in the atmosphere, precipitation, and ocean with increasing temperature.

interpret the ambient conditions. Changes in the ratio of ^{18}O to ^{16}O can only occur with the presence of water cycling, and therefore liquid water is required. Liquid water containing ^{18}O requires more energy to evaporate than the lighter ^{16}O water. Therefore, as ocean water temperature increases, more ^{18}O can be removed from the ocean water, decreasing the ^{18}O:^{16}O proportion in the ocean water while increasing the ^{18}O:^{16}O proportion in the atmosphere's water vapor (Figure 4.6). As air temperature decreases, water vapor containing the heavier ^{18}O condenses first, falling as precipitation, and the remaining water droplets are therefore enriched with ^{16}O. Since there is no record of water vapor preserved through time, precipitation stored as zircon crystals, ice (ice cores), rocks, coral ($CaCO_3$) or any other material that preserves the oxygen isotope record can be used to indicate the presence of liquid water and a water cycle, and provide a proxy record of relative temperature (e.g., warm or cold periods).

The reason that stable oxygen isotopes can be used to estimate when liquid water appeared on Earth is that evaporation and condensation are required for the ^{18}O:^{16}O ratio to change. Changes in ^{18}O:^{16}O preserved over time therefore indicate the presence of liquid water. Zircon crystals have been shown to provide a dateable reliable record of ^{18}O:^{16}O, and thus can be used as an indicator of interactions of groundwater with minerals such as zircon. Today, samples of magma containing zircon have $\delta^{18}O$ values less than 6.5‰. Similar values, all less than 6.5‰, have been found in zircon-containing igneous rocks from ancient volcanic calderas around the world, so the 6.5‰ $\delta^{18}O$ is the mantle-derived value at very high temperature formed without the influence of liquid water (Mojzsis et al., 2001; Valley, 2002). Values of $\delta^{18}O$ above 6.5‰, however, indicate alteration by liquid water at low temperatures and reworking of the original crustal mantle material. In other words, the crust cooled sufficiently for tectonic plates to form, and plate subduction occurred aided by the presence of surface liquid water. Samples from today's magma and past volcanic eruptions all have $\delta^{18}O$ less than 6.5‰. Zircon samples collected in Western Australia dated 3.91 to 4.28 billion years old, however, have $\delta^{18}O$ values as high as 15‰, far above the 6.5‰ mantle values, indicating that the crustal material was altered at low temperature under the presence of surface liquid water (Mojzsis et al., 2001; Valley et al., 2002). Oxygen isotope ratios measured in sedimentary rocks in Australia suggest that the Earth was likely entirely covered by liquid water 3.2 billion years ago (Johnson and Wing, 2020).

There are other independent sources of evidence of the presence of liquid water early in Earth's history. Rocks containing ripples, cross beds, or rounded stones, each indicative of liquid water action, found in southwestern Greenland have been dated to at least 3.85 billion years old (Nutman et al., 1997). Carbon-rich sedimentary rocks from West Greenland provide evidence of photosynthesis from highly productive aquatic plankton prior to 3.7 billion years ago (Rosing and Frei, 2004).

4.9 SUMMARY

Answers to the questions "When did liquid water first appear on Earth?" and "Does (or did) liquid water exist on other planets?" help to determine when life originated and whether life is (or was) possible on other planets. Hydrogen and oxygen are prevalent throughout the Universe, and some water may have been imported from space to Earth in the form of asteroids or comets. At the surface of the primordial Earth, surface temperatures would have been too high for condensation to occur for any water vapor released from outgassing from the interior. The "steam planet" period, when only water vapor existed, did not likely last too long on the geological timescale. As the surface eventually cooled and the lithosphere formed, liquid water began to appear at a time without terrestrial continental surfaces. In the oceans, single-cell anaerobic bacteria began to release oxygen as a product of photosynthesis; they evolved to aerobic bacteria, and oxygen eventually accumulated in the atmosphere during this Great Oxidation Event. In the meantime, the steady release of methane gas into the atmosphere coupled with the oxygen released through photosynthesis added to the liquid water formation. Since methane gas is a stronger greenhouse gas than carbon dioxide, the conversion of CH_4 to CO_2 resulted in sufficient cooling to induce the first global Ice Age, the Huronian Glaciation Event.

In today's atmosphere, hydrogen and oxygen gas thankfully do not bond to form water or an explosive amount of energy would be released. As stated by the Collision Theory, new molecules form when the reactants collide with sufficient energy (the activation energy) to break any existing bonds. The difference between the energy stored in chemical bonds in the reactants and the product is known as the change in enthalpy. Endothermic reactions require energy, thus cooling the immediate environment. An example of an endothermic reaction involving water is photosynthesis. In an exothermic reaction, energy is released, thus the environment warms. Common exothermic reactions that produce water include combustion and respiration, in addition to the formation of water itself.

The rate of chemical reactions, including those forming or requiring water, increases with temperature (energy) as given by the Arrhenius equation. An increase in the concentration of reactions increases the probability of collisions with other atoms or molecules, so the reaction rate increases with concentration. Increasing the surface area of the reactants also increases the chemical reaction rate.

Geological evidence can be used to estimate when liquid water first appeared on Earth. A method based on the rate of decay of oxygen isotopes gathered from mineral or rock samples is often used. Zircon crystals, the oldest terrestrial material known, preserve the oxygen isotope record well. Several oxygen isotope studies using zircon and other materials suggest that Earth was covered entirely by liquid water 3.20 billion years ago, not long after Earth became a planet 4.60 billion years ago. Other geologic evidence agrees with the zircon crystal studies, indicating that liquid water has predominated on Earth's surface for the majority of Earth's history, starting 0.5 to 1.0 billion years after Earth's formation. Perhaps Planet Earth should be called Planet Ocean instead.

4.10 QUESTIONS

4.1 What were the source(s) of hydrogen and oxygen on the primordial Earth?

4.2 The period when oxygen first appeared in Earth's atmosphere roughly 2.4 billion years ago is known as the Great Oxidation Event. What caused this increase in the oxygen concentration in the atmosphere, and how did this play a role in Earth's water cycle?

4.3 Combustion chemical reactions that produce water and release energy are common. Explain whether these chemical reactions are endothermic or exothermic, and why energy is released, and provide an example of how water is produced in these reactions.

4.4 Using the Arrhenius equation (Eq. 4.9) with the same pre-exponential (frequency factor) and activation energy used in Eq. (4.10), calculate the reaction rate at a temperature of 0 and 10 °C. How did the reaction rate change for this 10 °C increase in temperature?

4.5 How and why does increasing the concentration and proportions of reactants influence the rate of chemical reactions?

4.6 In 2008, a large explosion and resulting fire occurred in a sugar refinery in Georgia, United States, killing 14 people and injuring 40. A sugar dust explosion was cited as the cause of the tragedy. Why would sugar dust, but not the sugar itself, have contributed to causing this explosion? What could have been done to mitigate against it?

4.7 What did the Miller–Urey experiment illustrate in terms of the importance of water to the conditions making life on Earth possible?

4.8 What are the unique characteristics of zircon crystals that make them so important in helping identify the time far back in Earth's history when liquid water appeared on Earth?

4.9 How do stable isotopes of oxygen help to provide an indication of the presence of liquid water and a proxy for surface temperature?

4.10 Describe sources of evidence preserved in the geological record that help identify when liquid water appeared on Earth. Why is the presence of liquid water so important and closely tied to indicate when life first appeared on Earth or other planets?

REFERENCES

Arrhenius, S. (1869) 'On the influence of carbonic acid in the air upon the temperature of the ground', *Philosophical Magazine and Journal of Science*, 41, pp. 237–276. doi: 10.1002/cta.4490080404

Catling, D. C. and Zahnle, K. J. (2009) 'The planetary air leak', *Scientific American*, May, pp. 36–43.

Cloud, P. (1973) 'Paleoecological significance of the banded iron formation', *Economic Geology*, 68, pp. 1135–1143.

Criado-Reyes, J., Bizzarri, B. M., García-Ruiz, J. M., Saladino, R. and Di Mauro, E. (2021) 'The role of borosilicate glass in Miller–Urey experiment', *Scientific Reports*, 11, pp. 1–8. doi: 10.1038/s41598-021-00235-4

Feaga, L. M., A'Hearn, M. F., Sunshine, J. M., Groussin, O. and Farnham, T. L. (2007) 'Asymmetries in the distribution of H_2O and CO_2 in the inner coma of Comet 9P/Tempel 1 as observed by Deep Impact', *Icarus*, 190, pp. 345–356. doi: 10.1016/j.icarus.2007.04.009

Flament, N., Coltice, N., Rey, P. F. and Lyon, U. De. (2013) 'The evolution of the 87 Sr / 86 Sr of marine carbonates does not constrain continental growth', *Precambrian Research*, 229, pp. 177–188. doi: 10.1016/j.precamres.2011.10.009

Heisler, J., Tremaine, S. and Alcock, C. (1987) 'The frequency and intensity of comet showers from the Oort cloud', *Icarus*, 70, pp. 269–288. doi: 10.1016/0019-1035(87)90135-7

Johnson, B. W. and Wing, B. A. (2020) 'Limited Archaean continental emergence reflected in an early Archaean ^{18}O-enriched ocean', *Nature Geoscience*, 13, pp. 243–248. doi: 10.1038/s41561-020-0538-9

Lyons, T. W., Reinhard, C. T. and Planavsky, N. J. (2014) 'The rise of oxygen in Earth's early ocean and atmosphere', *Nature*, 506, pp. 307–315. doi: 10.1038/nature13068

Miller, S. L. (1953) 'A production of amino acids under possible primitive Earth conditions'. *Science*, 117, pp. 528–529. doi: 10.1126/science.117.3046.528

Mojzsis, S. J., Harrison, T. M. and Pidgeon, R. T. (2001) 'Oxygen-isotope evidence from ancient zircons for liquid water at the Earth's surface 4,300 Myr ago', *Nature*, 409, pp. 178–181. doi: 10.1038/35051557

Morbidelli, A., Chambers, J., Lunine, J. I. et al. (2000) 'Source regions and timescales for the delivery of water to the Earth', *Meteoritics and Planetary Science*, 35, pp. 1309–1320. doi: 10.1111/j.1945-5100.2000.tb01518.x

Morris, J. L., Puttick, M. N., Clark, J. W. et al. (2018) 'The timescale of early land plant evolution', *Proceedings of the National Academy of Sciences USA*, 115, pp. E2274–E2283. doi: 10.1073/pnas.1719588115

Nutman, A. P., Mojzsis, S. J. and Friend, C. R. L. (1997) 'Recognition of ≥3850 Ma water-lain sediments in West Greenland and their significance for the early Archaean Earth', *Geochimica et Cosmochimica Acta*, 61, pp. 2475–2484. doi: 10.1016/S0016-7037(97)00097-5

Piani, L., Marrocchi, Y., Rigaudier, T. et al. (2020) 'Earth' s water may have been inherited from material similar to enstatite chondrite meteorites', *Science*, 369, pp. 1110–1113.

Rosing, M. T. and Frei, R. (2004) 'U-rich Archaean sea-floor sediments from Greenland – indications of >3700 Ma oxygenic photosynthesis', *Earth and Planetary Science Letters*, 217, pp. 237–244. doi: 10.1016/S0012-821X(03)00609-5

Sleep, N. H., Zahnle, K. and Neuhoff, P. S. (2001) 'Initiation of clement surface conditions on the earliest Earth', *Proceedings of the National Academy of Sciences USA*, 98, pp. 3666–3672.

Tang, H. and Chen, Y. (2013) 'Global glaciations and atmospheric change at ca. 2.3 Ga', *Geoscience Frontiers*, 4, pp. 583–596. doi: 10.1016/j.gsf.2013.02.003

ten Kate, I. L. and Reuver, M. (2016) 'PALLAS: Planetary analogues laboratory for light, atmosphere, and surface simulations', *Netherlands Journal of Geosciences*, 95, pp. 183–189. doi: 10.1017/njg.2015.19

Tian, F., Toon, O. B., Pavlov, A. A. and De Sterck, H. (2005) 'A hydrogen-rich early Earth atmosphere', *Science*, 308, pp. 1014–1017. doi: 10.1126/science.1106983

Valley, J. W. (2002) 'Oxygen isotopes in zircon', *Reviews in Mineralogy and Geochemistry*, 53, pp. 343–385. doi: 10.2113/0530343

Valley, J. W., Peck, W. H., King, E. M. and Wilde, S. A. (2002) 'A cool early Earth', *Geology*, 30, pp. 351–354. doi: 10.1130/0091-7613(2002)030<0351:ACEE>2.0.CO;2

5 The Distribution of Liquid Water

Key Learning Objectives

After reading this chapter you will be able to:

1. Describe the current distribution of liquid water in oceans, lakes, wetlands, rivers, soils, and groundwater, and how these reservoirs of water are changing.
2. Explain why the majority of Earth's surface water is saline, and describe the effects of salinity on the properties of water that influence climate.
3. Discuss the importance, characteristics, and classification of major surface freshwater resources including lakes, wetlands, and rivers.
4. Recall how the quantity of water stored in soils and groundwater is determined.

5.1 Introduction

The dominant surface feature of Earth is liquid water, and the distribution of liquid water near the Earth's surface can be used as a template to describe the distribution and diversity of life, climate, and even surface topography. With Earth's dynamic water cycle, the distribution of liquid water today has little resemblance to the past or future. Earth's surface has been completely covered by liquid water, ice, or no water at all. Today, Earth's surface is covered by liquid water and ice with water vapor connecting these states through the water cycle. Water exists simultaneously in all three states and cycles between them. Before discussing the cycling of water, however, we will first discuss the distribution of liquid water, followed by frozen water in Chapter 6. Knowing the distribution and characteristics of surface water is not only important for water resources and management, but is also important to inform our understanding of changes in the water cycle. This understanding is critical, especially given current climatic and other environmental changes.

This chapter describes the current distribution of liquid water at or near the surface. The distribution of saline and fresh water in oceans, lakes, wetlands, streams and rivers, soils, and groundwater is described, with the terminology used to classify their many characteristics and formation processes. Since water stored as soil moisture provides a critical water supply for terrestrial vegetation and agriculture, aspects of its measurement as well as deeper groundwater supplies are included.

5.2 Oceans

All the water we see today likely originated from the collection of hydrogen and oxygen present during the planet's formation over four billion years ago. Then, surface temperatures were so high that water was all in vapor form. Only after sufficient surface cooling did condensation occur, precipitation form, and liquid water begin to accumulate and cover the entire planet. This was a time when the continents did not exist, other than as occasional volcanic island arcs protruding above shallow oceans. Once the continents formed, roughly two billion years after Earth's formation, coupling between the cycling of water over land and the oceans developed. Liquid water evaporated from the land and oceans, cooled and condensed in the atmosphere to form precipitation, and rivers formed, eroding the weathered landscape and returning dissolved and solid materials back to the oceans.

With time, water cycling created the conditions giving rise to today's oceans. The origin of the word **ocean** is from Greek mythology; Oceanus was one of the 12 Titans, children of Father Uranus (representing the sky) and Mother Gaia (the Earth). Today's oceans cover roughly 361 million km^2, 71% of the Earth's surface, with an average depth of 3,730 m (Cundy and Kershaw, 2013) and estimated volume of 1,338 million km^3 representing nearly 97% of all the water on Earth (Shiklomanov, 1993). All oceans contain dissolved salts at varying quantities (expressed as **salinity**), but oceans are not classified on the variations in salinity. Since all oceans are saline and contain 97% (by volume) of the world's water, roughly 97% of Earth's surface water is **saline**. Today's average ocean salinity is 35 grams of dissolved salt per liter (or kg) of water (35 parts per thousand, ppt or ‰), well above the 1 ppt threshold sometimes used to distinguish fresh from saline water: its dissolved salt content is greater than 1,000 parts per million (ppm), or more than 0.1%. Although oceans contain saline water, oceans, seas, and lakes are defined not by salinity but by size. An ocean is loosely defined as an open body of water far larger than a sea. A **sea** is smaller than an ocean but located within an ocean and partially enclosed by land. A **lake** is entirely enclosed by land.

With the majority of Earth's surface underwater, much of the ocean floor remains unexplored despite housing many of the surface's most prominent features. The **Mariana Trench** is the lowest point on Earth, 10,984 ± 25 m below current sea level (Gardner et al., 2014). Located in the western Pacific Ocean, the Mariana Trench, formed where the Pacific tectonic plate subducts beneath the overriding Philippine Sea Plate, is much deeper than Mount Everest is tall. Most of Earth's topographic features are under water, and it is said that more is known about the surface of Mars than Earth. This is based on the fact that the ocean floor that has been surveyed is mapped at a coarse 5-km spatial resolution, compared with Mars (20-m resolution), Venus (100-m resolution), or the Moon (100-m resolution) (Copley, 2014).

Where did the ocean salt come from? Salt that dissolves in water to form oceans originates from the land through the chemical weathering of the terrestrial surface. Once dissolved, the salinity changes, whether increasing with ocean evaporation, or decreasing with freshwater runoff or precipitation. Thus, ocean salinity is a result of the coupling between ocean and land through the properties of water and the water cycle. The origin of oceanic salt can be traced to the slight acidity of precipitation that aids terrestrial surface weathering. As a result of the chemical reaction between carbon dioxide in the air or soil pores and liquid water, carbonic acid (H_2CO_3) is formed:

$$CO_2(g) + H_2O(l) \rightarrow H_2CO_3(aq).$$

The carbonic acid then dissociates to a hydrogen ion (H^+) and a hydrogen carbonate ion (HCO_3^-), thereby increasing the water's acidity:

$$H_2CO_3 \rightarrow H^+ + HCO_3^-.$$

Atmospheric pollutants such as nitric oxide or sulfur dioxide from fossil fuel combustion can release even more hydrogen ions, decreasing the pH of precipitation to 3 or less. This is known as **acid rain**.

Unpolluted precipitation with a pH around 5.6 can chemically dissolve terrestrial minerals through **carbonation** (Chapter 9). Minerals dissolved through this weathering process, including sodium and chloride if present, then enter streams and rivers, ultimately draining into the oceans. Over time, the accumulation of the runoff from terrestrial surfaces into oceans has increased ocean salinity to its current concentration of roughly 35 g of dissolved salt per liter of water (35 ppt or 35,000 ppm). This combination of an abundant supply of sodium and chloride, weathering from terrestrial surfaces aided by acidic precipitation, runoff from rivers, and ocean evaporation resulted in 97 percent of the world's water being saline.

The salinity of Earth's surface waters has a major impact on climate. By forming a solution, the boiling- and freezing-point temperatures of saline water are raised and lowered (respectively), extending the temperature range where water remains liquid. At the ocean's current average salinity, the freezing point is roughly −2 °C (Chapter 3). This lower freezing point means that more of the oceans remain ice-free, aiding heat transfer between ocean and atmosphere. Ice-free water has a much lower **albedo** (the ratio of incident to reflected solar radiation; Chapter 8) than ice-covered water, permitting a much greater absorption of solar radiation. In addition, larger exchanges of energy and mass between the ocean and atmosphere in the forms of evaporation (the latent heat flux), convective heat loss (the sensible heat flux), and wind and waves occur over ice-free ocean surfaces.

Ocean salinity and temperature play a major role in global ocean circulation patterns, hence climate. The transport of energy from equatorial to polar regions is achieved primarily through ocean circulation. Vertical water temperature profiles are affected by the relationship between water temperature and density, with **thermal stratification** (lower-density warm water remains above higher-density cooler water) resulting in restricted vertical mixing when the water temperature exceeds the 3.98 °C temperature of maximum density. The density of water is also affected by the solute concentration. Saline water is denser than fresh water, and "sinks" beneath it, especially if that fresh water is also warmer. The combined influence of water temperature and salinity on the ocean's transport of energy is known as **thermohaline circulation** (Figure 5.1; temperature = *thermo*; salinity = *haline*) and has a significant influence on global and regional climate through ocean circulation (see Chapter 6, Section 6.3).

Ocean salinity also affects the water cycle through its influence on evaporation and precipitation formation. The addition of dissolved sodium and chloride ions lowers the **saturation vapor pressure** (the vapor pressure when the evaporation and condensation rates are equal;

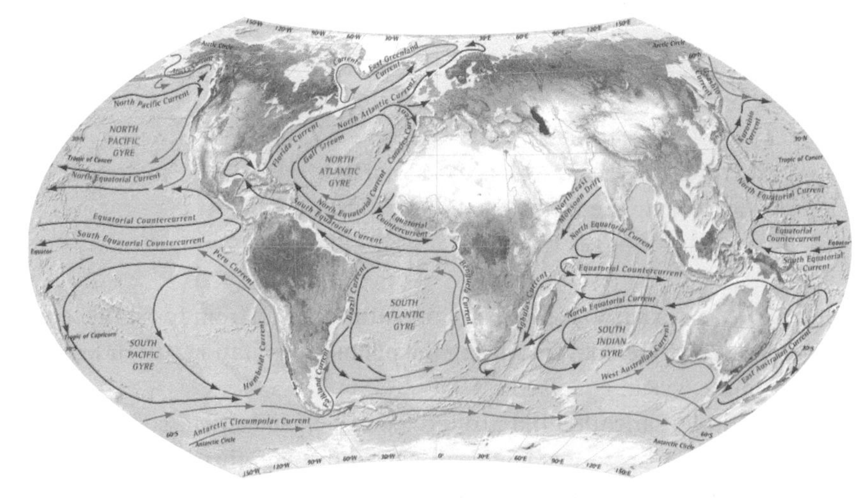

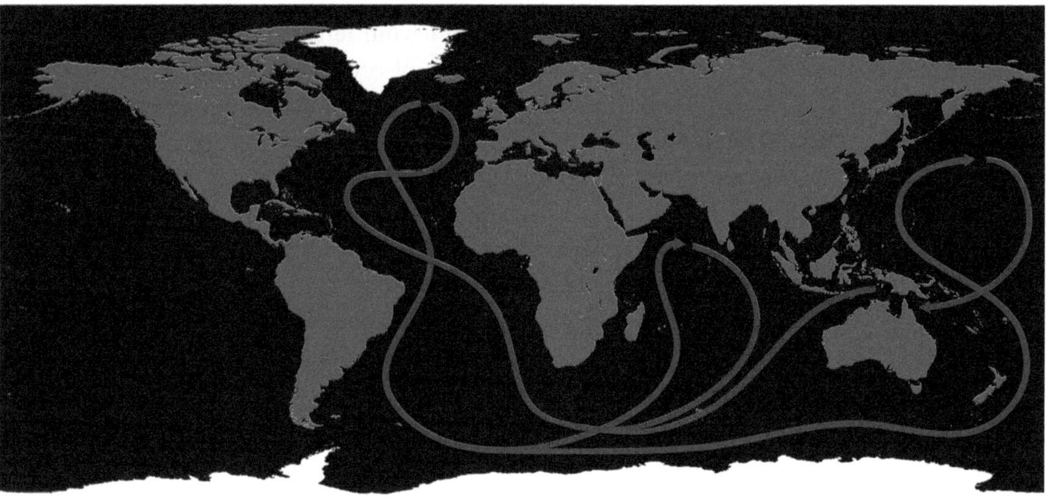

Figure 5.1 General global ocean circulation patterns. Top map shows circulation of warm surface (red) and cooler, deeper currents (blue), and the locations of the major rotating currents (gyres). Bottom map shows the deep thermohaline circulation of warm (red) and cold (blue) water. Credit (top): Dorling Kindersley, Getty Images. Credit (bottom): aristotoo, Getty Images.

Chapter 7) at the water surface. The lower saturation vapor pressure reduces the difference in vapor pressure gradient between the surface and the atmosphere, thus reducing evaporation (Chapter 7). The evaporation rate from the hypersaline Dead Sea (a lake) averaged 20 percent higher above a freshwater plume compared to the evaporation rate above the saltier open lake (Mor et al., 2018). Over time, evaporation from saline water tends to increase salinity, since only water molecules evaporate, not the solutes. Think of the Dead Sea in the Middle East or Great Salt Lake in North America, both located in regions with high evaporation rates and very little surface water input. Salt particles that are liberated from the ocean surface through wave action from storms and large waves or breaking waves along shorelines form the visible mist known as **sea spray**. Since water is the universal solvent, the salt liberated into the atmosphere through sea spray provides a strong source of **cloud condensation nuclei** required in the formation of precipitation (Chapter 7).

5.3 Lakes

Lakes contain most of the accessible surface liquid fresh water. The water contained in lakes is equal to 1.7 times the annual terrestrial precipitation, equivalent to a water depth of roughly 1.26 m if spread over the land area (Messager et al., 2016). **Fresh water**, with a salinity less than 1,000 ppm (0.1%) (although <500 ppm, 0.05%, is sometimes used), is required by nearly all organisms, yet makes up only 2.5% of the world's water, making it a crucial resource. Although lakes are usually fresh water, salinity does not define a lake. A **lake** is required to be surrounded by land (except for rivers or streams that flow in or out) and not part of an ocean. Great Salt Lake in North America is a classic example of this definition since it is quite saline (5–27%) but landlocked. **Ponds** are defined as smaller than a lake, but there is no universally accepted surface area to make the distinction. Depth is sometimes considered, with a pond being shallow enough to allow rooted aquatic vegetation growth and light penetration to the bottom. So, the definitions of oceans, seas, lakes, and ponds are all based on size and geography, not salinity. As an example of how confusing these definitions can be, consider the Caspian Sea. With a surface area of 371,000 km^2, the Caspian Sea is the world's largest body of water surrounded by land. Because it is landlocked, it should be defined as a lake. Owing to its large surface area and salinity (1.2%), however, it is commonly referred to as the Caspian Sea, not Lake Caspian.

Limnology, the study of lakes and all inland aquatic ecosystems, has provided detailed insight into the various types of lakes found in terms of their structure, aquatic biology, and ecological functions. Limnology as a field of study was established through research on Lake Geneva, which spans the Switzerland–France border, by **François-Alphonse Forel** (1841–1912), the "Father of Limnology".

Lakes are numerous in quantity yet collectively cover little of Earth. High-resolution satellite imagery estimates that there are 117 million lakes with a surface area greater than 2,000 m^2 worldwide, covering 3.7% of Earth's nonglaciated surface land area (Verpoorter et al., 2014). Of these, 27 million have a surface area larger than 10,000 m^2 (0.01 km^2), and 22 million are located between 60° N and 60° S latitudes. The highest concentration of lakes is in boreal and Arctic latitudes (45°–75° N), and 85% of all lakes are located at elevations less than 500 m above sea level (Verpoorter et al., 2014). This abundance of Northern Hemisphere high-latitude/low-elevation lakes reflects where most of the continental land mass currently exists, with surface drainage often restricted by impervious bedrock (e.g., the Canadian Continental Shield) and/or seasonally or permanently frozen ground (**permafrost**; Chapter 6) (Figure 5.2).

In regions largely covered by lakes, for example low-elevation boreal and Arctic regions such as northern Canada, or the North American Great Lakes region, lakes can have a significant influence on local climate and hydrology. In northern regions of Canada, lakes have a profound impact on regional climate and water balance (Rouse et al., 2005; 2008a;

Figure 5.2 Aerial photograph of a northern landscape with numerous small shallow lakes in the Canadian Continental Shield over the flat, poorly drained terrain. Photo credit: P. D. Blanken.

2008b). In summer, these northern lakes absorb solar radiation, storing heat and evaporating, keeping the surface cooler than the surrounding terrain. Further south over the Great Lakes, exposed ice-free water persisting late into the fall and winter often results in ample evaporation generating lake-enhanced precipitation, especially snow. In addition to causing major travel disruptions, the delayed meltwater runoff late in the spring in regions downwind of the Lakes with deep snow cover alters the local hydrology (Blanken et al., 2011; Kulie et al., 2021).

As well as potentially affecting regional weather and climate, lakes are also an important water reservoir providing freshwater resources for plants, animals, and people. In terms of water volume, the largest ten freshwater lakes store the majority (93%) of the global total (Table 5.1), highlighting the importance of individual lakes as important freshwater resources (70% of the total volume of water is contained in four lakes: Lakes Baikal, Tanganyika, Superior, and Malawi). The North American Great Lakes alone contain a combined 22,810 km^3, or 25% of the estimated 91,000 km^3 total fresh water contained in surface lakes, or roughly 20% of the world's total surface fresh water (i.e., excluding fresh water stored as groundwater).

5.3.1 Lake Formation

To understand how lakes form, we must consider two factors. The first is surface topography: the creation of an area lower than its surroundings in which water can be collected and

Table 5.1 Water volume and surface area of the largest freshwater lakes in the world (by volume)

Lake, continent	Volume (km^3)	Percent of global total volume (%)	Surface area (km^2)	Percent of global total surface area (%)
1. Baikal, Asia	23,600	25.9	31,500	0.63
2. Tanganyika, Africa	18,900	20.8	32,600	0.65
3. Superior, North America	12,100	14.2	82,100	1.64
4. Malawi, Africa	8,400	9.2	29,500	0.59
5. Vostok, Antarctica	~5,400	5.9	12,500	0.25
6. Michigan, North America	4,900	5.4	58,000	1.16
7. Huron, North America	3,540	3.9	59,600	1.19
8. Victoria, Africa	2,750	3.0	68,870	1.38
9. Great Bear, North America	2,236	2.5	31,000	0.62
10. Issyk-Kul, Asia	1,738	1.9	6,200	0.12

Percentages were determined using a total global freshwater lake volume of 91,000 km^3 (Shiklomanov, 1993) and a surface area of 5 million km^2 (Verpoorter et al., 2014).

retained. The second is the water balance (Chapter 1): input(s) of water into the lake must be at least equal to the output(s) to maintain a constant water level.

Ice sheets and glaciers often play a large role in lake formation. When ice sheets melt and glaciers recede, meltwater can collect in depressions left in the ground from the weight of the overlying chunks of ice, called **kettle lakes** (or kettle ponds). Rivers carrying meltwater (or any river) flowing over flat terrain meander and can eventually form **oxbow lakes** when the river channel is cut off from the main flow (Chapter 9). Small but numerous lakes known as **supraglacial lakes** can form on top of ice sheets when melting occurs, or under ice sheets or glaciers (**subglacial lakes**), such as Lake Vostok in the Antarctic, where the water's melting-point temperature is decreased because of the pressure of the overlying ice (Chapter 8). Under the Antarctic Ice Sheet, over 70 subglacial lakes have been identified with a combined estimated volume of 4,000–12,000 km^3 (Dowdeswell and Siegert, 2003) with the upper estimate comparable to the volume of water contained in Lake Superior (Table 5.1). The erosion and depression of the landscape from the former large North American **Laurentide Ice Sheet** with its meltwater is responsible for many of the thousands of small lakes throughout Canada, the midwestern and eastern United States, and the Great Lakes that supply drinking water for roughly 48 million people (Figure 5.3).

Lakes can form suddenly in depressions created by natural or human-made events. Following a large volcanic eruption, the void left, and the collapse of the surface due to the rapid drainage of magma, creates a large circular depression known as a **caldera**. This circular depression forms a natural closed basin that effectively collects rain and snow, ultimately forming a **volcanogenic** or **crater lake**. Crater Lake in Oregon, United States, and Lake Toba, the largest natural lake in Indonesia, which formed after the massive Toba eruption 74,000 years ago, are good examples of these.

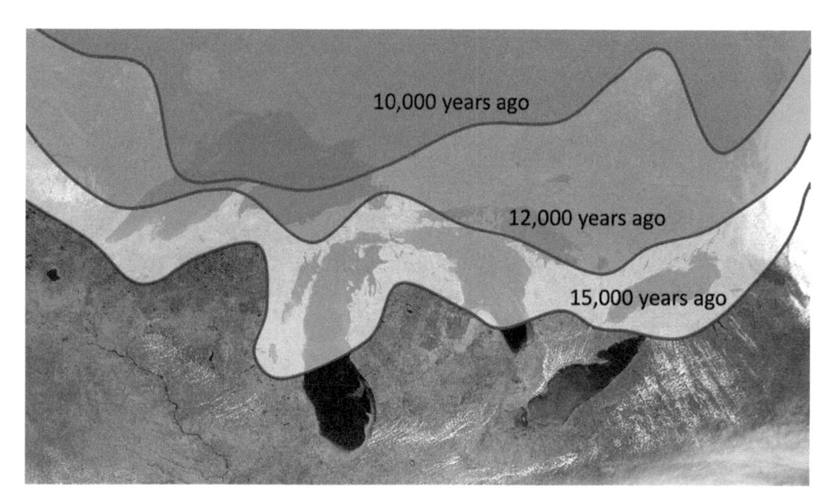

Figure 5.3 Approximate locations of the receding edge of the Laurentide Ice Sheet in the American Great Lakes Region. Meltwater provided by the receding ice sheet filled the Great Lakes basin with water, and the lakes changed in size and location as the ice advanced and retreated. Background image credit: SeaWiFS Project, NASA/Goddard Space Flight Center, and ORBIMAGE, acquired April 24, 2000. Ice sheet boundaries interpreted from Larson and Schaetzl (2001).

Volcanic eruptions can also create lakes when lava flows and dams rivers once cooled and solidified. The deep (259 m) subalpine Garibaldi Lake in British Columbia, Canada, formed in this manner. Lakes created by the natural damming of a river are known as **barrier lakes**. Landslides or rockslides in steep terrain, triggered by earthquakes or heavy rain events, can also form natural dams in rivers, creating lakes upstream. Examples include Quake Lake in Montana, United States, which formed in 1959 after a landslide triggered by an earthquake, or Sarez Lake in Tajikistan, formed in this way in 1911.

Perhaps the rarest natural lakes are **crater lakes** formed after a meteorite or asteroid impact. Such lakes are usually nearly circular and deep since that is the shape left by a large impact. Examples of such lakes include Pingualuit Lake in northern Quebec, Canada, 3.4 km wide and 396 m deep (Figure 5.4), or Manicouagan Reservoir located in central Quebec. The depression created by the large (estimated nearly 5 km diameter) asteroid that formed the Manicouagan Reservoir roughly 214 million years ago is in a region that has since experienced uplift, so today only the perimeter of the crater is filled with water, forming Manicouagan's unique circular rim shape.

Artificial lakes are those intentionally created by human activity through dam construction. The term **reservoir** applies to any lake, natural or artificial, that is used as a source of water. Artificial lakes are usually created for that purpose, hence the use of the common term reservoir. In addition to providing a source of drinking water, reservoirs also provide water for agricultural irrigation, hydroelectric power generation, flood mitigation, and/or recreational activities. Reservoirs are typically located in arid and semi-arid regions where runoff provided by spring-time snowmelt is captured to meet summer-time water

Figure 5.4 Pingualuit Lake, Quebec, Canada, a crater lake formed after a meteorite impact. Photo credit: Stocktrek Images via Getty Images.

demands. As the demand for water continues to increase with the increasing human population coupled with decreasing water supplies and higher temperatures, the volume of water stored in reservoirs has increased dramatically. Recent estimates indicate 16.7 million reservoirs with a surface area larger than 100 m^2 worldwide (2.8 million larger than 1,000 m^2) with a combined surface area of 305,723 km^2 and total volume of 8,069 km^3 (Lehner et al., 2011). This reservoir water volume is almost 9 percent of the total 91,000 km^3 of Earth's total fresh water contained in lakes. Nearly 8 percent of the world's rivers with an average flow above 1 $m^3\ s^{-1}$ are affected by the creation of dams to form these artificial lakes (Lehner et al., 2011).

5.3.2 Lake Classification

Lakes can be classified based on several characteristics: their formation (geology), hydrology, chemistry (nutrients or productivity), or temperature. As discussed in Chapter 3, liquid water density varies with temperature, reaching a maximum at 3.98 °C for fresh water. Therefore, water thermally stratifies at temperatures above 3.98 °C (Figure 5.5). This thermal density-based vertical stratification consists of a low-density **epilimnion layer** located above a colder high-density **hypolimnion layer**. The **thermocline**, the region of rapid temperature change, separates the epilimnion from the hypolimnion. Seasonally, the epilimnion layer may increase density with cooling and pass downward through the hypolimnion. This vertical mixing has both physical and biological importance since warm nutrient- and oxygen-rich water is mixed throughout the water column, providing energy, nutrients, and oxygen to the **benthic** (bottom-dwelling) organisms.

This temperature-induced thermal mixing is used for the classic classification of lakes summarized in Table 5.2. The basis of this classification is whether the lake mixes (Greek "miktos", **mictic**, meaning "mixed") or not (**amictic**), mixes once per year (**monomictic**), or several times per year (**polymictic**), all depending on whether the 3.98 °C temperature of maximum density is reached near the surface. This classification of lake type based on mixing characteristics results from lake depth and the unique density–temperature property of water, and is a useful classification for lake aquatic ecology studies.

Table 5.2 Lake classification (type) based on mixing classification from thermal structure

Lake type	Development of a seasonal ice cover?	Does surface temperature exceed 4 °C?	Thermal stratification period	Frequency of thermal mixing	Example
Amictic	Always ice-covered	No	None	Never	Lake Vostok, Antarctica
Cold monomictic	Yes	No	None	Never	Great Bear Lake, North America
Continuous cold polymictic	Yes	Yes	Daily	Daily	Lake Nipissing, North America
Discontinuous cold polymictic	Yes	Yes	Weeks	Irregular	Lake Winnipeg, North America
Dimictic	Yes	Yes	Months	Twice per year (start and end of ice cover)	North American Great Lakes
Warm monomictic	No	Yes	Months	Once per year	Lake Tanganyika
Discontinuous warm polymictic	No	Yes	Weeks	More than once per year	Loch Leven, Europe
Continuous warm polymictic	No	Yes	Hours	Daily	Lake Sonachi, Africa

Adapted from Lewis (1983).

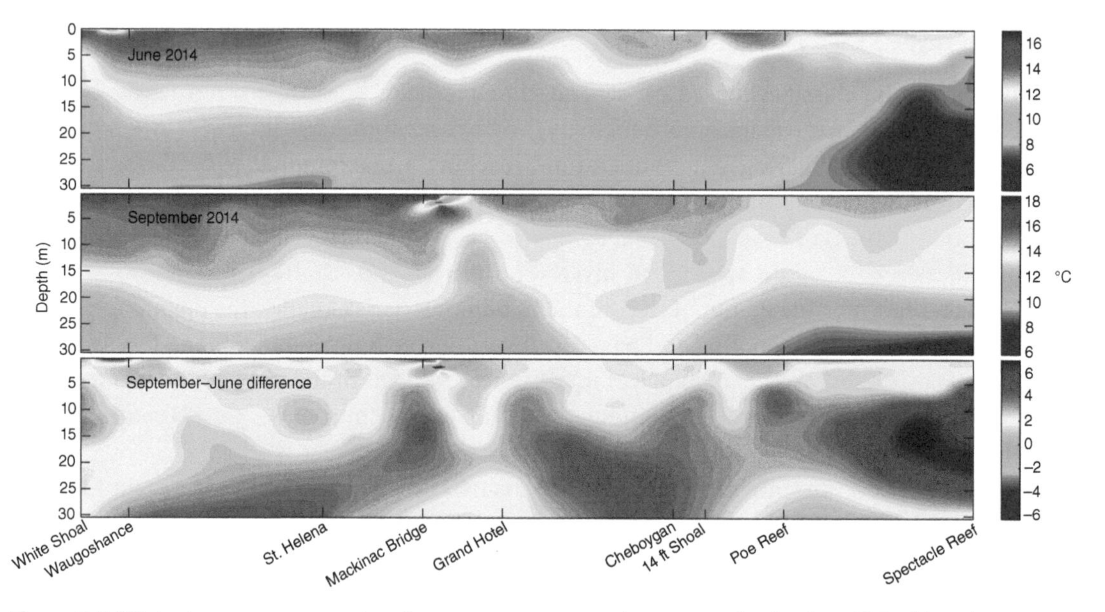

Figure 5.5 Water temperature structure from measurements taken across the Straits of Mackinac in Lakes Michigan and Huron in June and September 2009. Reproduced with permission from Barry and Blanken (2016).

5.4 Wetlands

Wetlands are land covered by water at some time during the year. They are a small portion of Earth's surface but play a large hydrologic and ecological role. With shallow, intermittent water coverage, wetlands often host unique aquatic and emergent vegetation. Coastal wetlands form **estuaries** (transition environments where freshwater river discharge mixes with ocean saltwater) or **marshes** (when frequently flooded by tidal saltwater) when adjacent to saline waters, or **riparian zones** when adjacent to freshwater rivers or lakes. While many estuaries and riparian wetlands are narrow and occur in small discontinuous segments, making them hard to detect, others are not. The broad, expansive wetlands found in flat areas that are poorly drained due to geology, topography, and/or permafrost can be massive. For example, wetlands in the central Amazon Basin have an estimated area of 250,000 km^2 (Richey et al., 2002), and the Hudson Bay Lowland in Canada, one of the three largest wetlands in the world, covers an even larger 320,000 km^2 (Roulet et al., 1994).

Natural wetlands cover a small (4%) fraction of the Earth's surface (Prigent et al., 2001) with a surface area of 29.83 million km^2 (Hu et al., 2017), and store an estimated 11,470 km^3 of fresh water, representing 0.03% of the world's total (Shiklomanov, 1993). Wetlands, however, host more than 40% of the world's plant and animal species, and provide traditional medicines for 80% of the world's population (Mitra, Wassmann and Vlek, 2003). The ecological services that wetlands provide are vastly disproportional to their surface area or freshwater volume, yet they face constant and continued degradation and loss. Studies report that 50% of the world's wetlands have been lost since 1900, and possibly 87% since 1700 (Davidson, 2014).

Wetland classification is based not only on hydrologic characteristics but also on ecosystem characteristics. Hydrologically, **wetlands** are defined as transitional areas between terrestrial and aquatic systems where the water table is usually at or near the surface (Cowardin and Golet, 1995). The **water table** is the depth where the soil is saturated with water. Wetland ecosystems can be divided into three main groups: **coastal**, **freshwater**, or **peatlands**. **Coastal wetlands** are those influenced by saline water though tides or floods and include **tidal saltmarshes** and **mangrove swamps**. **Freshwater marshes** and **swamps** are influenced by predominantly freshwater tides. Freshwater wetlands not influenced by tides or saline waters include freshwater marshes, swamps, and **riparian forests**. Lastly, **peatlands** are wetlands with deep, well-developed organic acidic soils (peat deposits). The low oxygen concentration in the saturated soils results in the slow and incomplete decomposition of organic matter and the accumulation of organic matter over time (referred to as **peat**) and acidic soils (Mitsch and Gosselink, 2015). Peatlands are the most widespread of wetlands worldwide, covering an estimated 4 million km^2 with 87% located within the boreal and Subarctic regions (Vitt, 2006).

Given the widespread dominance of peatlands (also known as **mires**), there is debate as to how they are best classified. The traditional classification is based on hydrology: a **bog** is a mire that receives most of its water from precipitation rather than groundwater or stream inflow, and the term **ombrotrophic bog** is used, meaning "rain-feed" (Greek). Since precipitation is already slightly acidic, and a bog lacks dissolved minerals and nutrients from groundwater inputs, bogs are acidic and nutrient-poor with vegetation that reflects these conditions. In contrast, a mire that receives most of its water from groundwater sources or stream inflow rather than precipitation is a **minerotrophic fen** or simply a **fen**. Owing to the additional inputs of nutrients and minerals from water flowing through soils and rocks, fens are nutrient-rich and tend to have soils far less acidic than bogs. Vegetation in a fen tends to be richer than vegetation found in a bog. Since the vegetation, soils, and ecology are so varied between bog and fen, debate on mire classification, whether based on water source, soil pH, soil fertility, or vegetation communities, continues (Hájek et al., 2006). Additional terms are used to further classify wetlands based on vegetation type and characteristics, with definitions that vary with region or country. On the basis of vegetation, a **marsh** is a wetland dominated by herbaceous (nonwoody) vegetation, and a **swamp** is a wetland dominated by woody tree and/or shrub vegetation (Holland, Whigham and Gopal, 1990).

5.5 Streams and Rivers

Streams and rivers connect landscapes, the continents to the oceans, and terrestrial to aquatic hydrology and processes. When soils are saturated from precipitation and/or the surface limits the vertical infiltration of water, surface runoff begins. If the topography is steep enough to induce lateral flow, water drops collect to form **rills**, rills connect to form **streams**, and streams connect to form **rivers**. All these overland flow units ultimately connect the land to the oceans through not just water but all that water transports: sediment, nutrients, organisms, and pollutants.

Since streams and rivers are frequently replenished by precipitation, they are mostly fresh water. Today, the world's rivers contain an estimated 2,120 km^3, a mere 0.006% of the world's total freshwater volume (Shiklomanov, 1993). This small volume does not imply that rivers do not serve vital hydrological and ecological functions. Indeed, streams and rivers and their associated riparian corridors are fragile freshwater ribbons that perform vital hydrologic, ecologic, and social services. River length is a better measure than water volume to highlight their importance.

Globally there are an estimated 17 million rivers, with an average length of 3.6 km (Lehner and Grill, 2013). An estimated 308,015 rivers are 10 km or longer, with the majority (91%) classified as short (10–100 km) and free-flowing (largely unobstructed by dams or diversions), with a combined length of 11,833,000 km (Grill et al., 2019). Of these, only 23% flow uninterrupted to the oceans, and only 37% have an unobstructed length longer than 1,000 km (Grill et al., 2019). Rivers collect surface water from an area, a drainage basin, that is defined by topography. Changes affecting water that occur in the drainage basin can therefore affect downstream regions as river channels merge. As a result, rivers are among the world's most degraded ecosystems (Best, 2019). There are several stressors placed on rivers resulting from human activity, such as damming to create reservoirs, water pollution from agricultural, urban, and industrial runoff (Chapter 15), sediment from runoff associated with changes in land use and land cover, invasive species introduction, and increased water temperature from climate change. These impacts have been so significant that in recent years the color of river and lake water has changed, especially near dams and urban areas (Gardner et al., 2020). Satellite remote sensing data show that one-third of larger rivers in the United States had significant color shifts in 2018 compared with 1984. Lakes appeared greener and rivers appeared yellower, indicating increased sediment loads in rivers and connected lakes and reservoirs, decreased water levels, and increased algal concentration.

5.6 Near-Surface Liquid Water

Fresh water located just beneath the land surface yet still accessible to plants and people represents an additional reservoir of fresh water. Fresh water stored in the near-surface unsaturated soil pore space (soil moisture) or occupying all pore spaces in the soil or rock (groundwater or aquifers) is not directly affected by conditions at the surface. When buffered against evaporative losses and potentially contaminated surface water inputs, water stored just beneath the surface often provides a clean and reliable water supply, especially during drought periods. Knowing how to calculate how much water is stored in soils is critical.

5.6.1 Soil Moisture

Water stored in soils is a vital resource for terrestrial vegetation (Chapters 11 and 12), so the quantification of **soil moisture** is important for plant ecology, especially agriculture. Soils contain mineral or organic solid materials with voids interspersed between containing air or water. The measure of the void or pore space in a soil is the **soil porosity** (f), the ratio of the **pore space volume** (V_f) to the **total soil volume** (V_t). The total volume of the soil is equal to the sum of the volumes of soil solids (V_s), air (V_a), and water (V_w) (Hillel, 2003) (Eq. 5.1):

$$f = \frac{V_f}{V_t} = \frac{V_a + V_w}{V_s + V_a + V_w}. \tag{5.1}$$

The density of the soil solids, **or particle density** (ρ_s), is equal to the soil solid mass (M_s) divided by the soil solid volume (Eq. 5.2):

$$\rho_s = \frac{M_s}{V_s}. \tag{5.2}$$

The **dry bulk density** ρ_b (noted as "dry" since the mass of water is excluded) is equal to the soil solid mass divided by the total soil sample volume consisting of the volumes of soil solids, air, and water (Eq. 5.3):

$$\rho_b = \frac{M_s}{V_t} = \frac{M_s}{V_s + V_a + V_w}. \tag{5.3}$$

When the dry bulk density and particle density are known, the soil porosity can be determined (Hillel, 2003) (Eq. 5.4):

$$f = \frac{\rho_s - \rho_b}{\rho_s}. \tag{5.4}$$

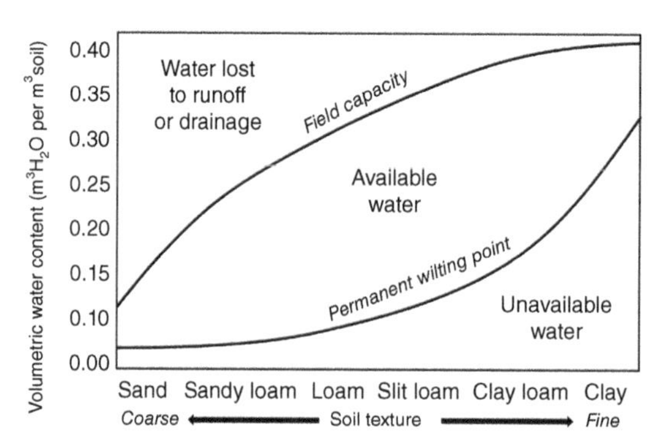

Figure 5.6 Soil water that is available to vegetation lies between the permanent wilting point and the field capacity, both of which vary with soil texture. Coarse-textured sandy soils lack water for plants since water easily drains. Fine-textured clay soils can retain more water, yet the water is tightly bound to clay particles and thus difficult for plants to extract.

Soil porosity varies from roughly 0.30 (30%) for mineral soils typically found in arid regions to 0.60 (60%) for organic soils typically found in wetlands. Therefore, somewhere between 30% and a maximum 60% of a volume of soil could be occupied by water, but rarely is the pore volume saturated. If the soil was perpetually saturated, it would be defined as **groundwater** with the uppermost saturated soil layer marking the top of the **water table**. The zone of soil that has a variable water content and is not always saturated is known as the **vadose zone**.

Soil water content can be measured either as the mass of water per mass of dry soil: the **mass** or **gravimetric water content**; or as the volume of water per volume of soil: the **volume** or **volumetric water content**. The mass water content (w) is the ratio of the mass of water (M_w) to the mass of the dry soil particles (M_s) (Hillel, 2003) (Eq. 5.5):

$$w = \frac{M_w}{M_s} \tag{5.5}$$

and the volume water content (θ) is the ratio of the volume of water to the soil's total volume (Eq. 5.6):

$$\theta = \frac{V_w}{V_t}. \tag{5.6}$$

If the soil were saturated, the soil's volume water content would equal its porosity. In the vadose zone, however, the soil is rarely saturated, and the water content varies from nearly completely dry to a value less than saturation. Soils are never completely dry since some water is tightly or **hygroscopically bound** in a thin layer to soil particles, especially clay particles that have a large surface area with a net negative charge to attract the bipolar water molecule. At the other extreme, excessive water will drain from soils owing to gravity. For vegetation, the range of water content available for plant growth is referred to as the **available water content** (Figure 5.6). This range lies between the two extremes of the soil being either too dry or too wet. At the dry end, the **permanent wilting point** is defined by the soil moisture content when water is so tightly bound to soil particles that the roots cannot supply enough energy to extract the water (Chapter 11). At the wet end, the **field capacity** is defined by the soil moisture content where excessive water drains under the force of gravity. When the soil is saturated, the plant's roots are not able to obtain a sufficient supply of oxygen for cellular respiration. The permanent wilting point varies with plant species, and the field capacity varies with **soil texture** (based on the mass fraction of sand, silt, and clay); thus, the available water content for plants is a complex variable that varies greatly in space and time.

Measuring soil properties such as density, porosity, or water content using the equations above is not difficult. All that is required is a small metal tin to collect a sample of in-situ soil. The volume of the tin can be calculated and its weight with the soil sample can be measured to provide the "wet" mass. After oven drying, typically for 24 hours at 105 °C, the soil sample is weighed again to provide the "dry" mass. The difference between the wet and dry masses equals the mass of water in the soil at the time the sample was collected. This method of measuring soil moisture is accurate but laborious and time-consuming. Therefore, several automatic measurement techniques and instruments have been developed to provide continuous in-situ soil moisture measurements with minimal site disturbance. A summary of several measurement techniques available to measure soil moisture is provided in Schmugge, Jackson and McKim (1980), or Susha Lekshmi, Singh and Shojaei Baghini (2014).

Regardless of how soil moisture is measured, the challenge lies not in the measurement but in how to quantify the large spatial and temporal variability. Soil moisture and porosity vary with soil texture. Soil texture varies widely across ecosystems and locally with slope, aspect (the direction the surface is facing), land use, and land cover. Soil moisture also varies vertically in each **soil horizon** that marks distinct layers of soil properties such as organic matter content, porosity, and rooting depth. In addition, soil moisture varies with precipitation amount, intensity, and type (e.g., rain or snow), and with the removal of water through surface runoff, drainage, evaporation, and transpiration. These processes controlling soil water inputs and outputs are complex and are revealed by how water content varies depending on whether the soil is wetting or drying, a general phenomenon referred to as **hysteresis**. Moreover, soil moisture varies across timescales both short (e.g., rain events, daily soil water evaporation and transpiration) and long (e.g., multiple-year droughts).

Remote sensing and satellite-based observations have mapped the changing spatial patterns in soil moisture, helping quantify the volume of water contained in soils (Figure 5.7). Globally, fresh water stored in soils is estimated at 16,500 km^3, a mere 0.05% of the world's total (Shiklomanov, 1993) and comparable to that of the second largest lake in the world (by volume), Lake Tanganyika (Table 5.1). As with the freshwater reservoirs discussed in this chapter, a small percentage of the world's total volume does not imply insignificance. Terrestrial plants require a reliable source of accessible liquid fresh water for photosynthesis as well as several other functions (Chapter 11). The vadose zone provides a simultaneous reservoir for water and oxygen, not just for vegetation, but for the billions of organisms that live just beneath the surface.

Despite its critical importance, soil moisture has been decreasing globally. Deng et al. (2020) found that the global average soil moisture was 0.271 m^3 of water per m^3 of soil (27.1%)

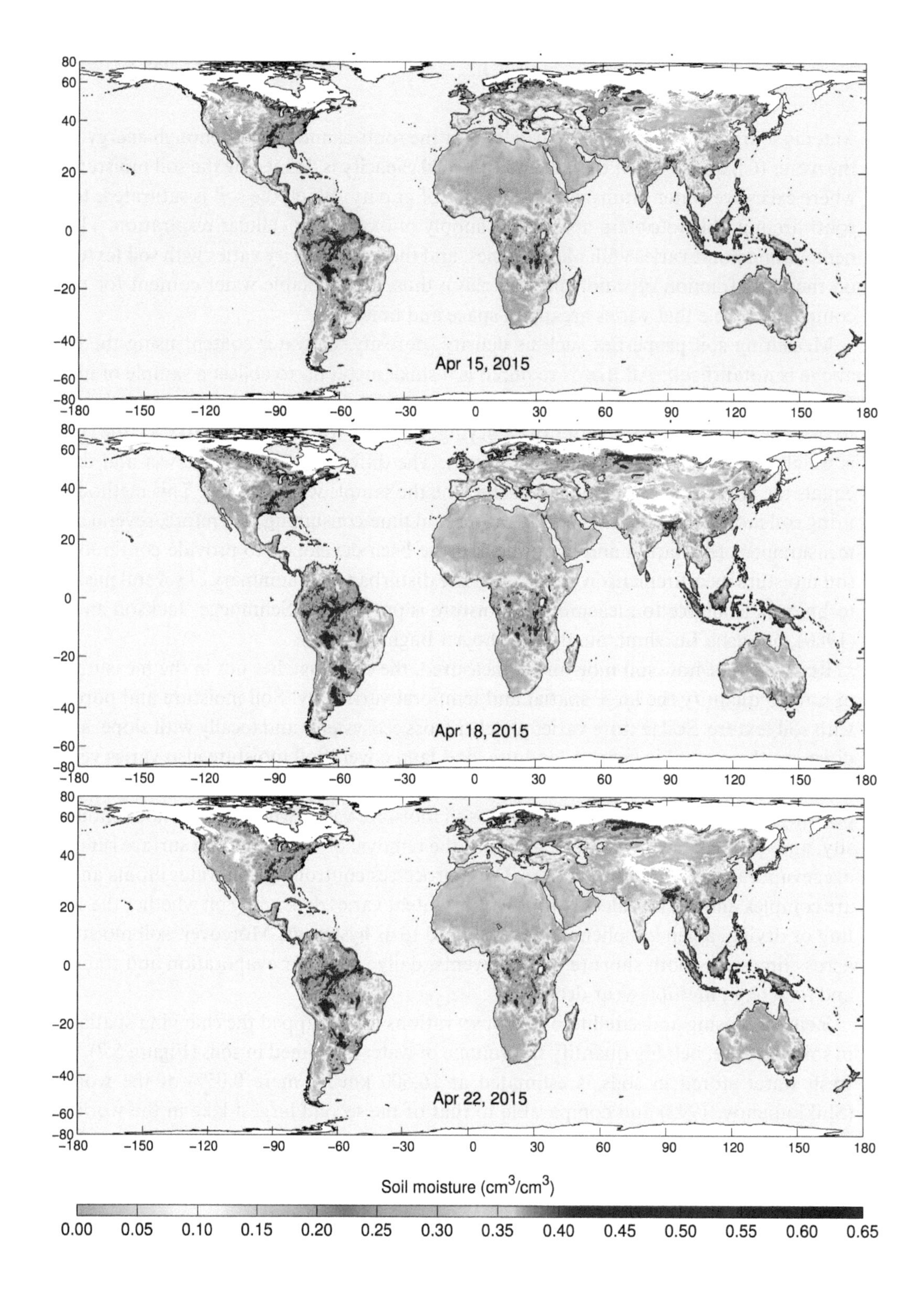

Figure 5.7 Global estimates of surface (top 5 cm) volumetric soil moisture on April 15, 18, and 22, 2015, created by NASA's Soil Moisture Active Passive (SMAP) observatory. Blue areas indicate high soil moisture content, yellow low soil moisture content, and white is snow or frozen ground.
Credit: NASA/JPL-Caltech.

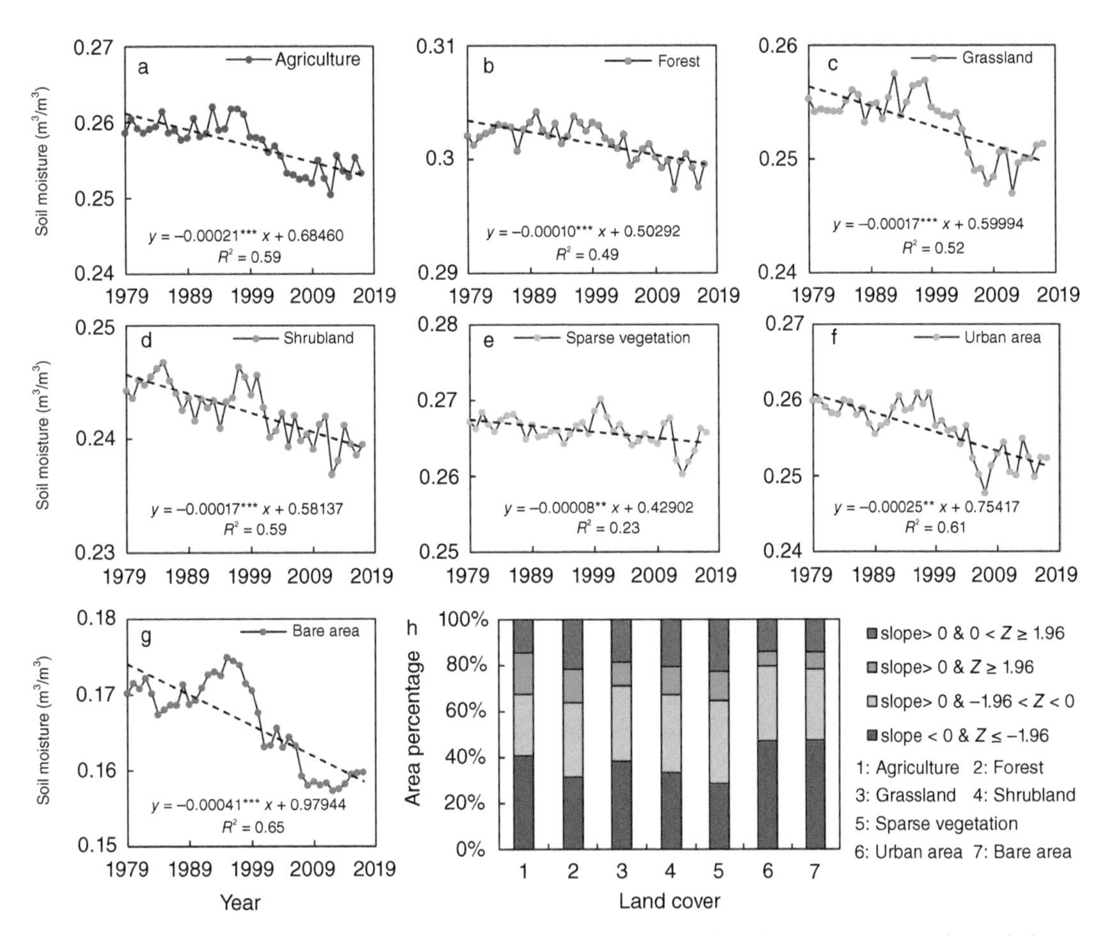

Figure 5.8 Trends in global soil moisture from different land uses and land cover types over the period 1979–2017. Reproduced with permission from Deng et al. (2020).

between 1979 and 2017. Decreasing soil moisture over this period was evident across a wide range in land uses and land covers (Figure 5.8) with 65% of this global drying trend induced by increasing temperatures. Such continued drying will have negative consequences for the global water cycle.

5.6.2 Groundwater

The largest reservoir of fresh liquid water is found beneath the surface. Water contained in soil and porous rock where all pore spaces are perpetually saturated is known as **groundwater**. There is an estimated 10,530,000 km^3 of fresh water (30.1% of the global fresh water total volume) and 12,870,000 km^3 of saline water stored as groundwater (Shiklomanov, 1993). This significant amount of fresh water, however, is neither easily accessible to vegetation, since plant roots (usually) will not grow in saturated soils, nor easily accessible for use as a resource for human activity, since pumps are usually required to extract the water to the surface. Since the water remains in contact with rocks and minerals for long periods, groundwater is often high in dissolved minerals (referred to as **hard water**), leaving an unpleasant taste and odor, and causing the corrosion of pipes and pumps. Regardless, groundwater is the world's most extracted raw material, currently being removed at a rate of 982 km^3 per year with roughly 70% used for agriculture (Margat and van der Gun, 2013).

How can we know how much water is stored as groundwater at depths that limit practical measurement other than at existing wells? Reports of groundwater withdrawals provide some localized information, but not at large scales. The physical properties of water can be used to help quantify this reservoir of water. The water molecule's high mass, coupled with its fluid motion in groundwater reservoirs, is large enough to cause variations in Earth's gravitational field that satellites can detect. Two identical satellites were launched in 2002 by NASA with the purpose of accurately mapping Earth's gravitational field. Called the **Gravity Recovery and Climate Experiment** or **GRACE**, these satellites followed one behind the other, 220 km apart, 500 km above Earth. When the leading satellite passed over a region of the Earth with a large mass and a resulting large gravitation field, such as ice sheet, ocean, or region with ample groundwater, its altitude and velocity decreased, resulting in a decreased distance between it and the trailing satellite. As the leading satellite came to the end of the region with a large mass and gravitational field, it ascended and accelerated while the trailing satellite slowed as it too passed over the same region. Measuring the changes in separation between the two satellites could therefore be used to map variations in Earth's gravitational field caused by changes in mass due to the presence of water.

Using data provided by GRACE, the magnitude of the stress on global groundwater has been quantified. An analysis of the 37 largest aquifer systems in the world showed that 21 aquifers were losing more water than they gained over the 2003–2013 study period (Richey et al., 2015). Eight of these aquifers were classified as "overstressed", the most overstressed located in the arid Arabian and Murzuk–Djado Basin, and in six of the eight aquifers, rangeland was the dominant land use (Richey et al., 2015).

The relocation of water from groundwater aquifers to surface reservoirs has been sufficient to slow the Earth's rotation, making the day length slightly longer. As the mass of a spinning object moves from the center to the periphery, the rotation slows, owing to conservation of angular momentum. Picture a figure skater changing their rotational velocity by moving their arms inward to accelerate, or outwards to decelerate. Liquid water has a similar effect when pumped from groundwater aquifers to the surface (analogous to the figure skater moving their arms outwards). For every 100 km^3 of water relocated to the surface, the daylength decreases by roughly 1 μs (microsecond; a millionth of a second) (Chao, 1995). With an estimated 10,000 km^3 of water moved to the surface and stored in constructed surface reservoirs, that is equivalent to an increase in daylength of 100 μs. Not a large number, but the fact that anthropogenic water relocation activity is having even a small impact on Earth's rotation is sobering.

5.7 SUMMARY

This chapter presented an inventory of the contemporary surface area and volume of surface liquid water. Figure 5.9 presents a graphical summary of these findings. Although 97% of all

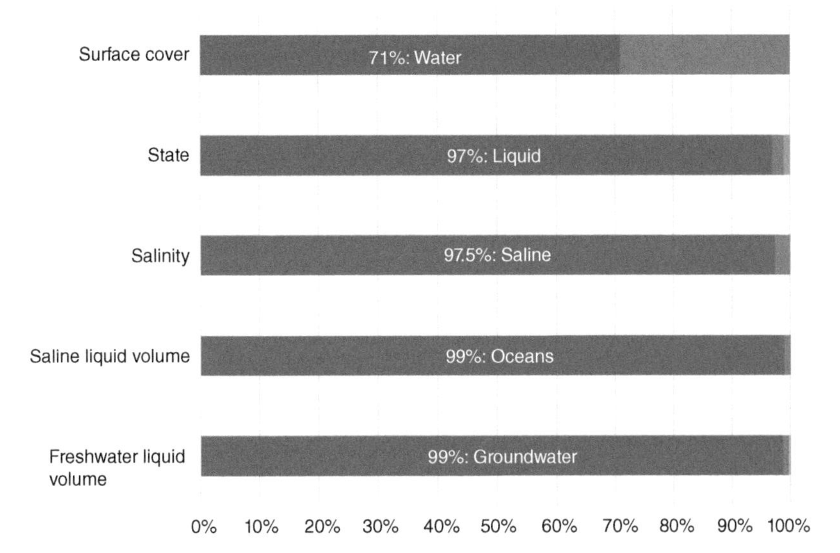

Figure 5.9 The percentage of Earth's surface area covered by water or land (top row). Water in the state of liquid, solid, gas (second row). Earth's liquid water that is saline or fresh (third row). Earth's total liquid saline water in oceans, groundwater, lakes (fourth row). Earth's total fresh water stored in groundwater, lakes, soils, wetlands, rivers (bottom row). Notes: Total surface area of Earth ~5.1×10^8 km^2. Total volume of water on Earth in all forms and locations, solid, liquid, and gas ~1.4×10^9 km^3. Total volume of liquid saline water on Earth ~1.35×10^9 km^3. Total volume of liquid fresh water as lakes, wetlands, rivers, soils, and groundwater ~10,651,090 km^3.

the water on Earth is liquid and covers 71% of Earth's surface, nearly all is in the form of saline ocean water. Over millennia, erosion from the land and evaporation from the oceans gradually resulted in the accumulation of salt in the oceans, with roughly 97.5% of today's water being saline. Through the ocean transport of energy by thermohaline circulation, saline water has a major effect on climate by increasing density and decreasing the freezing point relative to fresh water. Salinity also affects the rate of ocean evaporation by decreasing the surface saturation vapor pressure, and affects precipitation, by providing effective cloud condensation nuclei.

Only 2.5% of Earth's water is fresh, and nearly 99% of that fresh water is stored as groundwater and not easily accessible. This highlights the importance of surface liquid fresh water, where most resides in lakes. The definitions of many of the surface liquid water reservoirs such as lakes and wetlands are vague and vary with location. The numerous physical processes resulting in lake formation result in descriptions related to the formation process. Given the importance of the vertical mixing of water in lakes due to temperature-induced changes in density, limnologists often classify lakes based on the internal thermal structure. Wetlands, streams, and rivers have critical ecological and hydrological importance yet are subject to increasing stress due to climate change and changes in land use and land cover.

Terrestrial vegetation depends on the soil to provide the right combination of water, air, and nutrients. There is a delicate balance between having too much or not enough water in the soil that is accessible to vegetation. Estimates of global trends in soil moisture show that it is decreasing in most regions owing to increasing temperatures. Groundwater aquifers have also been depleted, and because of the lowering of regional water tables, stream and river base flow are expected to decrease.

This chapter's discussion of the contemporary surface area and volume of water stored in the liquid state paints a picture of irony. Earth is covered in liquid water, but nearly all of it is too saline to be of use for terrestrial plants or animals. Of the small fraction of water that is fresh, nearly all is stored beneath the surface and is not readily available. These facts highlight the importance of surface freshwater resources, but not all of it is in liquid form. Even though water in its solid state represents a small fraction of Earth's water volume, frozen water plays a major role in hydrology, climate, and ecology, as discussed in Chapter 6.

5.8 QUESTIONS

5.1 Explain the conditions that have resulted in 97.5 percent (by volume) of all the liquid water on Earth having a salinity greater than 1,000 parts per million. How does this saline water affect Earth's climate? Are there any conditions that could conceivably change to increase the portion of fresh water on Earth?

5.2 A newly discovered imaginary large body of water has a salinity of 1,100 ppm and is surrounded by land. Some, but not all, areas are shallow with rooted aquatic vegetation. Would this body of water be best classified as an ocean, sea, lake, pond, or wetlands, and why?

5.3 Great Salt Lake, a shallow saline lake located in Utah, United States, is rapidly decreasing in size, owing to excessive water use coupled with climate change. Consequently, large areas that were once submerged are being exposed. What are some of the major health and environmental concerns that will likely occur once the sediment is exposed, liberating dust across the region?

5.4 What are the differences between amictic, monomictic, and polymictic lakes? What are the advantages to using such a system to classify lakes?

5.5 Why are surface bodies of fresh water not classified by size? Thermal structure is used to classify lakes (e.g., Table 5.2), but a nutrient richness or productivity (trophic) classification could be used instead (e.g., oligotrophic or eutrophic lakes). Discuss the advantages and disadvantages of using one lake classification scheme over another.

5.6 The North American Great Lakes currently contain 20 percent of the world's surface fresh water and 25 percent of all the fresh water contained in surface lakes. Describe the processes that resulted in the formation of such a large portion of fresh water in this series of five connected lakes. Summarize the ecological, climatological, cultural, and/or economic impacts of the Lakes on the region, and how these impacts may change with changes in lake water quantity and/or quality.

5.7 Wetlands contain a minute (0.03 percent) fraction of the world's total volume of fresh water, yet there are numerous definitions and classifications associated with wetlands.

Discuss this apparent dichotomy between the small volume of water stored in wetlands and the wide attention paid to their form and function, as reflected by the many definitions and classifications.

5.8 Why is the length of rivers, rather than volume of water in rivers, a better metric to use to quantify their significance and function?

5.9 A soil has a solid mass of 400 g, a solid volume of 180 cm^3, and total volume of 280 cm^3. What is the soil's particle density, dry bulk density, and porosity? If this same soil has a mass and volume of 100 g and 58 cm^3 of liquid water (respectively), what are the gravimetric and volumetric water contents?

5.10 Recent changes in several of the reservoirs of liquid water such as lakes, rivers, soil moisture, and groundwater were discussed. For one or more of these, discuss reason(s) for any changes (trends), the implication(s) of these changes on the cycling of water, and possible action(s) to decrease the negative consequences of such changes.

REFERENCES

Barry, R. G. and Blanken, P. D. (2016) *Microclimate and Local Climate*. Cambridge University Press. doi: 10.1017/CBO9781316535981.

Best, J. (2019) 'Anthropogenic stresses on the world's big rivers', *Nature Geoscience*, 12, pp. 7–21. doi: 10.1038/s41561-018-0262-x.

Blanken, P. D., Spence, C., Hedstrom, N. and Lenters, J. D. (2011) 'Evaporation from Lake Superior: 1. Physical controls and processes', *Journal of Great Lakes Research*, 37. doi: 10.1016/j.jglr.2011.08.009.

Chao, F. (1995) 'Anthropogenic impact on global geodynamics due to reservoir water impoundment', *Geophysical Research Letters*, 22, pp. 3529–3532.

Copley, J. (2014) 'Just how little do we know about the ocean floor?', *The Conversation*, October 9 [online].

Cowardin, L. M. and Golet, F. C. (1995) 'US Fish and Wildlife Service 1979 wetland classification: A review', *Vegetatio*, 118, pp. 139–152. doi: 10.1007/BF00045196.

Cundy, A. and Kershaw, S. (2013) *Oceanography: An Earth Science Perspective*. Taylor and Francis. doi: 10.4324/9780203761083.

Davidson, N. C. (2014) 'How much wetland has the world lost? Long-term and recent trends in global wetland area', *Marine and Freshwater Research*, 65, pp. 934–941. doi: 10.1071/MF14173.

Deng, Y., Wang, S., Bai, X. et al. (2020) 'Variation trend of global soil moisture and its cause analysis', *Ecological Indicators*, 110, p. 105939. doi: 10.1016/j.ecolind.2019.105939.

Dowdeswell, J. A. and Siegert, M. J. (2003) 'The physiography of modern Antarctic subglacial lakes', *Global and Planetary Change*, 35, pp. 221–236. doi: 10.1016/S0921-8181(02)00128-5.

Gardner, J. R., Yang, X., Topp, S. N. et al. (2020) 'The color of rivers', *Geophysical Research Letters*, pp. 1–12. doi: 10.1029/2020gl088946.

Grill, G., Lehner, B., Thieme, M. et al. (2019) 'Mapping the world's free-flowing rivers', *Nature*, 569, pp. 215–221. doi: 10.1038/s41586-019-1111-9.

Hájek, M., Horsák, M., Hájiková, P. and Dítě, D. (2006) 'Habitat diversity of central European fens in relation to environmental gradients and an effort to standardise fen terminology in ecological studies', *Perspectives in Plant Ecology, Evolution and Systematics*, 8, pp. 97–114. doi: 10.1016/j.ppees.2006.08.002.

Hillel, D. (2003) *Introduction to Environmental Soil Physics*. Elsevier. doi: 10.1016/B978-0-12-348655-4.X5000-X.

Holland, M. M., Whigham, D. F. and Gopal, B. (1990) 'The Characteristics of Wetland Ecotones', in *The Ecology and Management of Aquatic-Terrestrial Ecotones. Man and the Biosphere Series*, Volume 4. Parthenon Publishing Group, pp. 171–198.

Hu, S., Niu, Z., Chen, Y., Li, L. and Zhang, H. (2017) 'Global wetlands: Potential distribution, wetland loss, and status', *Science of the Total Environment*, 586, pp. 319–327. doi: 10.1016/j.scitotenv.2017.02.001.

Kulie, M. S., Pettersen, C., Merrelli, A. J. et al. (2021) 'Snowfall in the Northern Great Lakes', *Bulletin of the American Meteorological Society*, 102, pp. 1317–1339.

Larson, G. and Schaetzl, R. (2001) 'Origin and evolution of the Great Lakes', *Journal of Great Lakes Research*, 27, pp. 518–546.

Lehner, B. and Grill, G. (2013) 'Global river hydrography and network routing: Baseline data and new approaches to study the world's large river systems', *Hydrological Processes*, 27, pp. 2171–2186. doi: 10.1002/hyp.9740.

Lehner, B., Reidy Liermann, C., Revenga, C. et al. (2011) 'High-resolution mapping of the world's reservoirs and dams for sustainable river-flow management', *Frontiers in Ecology and the Environment*, 9, pp. 494–502. doi: 10.1890/100125.

Lewis, W. M., Jr. (1983) 'A revised classification of lakes based on mixing', *Canadian Journal of Fisheries and Aquatic Sciences*, pp. 1779–1787.

Margat, J. and van der Gun, J. (2013) *Groundwater around the World* 1st ed. CRC Press. doi: 10.1201/b3977.

Messager, M. L., Lehner, B., Grill, G. et al. (2016) 'Estimating the volume and age of water stored in global lakes using a geo-statistical approach', *Nature Communications*, 7, pp. 1–11. doi: 10.1038/ncomms13603.

Mitra, S., Wassmann, R. and Vlek, P. L. G. (2003) 'Global inventory of wetlands and their role in the carbon cycle', *ZEF Discussion Paper on Development Policy, Bonn*, 64, p. 57.

Mitsch, W. J. and Gosselink, J. G. (2015) *Wetlands* 5th ed. Wiley.

Mor, Z., Assouline, S., Tanny, J., Lensky, I. M. and Lensky, N. G. (2018) 'Effect of water surface salinity on evaporation: The case of a diluted buoyant plume over the Dead Sea', *Water Resources Research*, 54, pp. 1460–1475. doi: 10.1002/2017WR021995.

Prigent, C., Matthews, E., Aires, F. and Rossow, W. B. (2001) 'Remote sensing of global wetland dynamics with multiple satellite data sets', *Geophysical Research Letters*, 28, pp. 4631–4634.

Richey, A. S., Thomas, B. F., Lo, M.-H. et al. (2015) 'Quantifying renewable groundwater stress with GRACE', *Water Resources Research*, 51, pp. 5217–5237. doi: 10.1002/2015WR017349.

Richey, J. E., Melack, J. M., Aufdenkampe, A. K., Ballester, V. M. and Hess, L. L. (2002) 'Outgassing from Amazonian rivers and wetlands as a large tropical source of atmospheric CO_2', *Nature*, 416, pp. 617–620. doi: 10.1038/416617a.

Roulet, N. T., Jano, A., Kelly, C. A. et al. (1994) 'Role of the Hudson Bay lowland as a source of atmospheric methane', *Journal of Geophysical Research*, 99, p. 1439. doi: 10.1029/93jd00261.

Rouse, W. R., Binyamin, J., Blanken, P. D. et al. (2008a) 'The influence of lakes on the regional energy and water balance of the central Mackenzie River Basin', in *Cold Region Atmospheric and Hydrologic Studies. The Mackenzie GEWEX Experience: Volume 1: Atmospheric Dynamics*. Springer, pp. 309–325. doi: 10.1007/978-3-540-73936-4_18.

Rouse, W. R., Blanken, P. D., Duguay, C. R. et al. (2008b) 'Climate–lake interactions', in *Cold Region Atmospheric and Hydrologic Studies. The Mackenzie GEWEX Experience*. Springer, pp. 139–160. doi: 10.1007/978-3-540-75136-6_8.

Rouse, W. R., Oswald, C. J., Binyamin, J. et al. (2005) 'The role of northern lakes in a regional energy balance', *Journal of Hydrometeorology*, 6. doi: 10.1175/JHM421.1.

Schmugge, T. J., Jackson, T. J. and McKim, H. L. (1980) 'Survey of methods for soil moisture determination.', *Water Resources Research*, 16, pp. 961–979.

Shiklomanov, I. A. (1993) 'World fresh water resources', in Gleick, P. H. (ed.) *Water in Crisis: A Guide to the World's Fresh Water Resources*. Oxford University Press, pp. 13–24.

Susha Lekshmi, S. U., Singh, D. N. and Shojaei Baghini, M. (2014) 'A critical review of soil moisture measurement', *Measurement: Journal of the International Measurement Confederation*, 54, pp. 92–105. doi: 10.1016/j.measurement.2014.04.007.

Verpoorter, C., Kutser, T., Seekell, D. A. and Tranvik, L. J. (2014) 'A global inventory of lakes based on high-resolution satellite imagery', *Geophysical Research Letters*, 41, pp. 6396–6402. doi: 10.1002/2014GL060641.

Vitt, D. H. (2006) 'Functional characteristics and indicators of boreal peatlands', in Wieder, R. K. and Vitt, D. H. (eds.) *Boreal Peatland Ecosystems*. Springer, pp. 9–24.

6 The Distribution of Ice

Key Learning Objectives

After reading this chapter, you will be able to:

1. Provide current estimates of the volume of frozen water stored as ice sheets, ice shelves, sea ice, glaciers, snow, frozen ground, and permafrost.
2. Explain how and why contemporary melting of ice has impacted and shifted the distribution of ice in the cryosphere and, consequently, had an impact on the climate of other regions.
3. Calculate the volume of liquid water contained in a snowpack.
4. Describe how contemporary changes in frozen ground are affecting the environment.

6.1 Introduction

The proportion of frozen water relative to the total volume of water on Earth is small, less than 2%, but has a significant impact on climate through the surface energy and radiation exchange, abiotic processes such as weathering and erosion, and biological processes such as the distribution and life cycles of plants and animals. The proportion of global fresh water contained in ice is nearly 69%, the largest of any reservoir, followed by groundwater at 30% (Shiklomanov, 1993). Together, then, frozen fresh water and groundwater currently store 99% of the world's fresh water. With such a large proportion of fresh water currently frozen, an understanding of the global distribution of ice is important in understanding freshwater resources.

The general term used to describe the region where water naturally exists in solid form is the **cryosphere**. *Cryo* is Greek, meaning icy cold, and it appears as a prefix in words such as **cryogenics** (the behavior of materials at low temperatures), or **cryoturbation** (the mixing of soils due to ice formation and melting). Given this broad definition of the cryosphere, naturally occurring frozen water can be further classified based on location and characteristics. This includes ice sheets and shelves, sea ice, glaciers, snow, and frozen ground, each of which is described in this chapter.

6.2 Ice Sheets and Shelves

An **ice sheet** exists when a land area is covered by more than 50,000 km^2 of ice (NSIDC, 2021). Currently there are two ice sheets on Earth, the **Greenland Ice Sheet** near the North Pole, and the **Antarctic Ice Sheet** at the South Pole. Combined, these contain over 99% of the world's frozen fresh water. This means that nearly 69% of the world's fresh water (in all states) is frozen in the Greenland and Antarctic ice sheets. The Greenland Ice Sheet covers 79% of Greenland, extending over an area of 1.7 million km^2 with an average thickness of about 1.5 km. The Antarctic Ice Sheet covers 98% of the Antarctic, spreading over an area of 14 million km^2 with an average thickness of about 2 km. It is the largest continuous mass of ice on Earth, containing about 61% of the world's fresh water. If the Antarctic Ice Sheet were to melt, the global sea level would rise by an astounding 58 m (Fretwell et al., 2012).

Ice sheets form when successive winter snow accumulation does not completely melt the following summer. Over time, the annual accumulation of snow compresses as the overlying mass increases, raising the snow's density from roughly 100 kg m^{-3} for fresh snow to roughly 900 kg m^{-3} for ice. Differences in density define the distinction between snow and ice, and other classifications such as **firn** (granular snow). The continued increase in mass and density can result in strain, deformation, and the movement or flow of the ice sheet from regions of high snow accumulation to lower coastal areas. If the ice sheet reaches the coast and floats over the ocean, it is referred to as an **ice shelf**.

For the mass of accumulated snow to exceed that lost to melting or sublimation, the winter must be long, the summer short, temperatures need to be below 0 °C for the majority of the year, and a reliable source of moisture to form precipitation must be available. The Arctic and Antarctic regions have met these requirements for centuries. Above the Arctic and Antarctic Circles (66.5° North and South latitudes, respectively), half the year is without sunlight, and during the short summer, despite constant sunlight, solar radiation inputs remain low because of the low solar altitude. The high **albedo** of the snow or ice-covered surface means that a large proportion of the incident solar radiation is reflected. The available surface energy must first be used to raise the temperature of the ice volume from below the freezing point to 0 °C (Chapter 8, Eq. 8.1).

The **mass balance** of an ice sheet is the difference between the mass inputs and outputs of frozen water. Conditions at the polar regions have for hundreds of thousands of years been suitable for the development of ice sheets, but major changes have been occurring in the mass balance of both the Greenland and Antarctic Ice Sheets with significant global consequences. An ice sheet accumulates mass when the annual snow accumulation exceeds the annual loss. Conversely, ice sheets lose mass when losses due to surface melting and ice loss from breaking ice shelves exceed snow accumulation, as in recent years. The term **ablation** is used for the total loss of mass regardless of process. The mass balance of the Greenland Ice Sheet was measured or reconstructed from ground-based measurements between 1972 and 2018 (Mouginot et al., 2019). Between 1972 and 1990 the ice sheet was nearly balanced (stable), with inputs from snow accumulation in the interior equal to outputs from glacial ice discharge into the oceans. Changes began in the 1980s when variation from the natural range increased and the loss of mass increased sixfold, resulting in sea level rise. Since 1972, the loss of mass from the Greenland Ice Sheet has raised sea levels by 13.7 mm, with half of that increase occurring during the last 8 years of the study (Mouginot et al., 2019).

Countless other studies confirm these findings. Using satellite measurements, Slater et al. (2021) found the Greenland Ice Sheet lost 3.8 trillion tonnes of ice between 1994 and 2017. Roughly half of this ice loss was associated with warm-summer induced increased meltwater runoff, and half due to increased glacier ocean discharge. Hanna et al. (2020) found that this loss of mass was more closely linked to increasing air temperatures, especially in coastal regions, than to decreases in snow accumulation rates.

The story of ice loss is similar in the Antarctic, yet for this ice sheet it is accelerated by **ice shelf collapse**, when large sections of floating ice break and collapse into the ocean. Ice shelf collapse accounted for the majority of mass loss from the Antarctic ice shelf and has increased 75 percent between 1996 and 2006 (Rignot et al., 2008). When ice sheets extend over water to form an ice shelf, they are susceptible to melting not only at the surface but also at the base (**basal melting**) from relatively warm ocean currents. In addition, **calving**, the breaking of large sections of ice along the edge of an ice shelf, promotes further lateral margin ice weakening. Recent large increases in air temperature in Antarctic coastal regions have further accelerated surface melting and hydraulic fracturing as surface meltwater flows vertically through the ice. Owing to these multiple mass-loss processes, it is estimated that the Antarctic has lost 8,667 ± 120 Gt of ice between 1994 and 2020 (more than double the total loss from the Greenland Ice Sheet) with roughly half due to reductions in extent and half due to decreases in thickness (Slater et al., 2020).

Antarctic Ice Sheet studies have shown that there are complex coupling and feedbacks between the ocean, ice, and atmosphere (i.e., all three states of water play a role). The ice-loss processes occurring at the margins of ice sheets are coupled to the loss of ice in regions far removed from the coast. In interior regions where the ice sheet is underlain by bedrock, mass loss is connected to and enhanced by the basal melting of the coastal ice shelves. Measurements and models have shown that a reduction in buttressing support derived from the thinning ice shelves can accelerate mass loss on the ice sheet itself (Pritchard et al., 2012) and therefore the entire ice sheet is sensitive to ocean temperature and currents. Snowfall, the only input in the mass balance, has increased over Antarctica by 45 percent compared with the pre-industrial period mean (Medley et al., 2018). As air temperature increases and more melting occurs, the atmosphere can contain more water vapor (see the saturation vapor pressure equation, Chapter 7, Eq. 7.2). This increased humidity and higher temperatures provide a likely explanation for the increased snowfall. The increase in snowfall, however, is not sufficient to offset the large ice mass loss that occurs primarily through ice shelf collapse. Models predict that with global warming of 2 °C above pre-industrial levels, the West Antarctic Ice Sheet will partially collapse, lose more than 70 percent of its present-day volume with warming between 6–9 °C, and completely melt with more than 10 °C warming (Garbe et al., 2020).

6.3 Sea Ice

Sea ice is defined as ice that forms and melts entirely over the ocean. Ice shelves, calving icebergs, and glaciers are generally not sea ice, since all originated on land. While some areas of the oceans are covered by ice year-round, most regions experience a seasonal pattern of winter ice formation and summer melting. Roughly 15 percent (25 million km^2) of the world's oceans are currently covered by sea ice at some time during the year, mostly in the polar regions (NSIDC, 2021).

Oceans in the polar regions contain dissolved and dissociated salts, lowering the water's freezing point and increasing its density (Chapter 3). The increased density transports the cold ocean surface water to deeper depths, affecting thermohaline ocean circulation (Chapter 5). Combined, the lower freezing point and increased density result in ice taking longer to form in polar ocean water than in fresher, warmer, water. The Arctic Ocean may not always have been saline, however. During ice ages, ice shelves may have been thick and of sufficient extent to block the inflow of saline water from the Atlantic Ocean. With isolation, the Arctic Ocean would gradually have become less saline with freshwater inputs from precipitation and melting snow and ice. There is evidence of a freshwater Arctic Ocean twice in the past 131,000 years (Geibert et al., 2021).

Sea ice has a strong influence on Earth's climate. Areas of open water absorb more solar radiation than adjacent ice-covered areas, resulting in further warming, greater melting of sea ice, increased exposure of open water, and so on. This positive **temperature–albedo feedback** amplifies warming but can also promote cooling; lower temperatures decrease the areas of exposed open water, the regional albedo increases with the increase in sea ice cover, temperatures decrease further due to reduced net solar radiation input, more water freezes, and so on. Currently, rapid warming at rates roughly double (or triple) the global average in the Arctic region is referred to as **Arctic amplification**, and is primarily due to the role of sea ice cover through this temperature–albedo positive feedback (Serreze and Barry, 2011).

The formation and melting of sea ice affect water temperature and density and therefore global ocean circulation and climate. Sea ice can dramatically influence local climate. Consider two locations at roughly the same latitude on either side of the Atlantic Ocean. On the far eastern coast of Canada lies Goose Bay in the Province of Newfoundland and Labrador, chosen in 1941 as a strategic location for an airport to service aircraft flying over the North Atlantic to help with World War II operations in Europe. Goose Bay's elevation is 12 m above sea level and its latitude is 53.30° N. The mean annual air temperature is 0.0 °C, ranging from −17.6 °C in January to 15.5 °C in July (a 33 °C annual range). At roughly the same latitude (53.20° N) and elevation (7 m) located 3,350 km across the North Atlantic Ocean lies Galway, Ireland. Galway's air temperature ranges from 5.0 °C in January to 15.0 °C in July (a 10 °C annual range), with an annual mean of 9.8 °C. Clearly the northern branch of the Gulf Stream current that veers directly towards Europe supplies warmth to Galway whereas Goose Bay feels the cold continental climate delivered by the prevailing westerly winds. This strong influence of the North Atlantic Ocean circulation pattern on regional climate can suddenly alter owing to changes induced by sea ice. Melting of Arctic sea ice and the export of fresh water likely played a key role in recent major climate shifts and future climate change. Increased freshwater input from melting Arctic ice could result in a slowing of the **Atlantic Meridional Overturning Circulation (AMOC)** and significant cooling and drying, especially across Europe (Jackson et al., 2015) (Figure 6.1).

Periods of abrupt cooling lasting hundreds of years have been linked to changes in the North Atlantic Ocean circulation. The **Little Ice Age** was a period of pronounced cooling across much of Northern Europe from around 1300 to 1850 CE. Famine, crop failure, and disease resulted in a population decrease of up to 50 percent. In London, the Thames River was completely frozen several times, allowing for so-called Ice Festivals to occur each winter. The slow growth rate of

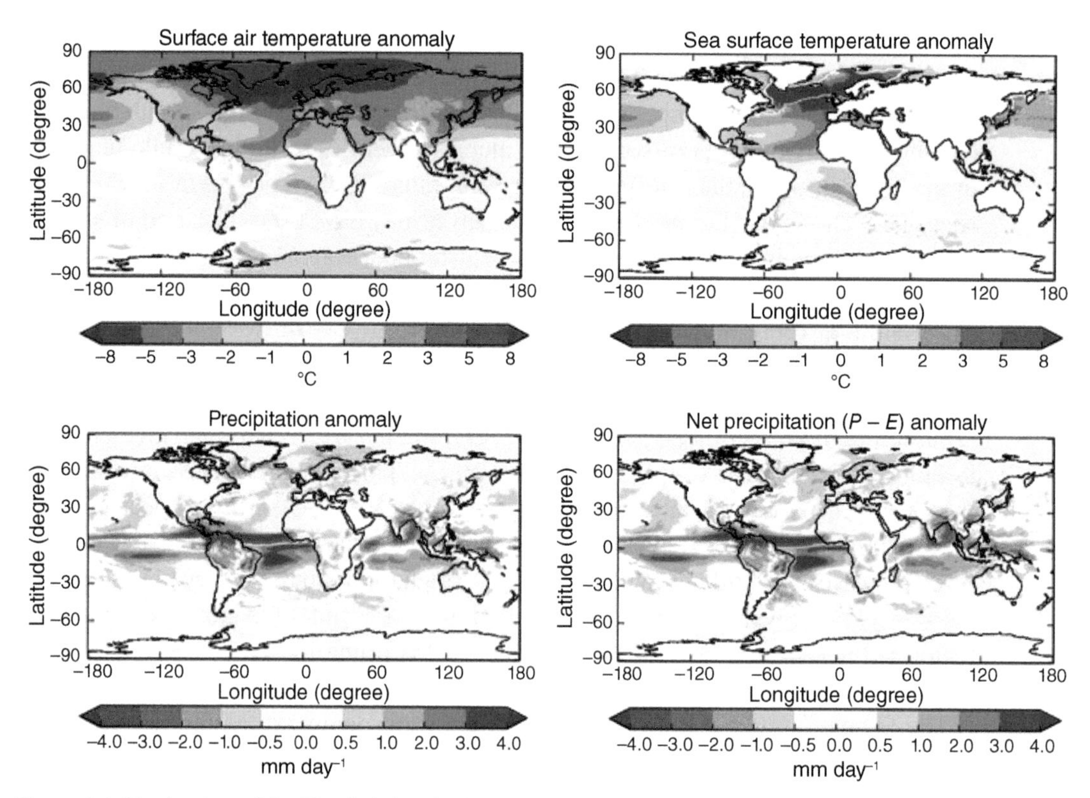

Figure 6.1 Freshening of the North Atlantic slows down the strength of ocean water circulation, resulting in a predicted decrease in surface air and sea temperatures (top left and right, respectively), decreased precipitation (bottom left) and drying (precipitation minus evaporation, bottom right) in the Northern Hemisphere and especially across Europe. Anomalies are relative to 30-year averages 60–90 years after the modeled freshwater perturbation. Reproduced with permission from Jackson et al. (2015).

trees resulted in high-density wood thought to provide the unique sound of Stradivarius violins. Prior to the Little Ice Age, the warm and wet conditions of the **Medieval Warm Period** from around 950 to 1250 CE also had a large societal impact. Warm and stable climate periods brought prosperity and stability whereas periods of high climate variability were associated with turmoil and conflict (Büntgen et al., 2011). Further in the past, the **Younger Dryas Period,** a period of cold and dry conditions across the northern latitudes around 12–13,000 years ago lasting roughly 1,000 years was also likely a consequence of changes in the position and strength of the Gulf Stream and the North Atlantic subpolar gyre due to the influx of cold fresh water into the Atlantic Ocean (Broecker et al., 1989). Somewhat ironically, the warming and melting at the end of the last Ice Age resulted in so much cold fresh water entering the Atlantic Ocean that a further thousand years of cooling resulted.

How does sea ice play a role in these abrupt and drastic changes in climate on either side of the Atlantic Ocean? In the North Atlantic Ocean, the counterclockwise circulation of southward-flowing cold water on the western side and northward-flowing warm water on the eastern side is known as the **North Atlantic Subpolar Gyre** (Chapter 5, Figure 5.1). The circulation and shape of this gyre is affected by the contrast in ocean water density, which in turn is affected by temperature and salinity. During periods of extensive Arctic sea ice, the ocean surface layers freshen as increased export of sea ice through the Fram Strait into the gyre eventually melts as it travels southwards. Salt is expelled as ice forms, hence multiyear sea ice consists of fresh water (Chapter 8). This additional import of fresh water during periods of extensive Arctic sea ice tends to weaken gyre circulation, resulting in a cooler climate in Northern Europe due to the reduction in northern heat transport on the western side of the gyre (Moreno-Chamarro et al., 2017). The period of anomalously large and abrupt export of Arctic sea ice during the fourteenth century that likely triggered abrupt climate changes such as the Little Ice Age is sometimes known as the **Great Sea Ice Anomaly**. Additional explanations for the increased Arctic sea ice extent during the fourteenth century include increased volcanic activity and intensity, and decreased solar irradiance, although several studies have shown that neither is required to initiate such large cooling events; cooling through sea ice feedbacks is sufficient (Miles, Andresen and Dylmer, 2020).

The Little Ice Age shows how important and sensitive regional climate is to sea ice conditions and the powerful temperature–albedo feedback, both ultimately connected to the physical properties of water. Therefore, how sea ice conditions are quantified requires strict definitions. Arctic sea ice varies seasonally, reaching maximum extent in March (September in the Antarctic) and minimum extent in September (March in the Antarctic). **Sea ice concentration** is defined as the relative area covered by ice compared with a reference area, and **sea ice area** is the total area covered by ice. **Sea ice extent** is defined as a region that has at least 15 percent ice concentration (NSIDC, 2021). Arctic sea ice extent varies annually from a minimum of roughly 6.3 million km^2 in September to a maximum 15.5 million km^2 in March. Seasonally opposite to the Arctic, the Antarctic sea ice extent varies annually from a minimum of roughly 2.8 million km^2 to a maximum of 18.5 million km^2 (1981–2010 median values; NSIDC, 2021). This annual variation in sea ice extent is large, yet superimposed on this annual trend is the recent long-term trend of decreasing sea ice extent.

Today's sea ice in both the Arctic and Antarctic is decreasing at an alarming rate. In the Arctic, the decline in sea ice area and extent has been accelerating. Over the 167 years from 1850–2017, the Arctic region lost sea ice area at a rate of nearly 1% (0.17 million km^2) per decade. Since 1948, however, the rate of sea ice area loss has more than doubled to 2.5% (0.46 million km^2) per decade, and the recent rate of loss (1979–2017) has increased to 6.2% (1.10 million km^2) per decade (Qiongqiong, 2021). If the current rate of loss continues, the Arctic Ocean will be free of summer sea ice by 2080 and free of ice entirely by 2160.

The sea ice story in the Antarctic is more complicated. Antarctic sea ice area has been slightly increasing since 1978 when satellite observations began. Some of this increase has been linked to changes in wind direction resulting in less ice divergence, but the evidence varies in part because of the immense size of the region (Holland, 2014). This increasing trend suddenly reversed in 2014 to a loss at a rate of 729,000 km^2 of sea ice per year between 2014 and 2017 (Parkinson, 2019), which has also been linked to changes in regional wind direction and an increase in wind speed (Wang et al., 2019). Although changes in regional winds appear to play a significant role, it remains unclear why these opposing trends in the Antarctic compared with the Arctic are occurring.

6.4 Glaciers

Valleys filled currently with rivers of ice, and the evidence on the landscape left after melting, offer dramatic visual clues to current and past climates. When the accumulation of snow on land exceeds ablation losses and then compresses to become ice, the ice may deform and flow downslope because of its weight. It is then classified as a **glacier**. Based on this definition, an ice sheet is a large glacier, but most often the term is used for smaller valley-confined ice (**valley glaciers** are common at ice sheet margins) or near mountain peaks (**cirque** or **alpine glaciers**). Glaciers contain ice but may also contain embedded sediment and rocks. In alpine regions, such **rock glaciers** can be found, consisting of angular fragments of frozen rock interspersed with ice.

Glacier formation requires an upper-elevation **accumulation zone** where air temperatures are consistently near or below freezing, which provides a reliable accumulation of enough snow exceeding any ablation losses to yield a positive mass balance and eventually glacier ice as the snow compresses. In the lower-elevation **ablation zone**, loss of snow and ice mass due to melting and/or sublimation or calving (if in coastal regions) exceeds gains, yielding a negative mass balance. Between the accumulation and ablation zones is the **equilibrium line**, the position (or elevation) where rates of accumulation and ablation are equal. A glacier is said to "advance" and move forward when accumulation exceeds ablation, and under such conditions the equilibrium line shifts to lower elevations, as in colder climates. Conversely, during warming periods, glacial retreat occurs as ablation exceeds accumulation, and the equilibrium line moves to higher elevations. The glacier does not recede in the sense that it flows uphill, but rather the glacier loses mass in the ablation zone.

Glaciers have a profound influence on landscape shaping. The large ice mass combined with increased flow velocity when confined in valleys results in **glacial scouring** of the underlying surface. Rocks can be plucked from the surface, embedded in the ice flow, scratch and scour the underlying surface as the ice advances, and be deposited far from their origin when the ice melts (these are called **glacial erratics**; Figure 6.2). Material scraped at the sides of the glacier can be deposited, forming a **lateral moraine**, or a **terminal moraine** when deposited at the glacier's snout.

Figure 6.2 A glacial erratic in Central Park, New York, deposited on top of scoured, striated bedrock after the retreat of the Laurentide Ice Sheet. Photo credit: M. Blanken.

Figure 6.3 Photographs of glaciers taken at the same location, dramatically showing the loss of glacial snow and ice. Top image: Qori Kalis Glacier, Peru, in July 1978 (left) and July 2011 (right). Bottom: Pedersen Glacier, Alaska, in the summer during the mid-1920s (left) and August 2005 (right). Photo credit: NASA Images of Change.

Glaciers are often small and located in remote inaccessible high-elevation and polar regions, and are therefore hard to inventory and monitor. Outside of Greenland and the Antarctic, recent glacier inventories are missing in 46 percent of the world's glacierized areas (Ohmura, 2009). Most of the estimated 95,000 glaciers that remain to be inventoried are in the Canadian Cordillera (54,000). Based on the 105,365 glaciers that have been inventoried, the best estimate of their combined surface area is 554,197.5 km^2 with a volume of 101,817.28 km^3 and a mean depth of 184 m (Ohmura, 2009). Most glaciers are found in polar regions, but some are found in high alpine regions such as in the Andes of South America. Since the start of the Industrial Revolution (ca. 1760), glaciers worldwide have been retreating and at a faster rate. Loss of glacial ice worldwide has been rapidly increasing, with estimates of 62 (1980s), 206 (1990s), 252 (2000s), and 327 Gt per year (2010s) (Slater et al., 2020). Photographs taken at the same location over time clearly corroborate the loss of glaciers worldwide (Figure 6.3).

The loss of glaciers means a loss of a proxy climate record since glacier ice cores provide climate information by the thickness of annual ice layers, air trapped within the ice, and oxygen isotope records (Chapter 4), especially in equatorial regions where climate proxy records are sparse. In the Andes, glaciers have been used to provide past precipitation information that can be used to help interpret anthropological questions such as whether drought caused the abandonment of the Inca citadel **Machu Picchu** in the sixteenth century (Wright, Witt and Zegarra, 1997). Glaciers in the Andes are retreating quickly, so there is an urgent need to retrieve ice cores (Fraser, 2019). In the aptly named **Glacier National Park** in Montana since the 1850s, rapid retreat at rates as high as 100 m per year have been associated with summer drought and low winter snowpack (Pederson et al., 2004). Over two-thirds of the estimated 150 glaciers in Glacier National Park that were present in 1850 were gone by 1980, and the Park may be glacier-free by 2030 (Hall and Fagre, 2003). Glacier retreat is usually a slow, gradual process but can also be sudden, with catastrophic consequences. Examples include: the Great Peruvian (or Ancash) Earthquake in 1970 that trigged the world's largest glacial avalanche, killing an estimated 70,000; the Mount Everest glacial ice avalanche in 2014, killing 16 and injuring 9; and the Himalayan Uttarakhand glacial ice avalanche and resulting flood in 2021, which killed an estimated 250 in northern India.

6.5 Snow

Snow is generically considered simply as frozen precipitation, where precipitation is defined as any water that has fallen from the atmosphere to the ground. Satellite-based estimates indicate snow covers on average 46 million km^2 (9%) of Earth's surface, 98% of that in the Northern Hemisphere (NSIDC, 2021). Snow's high albedo means that such snow-covered regions are closely coupled to temperature through the temperature–albedo positive feedback. Upon melting, the landscape changes from white to green as the rapid delivery of a winter's worth of precipitation triggers plant growth and many other forms of biological activity. In arid and semi-arid regions, water from the spring snowmelt provides a major source of fresh water required for the remainder of the year. Some 1.6 billion people, 22% of the global population, live in regions receiving water from a mountain snowpack (Immerzeel et al., 2020). Across 26% of Earth's land area, more than 50% of the runoff originates from seasonal snowmelt (Qin et al., 2020).

The term used to quantify the volume of liquid water contained in a snowpack (if all the snow melted) is the **Snow Water Equivalent**, or **SWE**. The liquid water content of snow (SWE) can be calculated from the depth of the snow (h_s) and the ratio of the density of the snow (ρ_s) to that of liquid water (ρ_w) (Eq. 6.1):

$$\mathrm{SWE} = h_s \frac{\rho_s}{\rho_w}. \tag{6.1}$$

Since variations in the density of water are small relative to changes in snow depth and density, ρ_w is usually taken as a constant 1 g cm^{-3}, and therefore Eq. (6.1) simplifies to SWE = $h_s \rho_s$ if ρ_s is also measured in g cm^{-3}. A 100-cm deep snowpack with an average snow density of 0.30 g cm^{-3} would have a liquid water content of 30 cm.

Throughout much of the mountainous regions in the Western United States and Alaska that are dependent on snow for water resources, over 850 **Snow Telemetry Sites (SNOTEL)** are used to provide near real-time data on the snowpack conditions (Figure 6.4). Often located in the remote, hard to access mountainous regions where snow falls, most SNOTEL sites consist of solar-powered sensors that transmit data via radio signals reflected off the ionized trails of

Figure 6.4 A SNOTEL site in a subalpine forest in the Rocky Mountains of Colorado roughly 3,000 m above sea level. Photo credit: P. D. Blanken.

meteors in a region located roughly 80 km above the Earth. This data transmission technique is known as **Meteor Burst Communication**. Along with meteorological measurements, snow depth is measured using an acoustic (sonic) sensor that emits ultrasonic sound pulses and measures the return interval. The weight of the overlying snow produces pressure on **snow pillows** (typically 3 m by 3 m), bladders filled with an antifreeze fluid. As the snow accumulates on the pillows, pressure is measured with a pressure transducer. Since pressure (P) is the force (F) per unit area (A), $P = F/A$, the pressure produced by the overlying snow can be written as $P = \rho_S g h_S$ (known as the **hydrostatic equation**) where g is the acceleration due to gravity ($g = 9.81$ m s^{-2}). Dividing the pressure measured by the snow pillow by the acceleration due to gravity is therefore equal to the snow water equivalent (Eq. 6.2):

$$\frac{P}{g} = \rho_S h_S = \text{SWE} \tag{6.2}$$

with SWE in units of the equivalent depth of liquid water.

A **snow course**, or transect of manual SWE measurements, is often used to spatially extrapolate the SNOTEL measurements over a wider area. From both SNOTEL and snow course measurements, it is clear that, in the western United States, large changes in the end-of-season snowpack (typically 1 April) are occurring. Between 1955 and 2016, nearly 92 percent of SNOTEL and snow course sites showed declining trends in SWE (Mote et al., 2018). The springtime pulse of meltwater released into streams is also changing, occurring 10–30 days earlier in the year 2020 than in 1948 (Stewart, Cayan and Dettinger, 2005). This decrease in the water provided by snow due to warming is only expected to continue (Pierce and Cayan, 2013), placing further stress on water resources in regions that depend on snow.

Across continental scales, changes in snow cover extent have been observed. In the Northern Hemisphere where 98 percent of snow cover is located, the spring **snow cover extent** (the area covered by snow) has decreased at a rate of 0.58 million km^2 per decade between 1967 and 2014 (Hernández-Henríquez et al., 2015). When the trends in snow cover extent are examined regionally and seasonally, the picture is more complex. Based on satellite data provided by the National Oceanic and Atmospheric Administration Climate Data Record (NOAA-CDR) there is a significant trend for earlier continental-scale disappearance of snow, but no change or even an increasing trend in the fall (October) snow cover extent over Eurasia (Brown and Derksen, 2013), North America, and the Northern Hemisphere as a whole (Rutgers University, 2021).

Why would snow cover extent be increasing in the fall when the climate is warming? In the spring, the disappearance of snow requires air temperatures above freezing, whereas in fall, snow formation requires air temperatures below freezing and precipitation to form. The increase in snow cover extent in the fall is unrealistic given that air temperatures in the adjacent Arctic have been increasing at a higher rate in the fall than spring (Brown and Derksen, 2013). In data ensembles of snow cover extent from both observations and models over 1981–2010, all showed a large decrease in the fall snow cover extent across the Northern Hemisphere except for the NOAA-CDR which showed a strong increasing trend (Mudryk et al., 2017). Improvements in the quality and quantity of the NOAA-CDR data over time may account for better detection, hence the apparent trend for increasing snow cover in the fall (C. Derksen, pers. comm, February 15, 2021). Mudryk et al. (2020) convincingly showed, however, that the Northern Hemisphere snow cover extent over 1981–2018 decreased in all months, exceeding losses of 50,000 km^2 per year in November, December, March, and May. It is essential to use multiple datasets to establish climate baselines for trends.

6.6 Frozen Ground and Permafrost

Ground ice can significantly modify and influence the environment in terms of temperature, drainage, and surface topography, and therefore several definitions are used. **Frozen ground** is soil or rock that contains ice (**ground ice**), thus the ground temperature is below the freezing point at some time. Ground that does not remain frozen year-long is called **seasonally frozen ground**. Ground that remains frozen for at least two consecutive years is called **permafrost**. The areal extent of the region underlain by permafrost further classifies the region as having **continuous permafrost** (90–100% of the area), **discontinuous permafrost** (50–90%), or **sporadic permafrost** (less than 50%). The layer above the permafrost that thaws seasonally is called the **active layer**, and this can be several tens of centimeters deep depending on the location.

The volume of water currently stored as ground ice is small with limited spatial distribution. There is an estimated 300,000 km^3 of water stored in frozen ground and permafrost, equivalent to less than 1% of the world's total volume of fresh water (Shiklomanov, 1993). Most permafrost is in North America over an area of 23 million km^2, nearly 25% of the Northern Hemisphere land area and 17% of the Earth's land area. Permafrost is found in alpine regions worldwide but is mostly discontinuous owing to the large spatial variation in the local climate, hence the extent is not well known. Permafrost has also been found beneath the polar oceans, but here too the extent is not well known. Most of today's permafrost formed during past glacial periods, since consistent air temperatures below the freezing point are required. Therefore, changes in permafrost extent or active layer depth are good indicators of past climate change rather than current conditions. The presence of frozen ground or permafrost does, however, play a key role in contemporary soil formation due to the volume expansion of water upon freezing (Chapter 3), and the recent degradation (melting) of permafrost has resulted in large changes in the landscape and subsequent feedbacks to the climate system.

Landscapes mostly in the Subarctic and Arctic regions of North America display a variety of unique landforms indicative of ground ice and permafrost. In flat areas underlain by permafrost, the drainage of liquid water is limited, resulting in many small shallow lakes across landscape (wetlands, Chapter 5). **Thermokarst**, **thaw**, or **tundra lakes** can form in depressions created by thawing permafrost (Chapter 10). Other examples of unique landforms include **ice wedge polygons**, **pingos**, and **hummock–hollow** terrain (Chapter 10). The thawing of permafrost due to the introduction of building materials such as roads, railroads, buildings or pipeline support piers that conduct heat into the permafrost has resulted in numerous examples of structural settling and even collapse.

Warming and thawing of ground ice in Subarctic and Arctic regions has released large quantities of greenhouse gases trapped within organic wetland soils. These water-saturated soils high in organic matter content due to the anoxic conditions are rich in carbon dioxide and methane, both powerful greenhouse gases. There is an estimated 1,600 billion tonnes of carbon stored in permafrost-containing soils, nearly double that contained in the atmosphere (Schuur et al., 2015). Release of these greenhouse gases through permafrost degradation is predicted to contribute an additional 0.13 to 1.69 °C of atmospheric warming by the year 2300 (MacDougall, Avis and Weaver, 2012). This is due to carbon positive feedback where the increasing CO_2 in the atmosphere from anthropogenic emissions creates warming, and that warming thaws permafrost, which in turn releases more greenhouse gases that enhance further warming.

6.7 SUMMARY

Most of the world's fresh water is currently stored as snow and ice, with the Greenland and Antarctic Ice Sheets storing 99 percent of the world's frozen fresh water and 69 percent of all fresh water. In addition to providing the major reservoir of fresh water, snow and ice play a major role in Earth's climate primarily through their highly reflective surface and the temperature–albedo positive feedback. With recent warming, the amplified increases in surface temperatures due to this temperature–albedo feedback have resulted in the loss of ice across the cryosphere. Dozens of studies have documented this loss in the Greenland and Antarctic ice sheets, ice shelves, and glaciers. Sea ice plays an important climate role through its influence on albedo and ocean circulation through modifications in ocean temperature and salinity. Rapid changes in sea ice conditions have been documented in both the Arctic and Antarctic, with the former expected to be ice-free in summer near the end of the twenty-first century.

Although glaciers do not contain a vast reservoir of fresh water, they scour and erode the landscape and offer a visible record of climate change. Precipitation in the form of snow is the ultimate source of ice for ice sheets and glaciers. Seasonal snow that does not remain year-long to become ice is an important water resource in many semi-arid and arid regions. In mountain landscapes the water content of the snowpack is often carefully measured, as the springtime runoff generated by the winter snowpack provides water resources through the rest of the year. The melting of permafrost has resulted in the release of trapped greenhouse gases such as methane that are powerful greenhouse gases. Throughout the cryosphere, the loss of snow and ice is resulting in dramatic changes with implications for water resources across the globe.

6.8 QUESTIONS

6.1 The mass balance of the Greenland and Antarctic ice sheets is affected by water in all three states: liquid, solid, and gas. Explain how water in each state affects both the accumulation and the ablation of mass in these massive polar ice sheets.

6.2 Describe why changes in ocean water temperature and salinity have such a large effect on ocean currents and therefore the climate in nearby coastal regions. How did past changes in the Atlantic Ocean affect weather and climate in Europe, and how might future changes do so?

6.3 The current rate of air temperature increase in the Arctic is two to three times the global average. Explain the connection between a decrease in sea ice area and air temperature increase. How would an increase in Arctic sea ice area affect air temperature?

6.4 Under current conditions, the Arctic Ocean will likely be ice-free in summer by 2080 and completely ice-free by 2160. Provide examples of how you imagine this major environmental change will affect the lives of the Indigenous peoples that live in the region.

6.5 Glaciers are relatively small, do not contain much water (compared with other global reservoirs), and are often isolated and hard to detect. However, glaciers can provide a wealth of information on past climates as well as current climate change. Give examples of how glaciers provide this important information on climate change.

6.6 Glaciers can collapse without warning, creating several hazards to those living nearby. Provide examples of recent glacier collapses, including reasons behind the collapse and the impact(s) resulting from the collapse. Are there any warning systems that could be designed to help mitigate this natural hazard?

6.7 The density of newly fallen snow can be 0.20 g cm^{-3} in cold, dry continental locations compared with 0.30 g cm^{-3} in mild, humid maritime locations. Calculate the water content contained in a 10-cm deep snowpack at each location. What depth of snow at the continental location is required to provide the water content provided by the snow at the maritime location?

6.8 The average person living in North America uses 475 liters of water per day. Calculate the depth of snow over a square meter of ground required to meet this daily demand, assuming a snow density of 0.25 g cm^{-3}.

6.9 Locations with permafrost, and those locations where permafrost is degrading (melting), can be hard to detect because of the high spatial and temporal variation across remote, inaccessible landscapes. Explore and describe some of the methods and techniques that have been used to detect and map the distribution of permafrost.

6.10 Under what conditions did most of the permafrost that is present today form, and what are the consequences to the water balance in regions where permafrost and other forms of ground ice are melting?

REFERENCES

Broecker, W. S., Kennett, J. P., Flower, B. P. et al. (1989) 'Routing of meltwater from the Laurentide Ice', *Nature*, 341, pp. 318–321.

Brown, R. D. and Derksen, C. (2013) 'Is Eurasian October snow cover extent increasing?', *Environmental Research Letters*, 8. doi: 10.1088/1748-9326/8/2/024006.

Büntgen, U., Tegel, W., Nicolussi, K. et al. (2011) '2500 years of European climate and human susceptibility', *Science*, 331, pp. 578–582.

Fraser, B. (2019) 'Daring scientists extract ice from Earth's highest tropical glacier', *Nature*, 573, pp. 171–172. doi: 10.1038/d41586-019-02566-9.

Fretwell, P., Pritchard, H. D., Vaughan, D. G. et al. (2012) 'Bedmap2: improved ice bed, surface and thickness datasets for Antarctica', *The Cryosphere Discussions*, 6, pp. 4305–4361. doi: 10.5194/tcd-6-4305-2012.

Garbe, J., Albrecht, T., Levermann, A., Donges, J. F. and Winkelmann, R. (2020) 'The hysteresis of the Antarctic Ice Sheet', *Nature*, 585, pp. 538–544. doi: 10.1038/s41586-020-2727-5.

Geibert, W., Matthiessen, J., Stimac, I., Wollenburg, J. and Stein, R. (2021) 'Glacial episodes of a freshwater Arctic Ocean covered by a thick ice shelf', *Nature*, 590, pp. 97–106. doi: 10.1038/s41586-021-03186-y.

Hall, M. H. P. and Fagre, D. B. (2003) 'Modeled climate-induced glacier change in Glacier National Park, 1850–2100', *BioScience*, 53, pp. 131–140. doi: 10.1641/0006-3568(2003)053[0131:MCIGCI]2.0.CO;2.

Hanna, E., Cappelen, J., Fettweis, X. et al. (2020) 'Greenland surface air temperature changes from 1981 to 2019 and implications for ice-sheet melt and mass-balance change', *International Journal of Climatology*, 41, pp. 1336–1352. doi: 10.1002/joc.6771.

Hernández-Henríquez, M. A., Déry, S. J. and Derksen, C. (2015) 'Polar amplification and elevation-dependence in trends of Northern Hemisphere snow cover extent, 1971–2014', *Environmental Research Letters*, 10, doi: 10.1088/1748-9326/10/4/044010.

Holland, P. R. (2014) 'The seasonality of Antarctic sea ice trends', *Geophysical Research Letters*, 41, pp. 4230–4237. doi: 10.1002/2014GL060172.

Immerzeel, W. W., Lutz, A. F., Andrade, M. et al. (2020) 'Importance and vulnerability of the world's water towers', *Nature*, 577, pp. 364–369. doi: 10.1038/s41586-019-1822-y.

Jackson, L. C., Kahana, R., Graham, T. et al. (2015) 'Global and European climate impacts of a slowdown of the AMOC in a high resolution GCM', *Climate Dynamics*, 45, pp. 3299–3316. doi: 10.1007/s00382-015-2540-2.

MacDougall, A. H., Avis, C. A. and Weaver, A. J. (2012) 'Significant contribution to climate warming from the permafrost carbon feedback', *Nature Geoscience*, 5, pp. 719–721. doi: 10.1038/ngeo1573.

Medley, B., McConnell, J. R., Neumann, T. A. et al. (2018) 'Temperature and snowfall in western Queen Maud Land increasing faster than climate model projections', *Geophysical Research Letters*, 45, pp. 1472–1480. doi: 10.1002/2017GL075992.

Miles, M. W., Andresen, C. S. and Dylmer, C. V. (2020) 'Evidence for extreme export of Arctic sea ice leading the abrupt onset of the Little Ice Age', *Science Advances*, 6. doi: 10.1126/sciadv.aba4320.

Moreno-Chamarro, E., Zanchettin, D., Lohmann, K. and Jungclaus, J. H. (2017) 'An abrupt weakening of the subpolar gyre as trigger of Little Ice Age-type episodes', *Climate Dynamics*, 48, pp. 727–744. doi: 10.1007/s00382-016-3106-7.

Mote, P. W., Li, S., Lettenmaier, D. P., Xiao, M. and Engel, R. (2018) 'Dramatic declines in snowpack in the western US', *npj Climate and Atmospheric Science*, 1. doi: 10.1038/s41612-018-0012-1.

Mouginot, J., Rignot, E., Bjork, A. A. and Wood, M. (2019) 'Forty-six years of Greenland Ice Sheet mass balance from 1972 to 2018', *Proceedings of the National Academy of Sciences USA*, 116, pp. 9239–9244. doi: 10.1073/pnas.1904242116.

Mudryk, L., Santolaria-Otin, M., Krinner, G. et al. (2020) 'Historical Northern Hemisphere snow cover trends and projected changes in the CMIP6 multi-model ensemble', *Cryosphere*, 14, pp. 2495–2514. doi: 10.5194/tc-14-2495-2020.

Mudryk, L. R., Kushner, P. J., Derksen, C. and Thackeray, C. (2017) 'Snow cover response to temperature in observational and climate model ensembles', *Geophysical Research Letters*, 44, pp. 919–926. doi: 10.1002/2016GL071789.

Ohmura, A. (2009) 'Completing the world glacier inventory', *Annals of Glaciology*, 50, pp. 144–148. doi: 10.3189/172756410790595840.

Parkinson, C. L. (2019) 'A 40-y record reveals gradual Antarctic sea ice increases followed by decreases at rates far exceeding the rates seen in the Arctic', *Proceedings of the National Academy of Sciences USA*, 116, pp. 14414–14423. doi: 10.1073/pnas.1906556116.

Pederson, G. T., Fagre, D. B., Gray, S. T. and Graumlich, L. J. (2004) 'Decadal-scale climate drivers for glacial dynamics in Glacier National Park, Montana, USA', *Geophysical Research Letters*, 31, pp. 2–5. doi: 10.1029/2004GL019770.

Pierce, D. W. and Cayan, D. R. (2013) 'The uneven response of different snow measures to human-induced climate warming', *Journal of Climate*, 26, pp. 4148–4167. doi: 10.1175/JCLI-D-12-00534.1.

Pritchard, H. D., Ligtenberg, S. R. M., Fricker, H. A. et al. (2012) 'Antarctic ice-sheet loss driven by basal melting of ice shelves', *Nature*, 484, pp. 502–505. doi: 10.1038/nature10968.

Qin, Y., Abatzoglou, J. T., Siebert, S. et al. (2020) 'Agricultural risks from changing snowmelt', *Nature Climate Change*, 10, pp. 459–465. doi: 10.1038/s41558-020-0746-8.

Qiongqiong, C., Wang, J., Beletsky, D. and Overland, J. E. (2021) 'Summer Arctic sea ice decline during 1850–2017 and the amplified Arctic warming during the recent decade', *Environmental Research Letters*, 16. doi: 10.1088/1748-9326/abdb5f.

Rignot, E., Bamber, J. L., van den Broeke, M. R. et al. (2008) 'Recent Antarctic ice mass loss from radar interferometry and regional climate modelling', *Nature Geoscience*, 1, pp. 106–110. doi: 10.1038/ngeo102.

Rutgers University (2021) Rutgers University Global Snow Lab. Available at: https://climate.rutgers.edu/snowcover/.

Schuur, E. A. G., McGuire, A. D., Schädel, C. et al. (2015) 'Climate change and the permafrost carbon feedback', *Nature*, 520, pp. 171–179. doi: 10.1038/nature14338.

Serreze, M. C. and Barry, R. G. (2011) 'Processes and impacts of Arctic amplification: A research synthesis', *Global and Planetary Change*, 77, pp. 85–96. doi: 10.1016/j.gloplacha.2011.03.004.

Shiklomanov, I. A. (1993) 'World fresh water resources', in Gleick, P. H. (ed.) *Water in Crisis: A Guide to the World's Fresh Water Resources*. Oxford University Press, pp. 13–24.

Slater, T., Lawrence, I. R., Otosaka, I. N. et al. (2021) 'Review article: Earth's ice imbalance', *The Cryosphere*, 15, pp. 233–246.

Stewart, I. T., Cayan, D. R. and Dettinger, M. D. (2005) 'Changes toward earlier streamflow timing across western North America', *Journal of Climate*, 18, pp. 1136–1155. doi: 10.1175/JCLI3321.1.

Wang, Z., Turner, J., Wu, Y. and Liu, C. (2019) 'Rapid decline of total Antarctic sea ice extent during 2014–16 controlled by wind-driven sea ice drift', *Journal of Climate*, 32, pp. 5381–5395. doi: 10.1175/JCLI-D-18-0635.1.

Wright, K. R., Witt, G. D. and Zegarra, A. V. (1997) 'Hydrogeology and paleohydrology of ancient Machu Picchu', *Ground Water*, pp. 660–666. doi: 10.1111/j.1745-6584.1997.tb00131.x.

7 Evaporation, Condensation, and Precipitation

Key Learning Objectives

After reading this chapter, you will be able to:

1. Summarize how the experiments by Dalton and Penman helped form our conceptual framework explaining the rate of evaporation.
2. Describe the controls of the evaporation rate and how it varies globally.
3. Explain the processes and importance of condensation.
4. Identify the requirements for precipitation formation, including cloud formation and the four main atmospheric cooling mechanisms.
5. Discuss the Bergeron, curvature (Kelvin), and solute (Raoult) effects, and their roles in the context of precipitation formation.

7.1 Introduction

Evaporation from liquid to gas is the largest flux in the water cycle, and condensation from vapor to liquid is required for precipitation to provide liquid water for evaporation. The importance of evaporation has been recognized for centuries, although it is challenging to measure and it varies spatially, hindering a complete understanding of its magnitude, distribution, and controls.

This chapter begins by examining the earliest studies that formed our fundamental understanding of evaporation. The concept of vapor pressure, the method used to measure it, and our knowledge of the importance of wind in controlling evaporation rates, were developed over 200 years ago. Field observations of evaporation from wet surfaces in the United Kingdom, coupled with analytic solutions used to calculate vapor pressures and the energy required to evaporate water, led to the development of simple equations to reliably predict evaporation from these well-watered surfaces. Subsequent research quantified the restrictions on evaporation from drier surfaces and the important role that vegetation plays in the water cycle. Using these studies as a foundation, today's remote sensing and models have been used to confirm and refine the important role of evaporation in the global water cycle.

Rain requires that evaporated water vapor condense back to a liquid. Based on the conservation of energy, condensation can be viewed as the opposite process of evaporation; hence many of the equations used to calculate evaporation can also be used to calculate condensation. This chapter shows how evaporation and condensation are coupled through vapor pressure and precipitation formation. Dew, frost, and cloud formation are discussed as examples of the condensation process. One of the most important requirements for condensation and cloud formation is the sufficient cooling of the atmosphere. Such cooling is primarily achieved through the vertical motion of air, and the four main mechanisms that result in vertical air motion (hence cooling) are described. To replenish the supply of surface liquid water that has been evaporated, water that has condensed in the atmosphere must be returned to the surface as precipitation, hence the processes involved in precipitation formation are described. Precipitation formation involves a delicate balance between the rates of evaporation and condensation from a water droplet that connect back to the properties of the water molecule. How these rates are determined is described.

7.2 Evaporation

Evaporation, the process of water changing states from liquid to gas, can be an important component of the local water balance, depending on the availability of surface liquid water and energy. Unlike rainfall or river flow, evaporation is hard to directly observe, yet is has long been recognized that understanding the rate and controls of evaporation are important to close the water balance. One of the first documented experiments leading to today's basic understanding was by **John Dalton** (1766–1844). One of Dalton's research objectives was to determine the maximum force that vapor could exert at different temperatures: what we refer to now as the **saturation vapor pressure**. Another was to "obtain a true theory of evaporation" since Dalton was suspicious of the **Chemical Solution Theory**, which stated that water and even metals simply dissolve into the atmosphere.

In his experiments, Dalton carefully measured the change in weight of small tin cylinders filled with various liquids under various conditions. A small flame under the tins heated the water, and evaporation was calculated by measuring the change in weight. Dalton varied the water temperature by moving the flame, and he measured the "...utmost force that certain vapours, as that from water, can exert at different temperatures" (p. 551, Dalton, 1802a) using several mercury-based barometer techniques. Dalton's techniques proved successful, and he was able to measure the saturation vapor pressure across a range of temperatures over several liquids: water, wine, ammonia, mercury, sulfuric acid. This enabled him to determine that the evaporation rate was proportional to the difference between the saturation vapor pressure and the vapor pressure in the surrounding air. By chance or design, Dalton also observed that evaporation was least when the doors and windows in his laboratory were closed, and the air was calm. With a moderate draft produced by the fire in his fireplace, evaporation increased, and increased further when the windows were open with strong winds outside. Clearly air movement played a role in evaporation, but Dalton provided only a qualitative description of the connection between wind and evaporation.

Dalton's experiments showed that rather than chemically dissolving in the atmosphere as solutions dissolve in water (as stated in the Chemical Solution Theory), the **evaporation rate** (E) depended on the difference between the surface **saturation vapor pressure** (e_s) and the atmospheric **vapor pressure** (e_a) with a functional dependence (f) on wind speed (u) (Eq. 7.1):

$$E = (e_s - e_a) f(u). \tag{7.1}$$

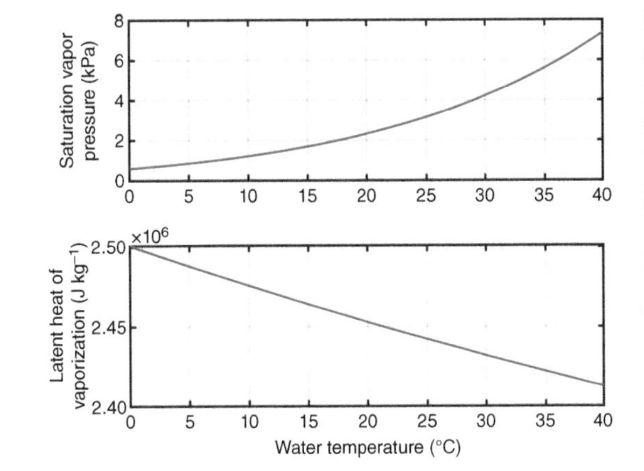

Figure 7.1 How the saturation vapor pressure (top panel) and the latent heat of vaporization (bottom panel) vary with temperature for a pure liquid water surface.

Dalton demonstrated that the saturation vapor pressure and therefore the evaporation rate were dependent on the temperature of the liquid. In warmer water, the increase in molecular motion made it easier for the bonds between water molecules at the surface to break and move into vapor form. The energy required to sever the bonds between molecules allowing a state change from liquid to gas is known as the **latent heat of vaporization** (λ: J kg^{-1}), and both λ and e_s vary with the water's surface temperature (Figure 7.1).

The latent heat of vaporization does not directly appear in Eq. (7.1), but this energy expresses itself through the saturation vapor pressure term as shown by Dalton's work. The **Clausius–Clapeyron relationship** (from **Rudolf Clausius**, 1822–1888, and **Benoit Paul Emile Clapeyron**, 1799–1864) shows how temperature and pressure are related. This relationship can be approximated by an equation of generic form given by **Teten's formula** (e.g., see Huang, 2018) (Eq. 7.2):

$$e_s = a \exp\left(\frac{bT}{c+T}\right). \tag{7.2}$$

Using coefficients $a = 0.61121$ kPa, $b = 17.502$, and $c = 240.97$ °C for pure liquid water as determined by Buck's analysis (1981), Eq. (7.2) becomes Eq. (7.3):

$$e_s = 0.61121 \exp\left(\frac{17.502\,T}{240.97+T}\right) \tag{7.3}$$

providing e_s in kPa with temperature (T) in °C. For air over ice, Eq. (7.2) can be used but with different coefficients: $a = 0.61115$ kPa, $b = 22.452$, $c = 272.55$ °C (Buck, 1981). Figure 7.1 (top panel) shows the relationship between the saturation vapor pressure and temperature calculated using Eq. (7.3), and how a small increase in temperature results in a large increase in the capability of the air to contain water vapor, especially as temperature increases. A 1 °C increase in air temperature from 1 to 2 °C results in a 0.05 kPa increase in e_s compared with a 0.26 kPa e_s increase from a 1 °C increase in air temperature from 31 to 32 °C. In both cases, a 1 °C increase in air temperature results in a 6–7 percent increase in the saturation vapor pressure. The fact that the amount of water vapor the air contains increases exponentially with temperature has implications for precipitation formation, the ability of humans to maintain a constant body temperature (Chapter 14), and climate change (Chapter 16).

The latent heat of vaporization decreases as water temperature increases (Figure 7.1, bottom panel). This is because temperature is a measure of kinetic energy, and in warmer water the molecules have greater motion, thus requiring less energy to break the connecting hydrogen bonds. The variation of λ with temperature (T) in water can be approximated using Eq. (7.4):

$$\lambda = 1.91846 \times 10^6 \left[\frac{T}{(T - 33.91)} \right]^2 \tag{7.4}$$

with λ in J kg^{-1} when T is provided in kelvin (K) (Henderson-Sellers, 1984). Using Eq. (7.4), for example, a lake with surface temperature of 10 °C (10 °C + 273.15 = 283.15 K) requires 2.48 million joules to evaporate a kilogram of water. If the lake's surface temperature increased to 20 °C (293.15 K), λ decreases to 2.45 million J kg^{-1}, but still a large amount of energy. Although evaporation and boiling can be confused since both are changes of state from liquid to gas, **boiling** occurs only when boiling-point temperature is reached, with the vapor forming throughout the liquid. In contrast, evaporation may occur at any temperature, with the vapor forming only at the water surface.

To put the energy required to evaporate water in perspective, compare it to the energy required to move a large, heavy object such as a piano. A piano with a mass of 300 kg might require a force of 400 N (a newton is a kg m s^{-2}) to be pushed. The work required would be 400 kg m s^{-2} multiplied by the distance pushed (say 3 m), equal to 1,200 kg m^2 s^{-2} (joules). Dividing work by the mass of the piano (1,200 J / 300 kg) gives 4 J kg^{-1}, a tiny fraction of the roughly 2.5 million J kg^{-1} required to evaporate water. Since the density of liquid water is approximately 1,000 kg m^{-3}, a kilogram of water has a volume of roughly 0.001 m^3 (one liter), the volume of a typical water bottle. To evaporate this liter of water requires the equivalent amount of energy to move the 300-kg piano with a force of 400 N a distance of 6.25 km ($d = W/F$ = 2.5 million kg m^2 s^{-2} / 400 kg m s^{-2} = 6,250 m).

Another way to visualize the magnitude of λ is to express it in terms of a common household lightbulb. A 10-W LED lightbulb uses energy at a rate of 10 joules per second and would need to be on for 69 hours to use the equivalent amount of energy required to evaporate one liter of water (2.5 million J / 10 J s^{-1} = 250,000 s = 69 hours).

7.2.1 Controls on the Evaporation Rate

Dalton was also intrigued by the balance between water inputs and outputs over natural surfaces (Dalton, 1802b). He described one of the first water balance studies comparing water inputs from rain and dew to water outputs from runoff and evaporation over a year in England and Wales. Dalton appreciated the complexity of the spatial variation in evaporation across the landscape as it affected the region's water balance. He wrote:

> Upon looking over the surface of any country, three principal varieties of surface present themselves to view, as far as respects evaporation, namely, *water*, ground covered with grass and other vegetables, and bare soil. The difficulties that occur in attempts to find the quantity of water evaporated in those three

Figure 7.2 Penman's evaporation field experiment at Rothamsted Experimental Station in June 1944. The evaporation tanks are visible in the foreground. Reproduced with permission from Penman (1948).

cases, are perhaps the principal reason why our knowledge of this head is so imperfect. (Dalton, 1802b, pp. 357–358)

To measure evaporation under ambient outside conditions, Dalton inserted a 10-inch diameter, 3-foot-deep cylinder into the ground and recorded the difference in water entering the top and leaving the bottom over the course of a year, with the difference taken as an approximation of evaporation. Based on the available data and his calculations, he concluded that inputs of water by rain and dew were "...equivalent to the quantity of water carried off by evaporation and by the rivers" (Dalton, 1802b).

In England and Wales, where rainfall is plentiful, most surfaces have an ample supply of water. Nearly 150 years after Dalton's research, field observations of evaporation were performed at Rothamsted Research (formerly the Rothamsted Experimental Station, located just north of London) using methods similar to Dalton's. At Rothamsted, **Howard Penman** (1909–1984) developed empirical relationships to predict evaporation from open water and well-watered bare soil and grass (Penman, 1948) (Figure 7.2). Penman's observations reinforced the idea that in addition to having a source of liquid water and a supply of energy to provide the latent heat of vaporization, there must also be some mechanism to remove water vapor for continued evaporation to occur as represented by Dalton's wind function $f(u)$. Penman's measurements and resulting analysis provided his best empirical estimate of the evaporation rate from the tanks filled with only water (E_0 in mm per day; Eq. (7.5)):

$$E_0 = 0.35 \times \left(1 + 9.8 \times 10^{-3} u_2\right)\left(e_s - e_a\right) \quad (7.5)$$

where u_2 is the horizontal wind speed measured in "miles per day" at a height of 2 m above the surface, and e_s and e_a are the saturation and ambient vapor pressures (respectively) in units of mm of mercury. Although empirical, Eq. (7.5) follows the same general form as Dalton's equation (Eq. 7.1), with the evaporation rate from wet surfaces equal to the vapor pressure gradient multiplied by a wind function. This simple equation worked so well in predicting evaporation from well-watered surfaces that it is now used in countless applications.

These classic studies by Dalton and Penman (and others) summarized the primary evaporation requirements and controls. To borrow the economic terms of supply and demand, the first and most obvious requirement is a supply of liquid water, sometimes referred to as the **source strength**. In both Dalton's and Penman's studies, the surfaces had no limit on the availability of water for evaporation. Above such surfaces the relative humidity is near 100 percent, so the vapor pressure is equal to the saturation vapor pressure calculated at the surface temperature using Eq. (7.3). In Chapters 11 and 12 the active role that vegetation plays in controlling and limiting the supply of water to the atmosphere is discussed.

The atmosphere's demand or **sink strength** for water vapor is represented by the dryness of the air above the surface. The drier the air, or specifically, the greater the difference between the humidity at the surface and the air above, the greater the evaporation rate. The difference between the surface and atmospheric humidity is represented by the $(e_s - e_a)$ term in Eqs. (7.1) and (7.5). The wind function is connected to the "sink strength" term as follows: Imagine a glass of water sitting on a table, freely evaporating water into the air above. If a cover were placed over the glass, evaporation would eventually decrease since the vapor pressure in the trapped air would ultimately equal the saturation vapor pressure. Removing the cover, allowing dry air to enter, would allow evaporation to continue. The faster dry air is supplied (i.e., wind speed), the greater the evaporation rate.

As discussed, energy is required to evaporate water, so where does this energy appear in Eq. (7.1)? There is no direct expression for the latent heat of vaporization in Dalton's 1801 essay (Dalton, 1802a). Dalton's finding on the relationship between evaporation, saturation vapor pressure and temperature intrinsically contains the latent heat of vaporization; the saturation vapor pressure varies directly with temperature, and temperature determines the latent heat of vaporization (Figure 7.1).

To summarize, the fundamental controls on evaporation are the availability of liquid water at the surface, the availability of energy to evaporate this water, and the availability of sufficiently dry air above to maintain evaporation. The first and foremost requirement, that of liquid water at the surface, has strong biological controls with important consequences for plants (Chapters 11 and 12), insects and reptiles (Chapter 13), and people (Chapter 14).

7.2.2 Global Evaporation Patterns

As noted by Dalton, evaporation occurs from several sources, including open water, bare soil, and vegetation, so all need to be considered in global evaporation estimates. Comparable units should be used to allow for an equal comparison. Evaporation can be expressed as the mass of water vapor passing through a square meter of surface area over some time interval (kg m^{-2} s^{-1}). These units are useful in water balance studies when evaporation in millimeters per day (determined by dividing kg m^{-2} s^{-1} by the density of water, then converting meters to millimeters and seconds to a day) can be compared to precipitation, also in mm per day. Alternatively, evaporation can be expressed as the equivalent energy required to convert a mass of water from liquid to gas by means of the latent heat of vaporization (the latent heat flux, W m^{-2}), convenient units for energy balance studies. For example, solar radiation measured in W m^{-2} can be easily compared to W m^{-2} of evaporation.

Evaporation unit conversion can be confusing. An evaporation rate (E) of 6 mm per day (equivalent to 6.94×10^{-5} kg m^{-2} s^{-1} after multiplying by the density of water, converting days to seconds, and millimeters to meters), when multiplied by λ at an air temperature of 20 °C (293 K), gives an equivalent λE of 170 J m^{-2} s^{-1} (W m^{-2}):

$$E = 6\frac{\text{mm}}{\text{day}} = 6.94 \times 10^{-5}\,\frac{\text{kg}}{\text{m}^2\text{s}}$$

$$\lambda = 1.91846 \times 10^6 \left[\frac{293\,\text{K}}{(293\,\text{K} - 33.91)}\right]^2 = 2.45 \times 10^6\,\frac{\text{J}}{\text{kg}}$$

$$\lambda E = \lambda \times E = \left(2.45 \times 10^6\,\frac{\text{J}}{\text{kg}}\right) \times \left(6.94 \times 10^{-5}\,\frac{\text{kg}}{\text{m}^2\text{s}}\right) = 170.38\frac{\text{J}}{\text{m}^2\text{s}} = 170.38\frac{\text{W}}{\text{m}^2}.$$

In water balance studies, precipitation is often compared to evaporation. To convert from kg m^{-2} s^{-1} to mm per day, E is divided by the density of water, meters are converted to millimeters, and seconds are converted to days:

$$\left(\frac{\text{kg}}{\text{m}^2\text{s}}\right) \times \left(\frac{\text{m}^3}{1{,}000\,\text{kg}}\right) \times \left(\frac{60\,\text{s}}{\text{min}}\right) \times \left(\frac{60\,\text{min}}{\text{hour}}\right) \times \left(\frac{24\,\text{hour}}{\text{day}}\right) \times \left(\frac{1{,}000\,\text{mm}}{\text{m}}\right) = \frac{\text{mm}}{\text{day}}.$$

For example, an evaporation rate of 6.94×10^{-5} kg m^{-2} s^{-1} is equal to 6 mm per day:

$$\left(\frac{6.94\times10^{-5}\,\text{kg}}{\text{m}^2\text{s}}\right) \times \left(\frac{\text{m}^3}{1{,}000\,\text{kg}}\right) \times \left(\frac{60\text{s}}{\text{min}}\right) \times \left(\frac{60\,\text{min}}{\text{hour}}\right) \times \left(\frac{24\,\text{hour}}{\text{day}}\right) \times \left(\frac{1{,}000\,\text{mm}}{\text{m}}\right) = 6\frac{\text{mm}}{\text{day}}.$$

Evaporation measurements remain sparse, so models and remote sensing data are often used to estimate evaporation at large spatial scales. Using a global evaporation model, GLEAM (Global Land-surface Evaporation: the Amsterdam Methodology), primarily driven by satellite-based remote sensing data, Miralles et al. (2011) produced maps of terrestrial evaporation and precipitation (mm per year). They estimated that the annual land evaporation was 67.9×10^3 km^3 when averaged over 2003–2007, the vast majority (80%) from **transpiration**, water that has evaporated from inside the leaves of vascular plants (Chapter 12), and 11% from the evaporation of precipitation intercepted by vegetation, indicating that 91% is directly influenced by vegetation (Figure 7.3). Of the remaining 9%, 7% was bare soil evaporation and 2% was snow sublimation. The factors limiting evaporation from the land surface were the available energy and the available water dictated by the volume of precipitation. Overall, 58% of precipitation falling on land surface was evaporated, leaving 42% as runoff into oceans (Miralles et al., 2011).

Ocean evaporation measurements are practically nonexistent and are therefore estimated using models driven by satellite-provided data. One such estimate calculated an annual global ocean average of 112 cm per year over 1958–2005, with large annual and spatial variation (Figure 7.4) (Yu, 2007). Multiplying that evaporation rate by the surface area of the world's oceans (361,132,000 km^2) gives 404.5×10^3 km^3 per year, nearly six times the estimated volume of water evaporated from the land surface. Adding the estimated global land evaporation of 67.9×10^3 km^3 to the ocean evaporation of 404.5×10^3 km^3 gives a total global evaporation of 472.4×10^3 km^3, with ocean evaporation accounting for 86 percent of the global total (404.5×10^3 km^3 / 472.4×10^3 km^3). Yu (2007) observed a shift from a decreasing trend in

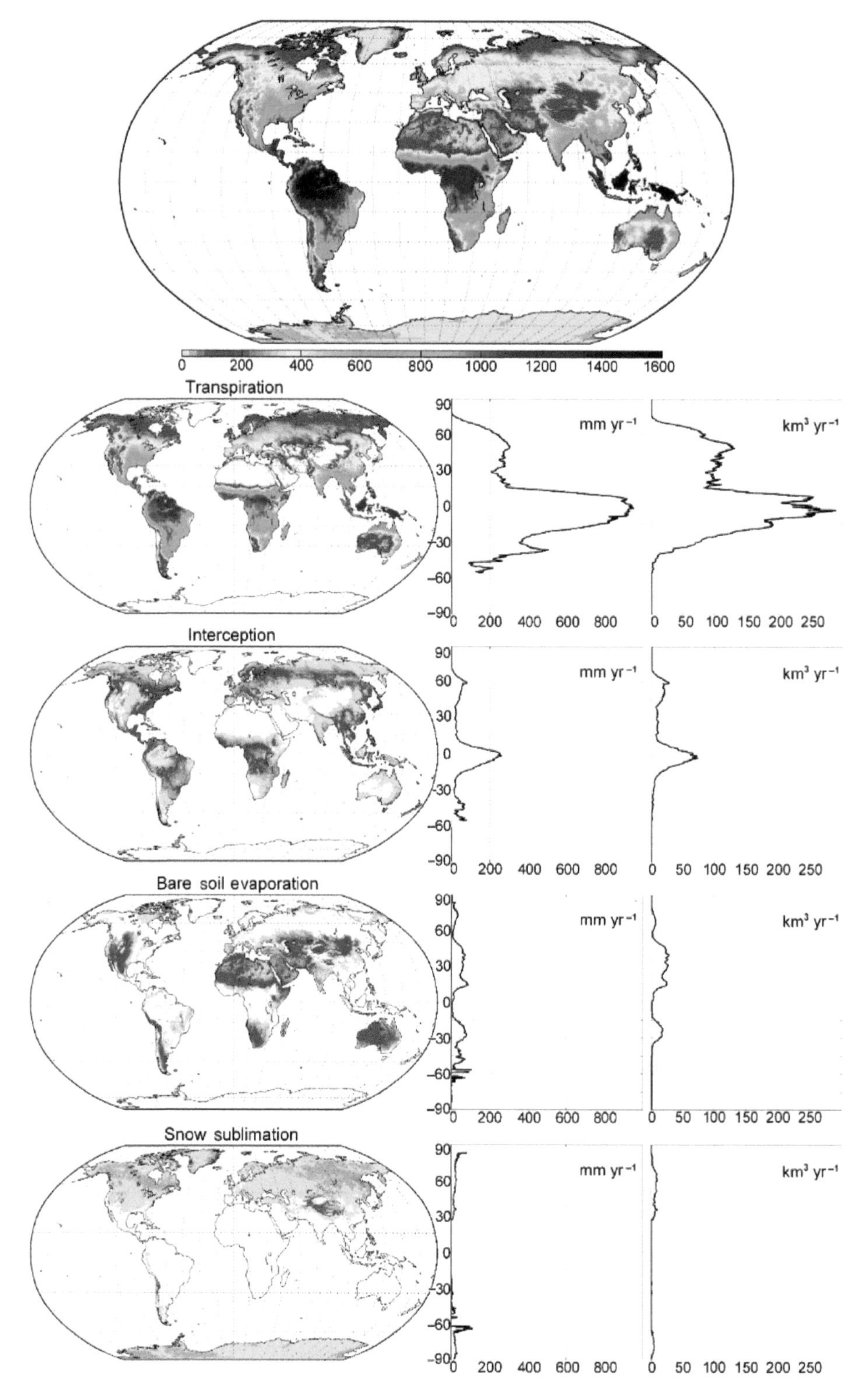

Figure 7.3 Land evaporation (top map) with contributions from transpiration, evaporation of intercepted water, soil evaporation, and snow sublimation (bottom maps and latitudinal profiles) in mm per year over 2003–2007. Reproduced from Miralles et al. (2011) under Creative Commons Attribution 4.0 International License (CC BY 4.0) https://creativecommons.org/licenses/by/4.0/.

ocean evaporation to an increasing trend in 1977–1978, and attributed the decreased evaporation to variability in the air–sea humidity gradient, and increased evaporation to increased wind speed. This increase in wind speed acted to maintain a strong air–sea vapor humidity gradient, thus increasing evaporation.

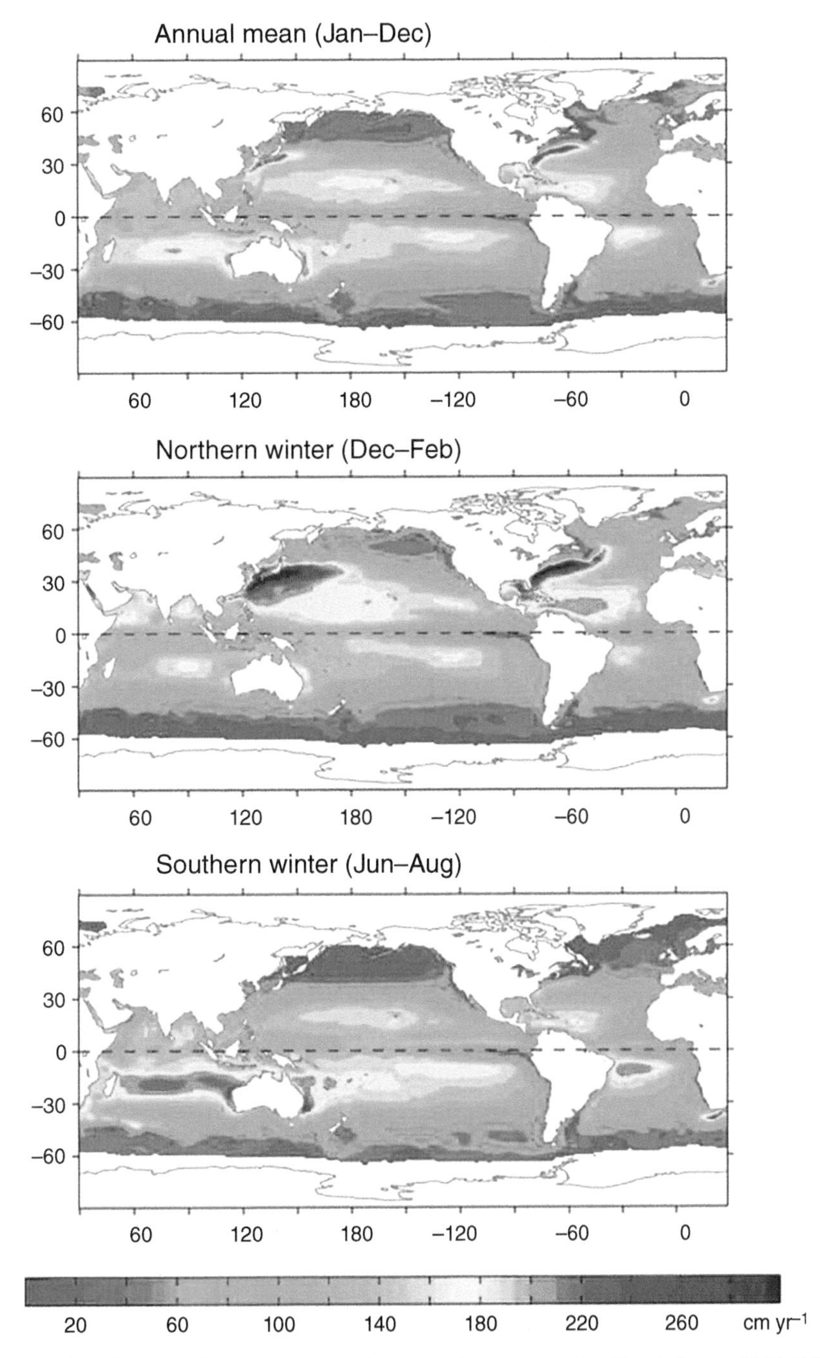

Figure 7.4 Annual and seasonal ocean evaporation rates (cm per year) estimated over 1958–2005. Reproduced with permission from Yu (2007). © American Meteorological Society.

7.3 Condensation

Condensation is the change of state from a gas to a liquid. As the counterpart to evaporation, energy and mass are conserved when water vapor returns to its liquid form and energy is released back to the surrounding environment. The processes represented in the equations that determine the evaporation rate are similar to those used to calculate the condensation rate. If the vapor pressure in the air exceeds that at the surface, the difference will be negative, hence E calculated using Eq. (7.1) would be a negative number, implying condensation on the surface. Condensation also occurs in the atmosphere when water condenses on small particles to form clouds, releasing heat in the process, and, if conditions are right, forming precipitation. Thus, the condensation of water has major consequences for the redistribution of energy and water in both space and time.

To **condense** literally means to compress and increase density. A volume of air containing water vapor, when compressed (meaning volume decreases and pressure increases), brings the individual water molecules closer together until they bond to form a liquid. An increase in pressure alone can result in condensation, but the velocity of the individual water molecules, as measured by their average kinetic energy (temperature) also plays a role. When pressure is constant, condensation can occur as temperature decreases. As the temperature decreases, the water molecules slow and therefore have a greater probability of colliding and bonding to form a liquid. This relationship between the state of water as a function of pressure and temperature is shown by the state diagram (Chapter 3), with the boundary between gas and liquid given by the saturation vapor pressure equation (Eq. 7.3). Even when evaporation is large, some water molecules will return to the evaporating surface, condensing back to liquid. Under typical outside conditions, dry air constantly replaces moist, so evaporation continues, as observed by Dalton. If, however, the humidity in the air above the evaporating surface continues to increase, Eq. (7.1) predicts that evaporation will cease when there is no difference in vapor pressure. Although under these conditions there would indeed be no change in the water level surface, evaporation is occurring, but the evaporation rate is equal to the condensation rate. The vapor pressure measured when the evaporation and condensation rates are equal is the saturation vapor pressure, and this increases with temperature since evaporation also increases with temperature. The air is "saturated" in the sense that the evaporation of n molecules requires n molecules to condense back to the water surface.

Since condensation literally refers to an increase in density and the saturation vapor pressure defines the pressure when the rates of evaporation and condensation are equal, it is often convenient to convert units of vapor pressure to units of vapor density. Vapor pressure (e_a; Pa = J m^{-3}) can be converted to vapor density (ρ_v; g m^{-3}) using the **Ideal (or Universal) Gas Law**. This was originally stated by **Benoit Paul Emile Clapeyron** (1799–1864) for gases that are "ideal" in the sense that the molecules are small with negligible volume and move randomly with elastic collisions between them. The Ideal Gas Law states that the pressure (P), volume (V), temperature (T), and quantity of the gas in moles (n) are related (Eq. 7.6):

$$PV = nRT \tag{7.6}$$

where R is the **ideal (universal) gas constant** 8.314 J K^{-1} mol^{-1} independent of the gas, determined by multiplying Avogadro's constant (N_A) by the Boltzmann constant (k). In reality, no gas behaves as an ideal gas. Therefore, the ideal gas constant can be replaced with a **gas-specific constant,** equal to R divided by the molar mass of the gas (M; g mol^{-1}). Using the 28.96 g mol^{-1} mean molar mass of dry air (M_d), the gas constant for dry air (R_d; J kg^{-1} K^{-1}) is:

$$R_d = \frac{R}{M_d} = \frac{8.314\,\text{J mol}^{-1}\text{K}^{-1}}{28.96\text{ g mol}^{-1}} \times \frac{1{,}000\text{ g}}{\text{kg}} = 287.09\frac{\text{J}}{\text{kg K}}.$$

Using the 18 g mol^{-1} molar mass of water vapor (M_w), the gas constant for water vapor (R_v; J kg^{-1} K^{-1}) is:

$$R_v = \frac{R}{M_v} = \frac{8.314\,\text{J mol}^{-1}\text{K}^{-1}}{18\text{ g mol}^{-1}} \times \frac{1{,}000\,\text{g}}{\text{kg}} = 461.89\frac{\text{J}}{\text{kg K}}.$$

Using the water-specific gas constant, temperature, and the Ideal Gas Law, vapor pressure can be converted to vapor density (ρ_v). For example, a vapor pressure of 1 kPa (1,000 Pa or 1,000 J m^{-3}) at an air temperature (T) of 20 °C (293 K) is equivalent to $\rho_v = 7.4$ g m^{-3}:

$$\rho_v = \frac{e_a}{R_v T} = \frac{1{,}000\text{ J m}^{-3}}{\left(461.89\text{ J kg}^{-1}\text{K}^{-1}\right)\left(293\text{ K}\right)} = 7.4\text{ g m}^{-3}.$$

7.3.1 Dew

Dew forms when water condenses from the air onto the ground or surface vegetation and remains liquid. **Frost** forms when the condensed water freezes on the surface. Near open water, damp soils, and other wet surfaces, the air in contact with the surface is close to being saturated with water vapor. Since the humidity in the air above cannot exceed saturation, the only way to achieve a higher vapor pressure in the air than at the surface (known as a **vapor pressure inversion**) is through a decrease in surface temperature. Over the course of a day, the minimum surface temperature is typically reached just prior to sunrise. Therefore, this is the time when **dewfall** is common (Monteith, 1957).

Since vapor pressure inversions are relatively rare and of short duration, the addition of surface water by dewfall is small in most regions. In some locations such as deserts, however, dewfall can be a significant source of precipitation. By calculating dewfall from periods at night when the latent heat flux was negligible, Hao et al. (2012) found that dewfall accounted for up to 50 percent of the annual rainfall in a riparian *Populus* forest located in the arid Taklimakan region in China. There are discussions of the potential to artificially collect or harvest water from the atmosphere to supplement rainfall in arid regions (Vuollekoski et al., 2015).

7.3.2 Clouds

Clouds consist of small, condensed water vapor droplets, and/or ice crystals if the air temperature is below the freezing point, that remain suspended in the atmosphere. Clouds are not water vapor, as water vapor is invisible to the human eye (Chapter 2). Condensation and cloud formation requires not only particulate matter suspended in the atmosphere for water vapor to condense onto (**cloud condensation nuclei**; Section 7.4.1), but also sufficient humidity and decrease in temperature for the saturation vapor pressure to be reached. The required temperature for condensation is aptly named the dew-point temperature.

The **dew-point temperature** (T_d) is the temperature at which air at constant pressure will attain 100% relative humidity, meaning the vapor and saturation vapor pressures are equal. The dew-point temperature can be calculated by solving the saturation vapor pressure equation (Eq. 7.3) for temperature. Solving for T gives the temperature for saturation, the dew-point temperature T_d (°C), since the saturation vapor pressure at the dew-point temperature $e_s(T_d)$ is equal to the ambient vapor pressure e_a (Eq. 7.7):

$$T_d = \frac{c \ln\left(e_a / a\right)}{b - \ln\left(e_a / a\right)}. \tag{7.7}$$

Using Buck's (1981) coefficients a = 0.61121 kPa, b = 17.502, and c = 240.97 °C in Eq. (7.3) gives e_s = 1.23 kPa at a temperature of 20 °C:

$$e_s = 0.61121\,\text{kPa} \times \exp\left(\frac{17.502 \times 20\,°\text{C}}{240.97\,°\text{C} + 20\,°\text{C}}\right) = 2.34\,\text{kPa}.$$

If the relative humidity is 50%, then the vapor pressure would be 1.17 kPa:

$$e_a = 0.50 \times 2.34\,\text{kPa} = 1.17\,\text{kPa}.$$

Equation (7.7) can then be used to calculate the equivalent dew-point temperature at 1.17 kPa, equal to 9.27 °C:

$$T_d = \frac{240.97\,°\text{C} \times \ln\left(1.17\,\text{kPa} / 0.61121\,\text{kPa}\right)}{17.502 - \ln\left(1.17\,\text{kPa} / 0.61121\,\text{kPa}\right)} = 9.27\,°\text{C}.$$

The difference between the air and dew-point temperature is the **dew-point spread**, and the greater the dew-point spread, the drier the air. The dew-point temperature is often negative in dry air (e.g., T_a and T_d of 12 and −10 °C, respectively, give a relative humidity of 21%). A high dew-point temperature is indicative of a humid atmosphere, which has implications for human comfort and thermal stress (Chapters 14 and 16).

There are two ways for the water vapor contained in an air parcel to condense into liquid form, assuming condensation nuclei are present. One is to increase humidity (vapor pressure) through an increase in evaporation or by the intrusion of a higher-humidity air mass. The other is to decrease the air temperature (T_a), which would decrease the saturation vapor pressure and the dew-point spread. There is usually a combination of an increase in humidity and a decrease in air temperature resulting in $T_a = T_d$ and the relative humidity reaching 100%. Consider the land adjacent to a large body of water such as the North American Great Lakes. In early summer (June, July), the land surface is quite warm whereas the lakes are still cool. The warmer air over the land has a higher saturation vapor pressure than cooler air over the lakes (Eq. 7.3), so when the air over the land surface has an ample source of moisture such as after rain, humidity over the land will be higher than over the lake. When warm humid air from the land moves horizontally (horizontal air flow is **advection**; vertical air flow is **convection**) over the adjacent

Figure 7.5 Advection fog near the Mackinac Bridge at the confluence of Lakes Michigan and Huron (foreground). Photo credit: P. D. Blanken.

cooler, drier air over the lake, the air will cool to its dew-point temperature, and dense **advective fog** can suddenly form (Figure 7.5). Under these circumstances, water has been transported horizontally through the atmosphere from the land to the lake, and the latent heat of vaporization that evaporated the water over the land has been released over the lake when the water vapor condensed; mass and energy have been transferred in space and time.

Even without an increase in humidity, condensation can be achieved through a decrease in temperature. In the atmosphere there is a sharp decrease in temperature with altitude at a rate of a decrease of roughly 1 °C for every 100 m increase in altitude. This is known as a lapse rate, more specifically the **dry adiabatic lapse rate** (DALR). This lapse rate is specific for dry air, meaning that T_d is less than T_a and condensation does not occur. This is significant because if condensation does occur, the energy equal to the latent heat of vaporization is released; roughly 2.5 million joules for every kilogram of water that condenses. This release of heat results in the surrounding air warming, so the temperature decreases at a rate less than 1 °C per 100 m, with the exact rate dependent on humidity (typically 0.6–0.9 °C m^{-1}). This is known as the **moist** or **wet adiabatic lapse rate** (MALR). In either the dry or moist conditions, the term **adiabatic** is used to describe the conditions within a parcel of air without heat or mass exchange with the surrounding air. When a parcel of rising air condenses, the heat released often results in a highly unstable air mass with rapid vertical ascent, hence further cooling and condensation. The rapid formation and growth of summertime thunderstorms is an example of this.

Based on the conservation of energy, the latent heat of vaporization required to evaporate water at one location (typically the surface) and time is released at a different location and time where and when condensation occurs. An approximate calculation of the height of the **cloud condensation level** (CCL) (the cloud base) can be used to illustrate the difference between the location of evaporation at the surface and the altitude of condensation above the surface (Eq. 7.8). If a parcel of air near the surface has an air temperature of 30 °C and a dew-point temperature of 5 °C (warm and dry air), the air needs to ascend to an altitude of 2,500 m to cool to the dew-point temperature for saturation and condensation to occur:

$$\text{CCL} = \left(T_a - T_d\right) \times \text{DALR} = \left(30\ ^\circ\text{C} - 5\ ^\circ\text{C}\right) \times \frac{100\ \text{m}}{1\ ^\circ\text{C}} = 2{,}500\ \text{m}. \tag{7.8}$$

If the humidity is sufficient for condensation to continue, heat will be released, and the air will continue to rise until all the water vapor has condensed or the temperature inversion high in the tropopause is

Figure 7.6 A large anvil-shaped cloud formed during a summer thunderstorm. Photo credit: P. D. Blanken.

reached, indicated in large thunderstorms by the anvil-like shape at their top (Figure 7.6). Since the latent heat of vaporization is large, so is the heat released with condensation, as demonstrated by the intense energy associated with thunderstorms and, on a grander scale, hurricanes.

7.3.3 Atmospheric Cooling Mechanisms

The vertical movement of air is an effective way for air to cool and reach the dew-point temperature. There are four main mechanisms resulting in this vertical air motion: convection, convergence, frontal, and orographic. **Convection** is defined as the vertical motion of a parcel of air through the surrounding atmosphere due to a decrease in pressure relative to the surrounding air. This is the bulk movement of air with the decrease in pressure associated with an expansion of the air parcel and is a result of surface heating. With expansion, the air parcel becomes unstable and buoyant, and rises. These parcels of rising air, or thermals, are common on summer days with initially calm conditions and strong surface heating.

When air rises from a warm surface, air must replace that displaced air, often resulting in strong surface winds. This is why hot regions such as deserts are often windy, and why strong surface winds develop in addition to strong vertical winds near thunderstorms. These circulations of rising air that eventually cool and descend are called **thermal** or **convective cells**. Thermal cells can occur at small scales, such as small turbulent eddies resulting from the heating of leaves on a tree, medium scales, such as thunderstorms, or large **Hadley circulation cells** that encircle the globe.

With a large, organized region of convection, surface air flowing into the region from opposing directions results in the **convergence** of air and further vertical air motion. On the global scale, persistent solar heating of the Earth's surface at the Equator year-round results in high surface temperatures and convection. Because of the **Coriolis effect** (quantified by **Gaspard-Gustave de Coriolis**, 1792–1843), once cooled, the air descending at roughly 30° latitude north and south of the Equator diverts eastward in the Northern Hemisphere and westward in the Southern Hemisphere. At the surface near the Equator, air is drawn in to replace the ascending air, again diverting to the right or left of its path because of the Coriolis effect in the Northern or Southern Hemisphere, respectively. These persistent surface winds are known as the **trade winds**. Since the equatorial surface heating persists year-long, so do the Hadley circulation cells (described by the lawyer and amateur meteorologist **George Hadley**, 1685–1768). This region is aptly named the **Intertropical Convergence Zone** (ITCZ) as surface air converges from all directions between the Tropics of Cancer and Capricorn. The ITCZ is easily seen in most

Table 7.1 **Molar mass and volumes for constituents of dry air at standard conditions**

Constituent	Molar mass (M: g mol^{-1})	Molecule in air	Molar mass (M: g mol^{-1})	Mole fraction in air (V: mol element per mol air)	Molar mass × mole fraction in air (g element per mol air)
Nitrogen: N	14.0067	N_2	28.0134	0.7808	21.8729
Oxygen: O	15.9994	O_2	31.9988	0.2095	6.7037
Argon: Ar	39.9480	Ar	39.9480	0.0093	0.3715
All trace gases excluding H_2O		CO_2, Ne, He, CH_4, Kr, H_2, Xe	41.5000	0.0004	0.0166
Dry air				***1.0000***	***28.9647***

satellite images by the band of high cumulus and cumulonimbus clouds that encircles the globe near the Equator.

Air also rises because of horizontal differences in air temperature or humidity commonly associated with meteorological fronts, hence the term **frontal lifting**. These are referred to as a **warm front** when warmer air advances into colder air, a **cold front** when colder air advances into warmer air, a **stationary front** when there is no movement of either air mass, or an **occluded front** when the faster-moving cold front displaces all the surface warmer air aloft. The term **dry line** or **dew-point front** is used in reference to humidity, marking the location with a sharp contrast between dry and humid air. Differences in air density resulting from differences in temperature or humidity lift and cool the air, making condensation more likely. In a warm front, the warmer, lower-density air has difficulty displacing the colder surface air and therefore rises over it, often resulting in a wide area of gentle precipitation. In a cold front, colder, higher-density air easily displaces warmer surface air, resulting in a narrow area of intense precipitation. Widespread precipitation occurs with an occluded front when the cold air has displaced all the warm air aloft, and a stationary front usually has no precipitation since there is no horizontal, hence vertical, air movement.

The dry line or dew-point front is relatively little known but has relevance to precipitation and demonstrates a particularly interesting aspect of water vapor. Severe storms with intense precipitation often develop along the boundary between dry and humid air. At similar temperatures, dry air is denser than moist air, so the colloquial “hot and heavy humid air” is false. To prove this requires looking at the molar mass of the atmosphere. The average molar mass of a gas mixture (M_{total}) is equal to the sum of the molar mass (M_i: g mol^{-1}) multiplied by the mole fraction (V_i: mols) of each gas i relative to the total moles of all gases in the mixture (mol mol^{-1}) for all individual gas components from 1 to n, where n is the total number of gases in the mixture (Eq. 7.9):

$$M_{total} = M_1V_1 + M_2V_2 + \ldots + M_nV_n. \tag{7.9}$$

The molar mass for dry air can be calculated using Eq. (7.9) and the data provided in Table 7.1. At standard conditions, 1 mol of a gas will occupy a volume of 22.4 liters. With this standard volume, if the mass increases, so will the density.

Table 7.2 **Molar mass and volumes for constituents of moist air at standard conditions**

Constituent	Molar mass (M: g mol^{-1})	Molecule in air	Molar mass (M: g mol^{-1})	Mole fraction in air (V: mol element per mol air)	Molar mass × mole fraction in air (g element per mol air)
Nitrogen: N	14.0067	N_2	28.0134	0.7808	21.8729
Oxygen: O	15.9994	O_2	31.9988	0.2095	6.7037
Argon: Ar	39.9480	Ar	39.9480	0.0093	0.3715
Water vapor: $H_2 + O$	2(1.00784) + 15.9994	H_2O	18.0151	0.0004	0.0072
Moist air				***1.0000***	***28.9553***

For dry air with no water vapor, from Eq. (7.9) using the data in Table 7.1 the sum of the molar mass multiplied by the mole fraction in dry air for each of the molecules is 28.9647 grams per mole in a volume of 22.4 liters with the vast majority (98.66 percent) from nitrogen and oxygen gas.

For comparison, the same calculation for air that contains water vapor can be performed using Eq. (7.9) and the data provided in Table 7.2. To illustrate, water vapor is now included to replace all the trace gas elements. With this addition, the total molecular mass of the moist air decreases to 28.9553 grams per mole in a volume of 22.4 liters. This difference arises because the sum of molar mass of the trace gas molecules (41.5 g mol^{-1}) is much larger than the molar mass of water vapor (18 g mol^{-1}), so the addition of water vapor will always decrease the air's total molecular mass. With the same volume of air (i.e., similar temperature and pressure), dry air will have a higher density than moist air.

This one-hundredth of a gram difference in molecular mass in air is sufficient to affect the convective initiation required to trigger severe storms. Although this difference in molecular mass and density between dry and moist air may seem trivial, a slight difference can have major consequences. Especially in the central United States in the late spring and early summer, dry air often advances into humid air after the passage of a warm front. At the surface, the higher-density dry air wedges under the lower-density moist air, vertically displacing the moist air where cooling and condensation may occur, thus forming severe thunderstorms.

The final atmospheric lifting mechanism is **orographic lifting**, where topography forces air to rise. A classic example of orographic lifting occurs when a high mountain range is oriented perpendicular to the prevailing wind direction in a coastal region. Such an arrangement is common given that geological subduction zones occur when an oceanic tectonic plate subducts under a continental plate, giving rise to volcanic mountain formations such as along the entire west coast of North and South America. Owing to the presence of the thermally driven Hadley cells near the Equator coupled with the Coriolis effect, the prevailing winds are from the east between latitudes 30° S and 30° N, and from the west between latitudes 30° and 60° North and South. Therefore, along the west coast of North and South America the prevailing winds blow primarily over thousands of kilometers of ocean or land, depending

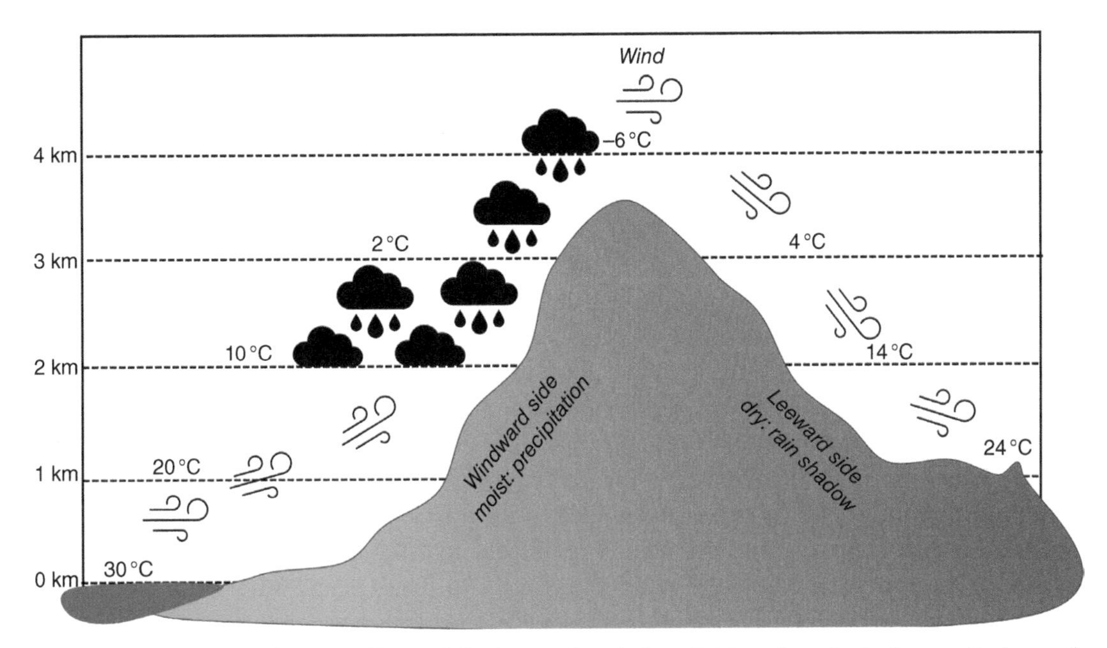

Figure 7.7 Formation of orographic precipitation on the windward side and a rain shadow on the leeward side of a topographic barrier. Prevailing wind direction is from left to right. Example air temperatures shown are calculated with an initial sea-level temperature of 30 °C, a dew-point temperature of 10 °C, a DALR of 10 °C per km, and an MALR of 8 °C per km. Note the higher air temperatures on the leeward side.

on the latitude, before the air is forced to rise over the American Cordillera. The associated precipitation pattern created because of orographic lifting is referred to as **orographic precipitation** on the wet windward side of the topographic barrier and a **rain shadow** on the dry leeward side.

Orographic precipitation is dramatic. Consider locations along the west coast of North or South America near 50° N or S latitudes. In North America, your location would be about mid-way across Vancouver Island, Canada, with a maximum elevation of about 1,500 m above sea level. On the west coast of the Island the mean annual precipitation is about 10,500 mm per year. On the east side of the island, only 100 km away, the mean annual precipitation is about 1,000 mm (Moore et al., 2010). This pattern repeats with high precipitation on the western side and low precipitation on the eastern side across the western coast of mainland British Columbia. A similar pattern exists in South America at 50° S along the Patagonia Archipelago. Excessive rain falls on the western side of the Andes with desert conditions occurring just to the east in the Patagonian region. In contrast, locations near 25° N or S latitudes experience a reversal in this precipitation pattern, with wet conditions east of the mountains and desert-like conditions on the ocean side.

This orographic precipitation pattern can be explained with an illustration (Figure 7.7). As prevailing winds force air to rise over topography, the air initially cools following the DALR of 1 °C per 100 m. Once the dew-point temperature is reached, condensation and precipitation are likely, and the lapse rate switches to the MALR resulting in wet conditions on the windward side. Drier air descending on the leeward side will warm following the DALR with the increasing air temperature drying the air further, resulting in a lack of precipitation and the rain shadow.

7.4 Precipitation Formation

Precipitation, defined as water from the atmosphere that reaches the Earth's surface, returns evaporated water back to the Earth's surface. After evaporation, precipitation is the largest flux in the global water cycle. Precipitation formation is a delicate process directly dependent on four properties of water. First, water vapor needs a surface to condense onto, a **condensation nucleus**, and fresh water becomes a solution as the water dissolves the material on which it condensed. Second, the lowering of the freezing point means that the solution can remain in liquid form while at the same time surrounded by ice crystals. This mixture of liquid and frozen water results in vapor pressure gradients forming due to differences in the saturation vapor pressure over liquid water compared with any adjacent ice. Third, small water droplets can form due to the large surface tension of water, and the vapor pressure around the droplet changes with the size of the droplet. And fourth, the vapor pressure around the droplet changes as the droplet changes volume owing to changes in the solute concentration. Precipitation requires a delicate balance between the rates of evaporation and condensation, all mediated by the unique properties of water.

7.4.1 Cloud Condensation Nuclei

For water vapor to condense in the atmosphere, small, suspended solid particles, **cloud condensation nuclei** (CCN), are required. For a solid particle to remain suspended in the atmosphere, it must be small and light, and typically have a diameter of 0.2 μm, much smaller than a raindrop's typical diameter of 2,000 μm. The term **aerosol** is used to describe the mixture of air containing particles with a diameter less than 1 μm. Another characteristic of CCN is that the material is **hygroscopic**, attracting and absorbing the water to its surface due to its slight electrical charge. Ionic compounds such as sodium chloride (NaCl) are readily attracted to water vapor molecules, forming a solution once dissolved after water condenses on the CCN.

With nearly 70 percent of the Earth's surface covered by saline oceans, ample condensation nuclei in coastal areas and over the open ocean are produced by waves. **Sea spray**, often visible from the mist produced by waves breaking along a coastline, can liberate salt-rich condensation nuclei into the atmosphere (Figure 7.8). Dimethyl sulfide, produced by marine biogenic activity and released into the atmosphere, is a key contributor to the production of CCN (Croft et al., 2021). Other sources of natural CCN include wind-blown dust, fires, and volcanic eruptions. Anthropogenic CCN are produced from fossil fuel combustion or intentional cloud-seeding operations. Organic molecules such as carbon can form water-soluble polar compounds that serve as CCN or ice nuclei (IN) when temperatures are below the freezing point. Even bacteria originating from plant surfaces and soil have been found in clouds, indicating their ability to serve as CCN or IN (Sun and Ariya, 2006).

Figure 7.8 Sea spray serving as effective cloud condensation nuclei near Dana Point, California, United States. Photo credit: P. D. Blanken.

7.4.2 The Bergeron Process

Water simultaneously existing in all three states is common, and the cycling between these states in clouds determines if precipitation will form. Since air temperature decreases rapidly with altitude, air temperatures are usually below 0 °C in clouds, meaning ice crystals are common even in warm regions. At the same time, condensed water creates a solution when dissolving the CCN, lowering the freezing point. Supercooled liquid water droplets can coexist with ice crystals in clouds, referred to as a **mixed-phase cloud**. In such clouds, a unique interplay develops between water droplets and ice crystals due to differences in vapor pressure over each surface. The analytic approximation to the Clausius–Clapeyron relationship (Buck, 1981) for a flat surface of pure liquid water (Eq. 7.3, repeated here) is:

$$e_s = 0.61121 \exp\left(\frac{17.502\,T}{240.97 + T}\right).$$

The calculation of the saturation vapor pressure over a pure flat ice surface ($e_{s(ice)}$) at temperature T (°C), however, requires different values for coefficients a, b, and c (Eq. 7.10) (Buck, 1981):

$$e_{s(ice)} = 0.61115 \exp\left(\frac{22.452\,T}{272.55 + T}\right). \tag{7.10}$$

When the saturation vapor pressures over liquid water (Eq. 7.3) and ice (Eq. 7.10) are plotted together, it is apparent that regardless of the water temperature, $e_{s(ice)}$ is always less than $e_{s(liquid)}$ (Figure 7.9).

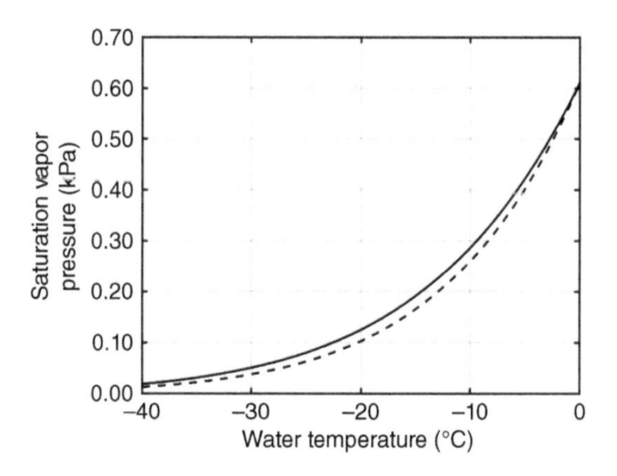

Figure 7.9 The saturation vapor pressure as a function of water temperature calculated above pure flat liquid water using Eq. (7.3) (solid line) and above pure flat ice using Eq. (7.10) (dashed line).

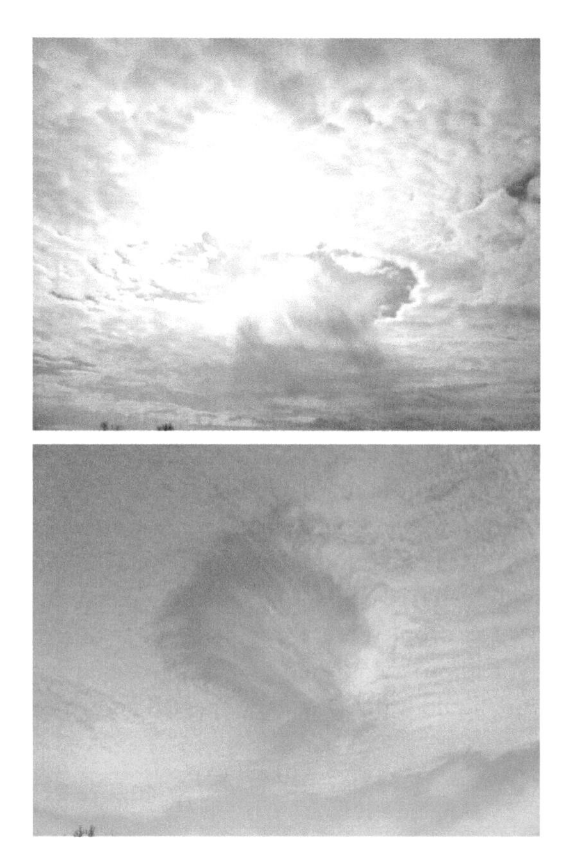

Figure 7.10 Hole-punch or fall-streak clouds illustrating the Bergeron process. Photo credits: P. D. Blanken.

In a cloud with high relative humidity, the vapor pressure is always higher near the liquid water than the ice crystal (Figure 7.9). Thus, water from the liquid droplet evaporates and deposits on the ice crystal, the droplet shrinks, and the ice crystal grows. This effect can be seen in **hole-punch clouds** or **fall-streak clouds** (Figure 7.10). This same effect also occurs at the base (**depth hoar**) or surface (**surface hoar**) of a snowpack where temperature gradients create vapor pressure gradients and ice crystal growth.

This process resulting in ice crystal growth is known as the **Bergeron process**, or more formally the **Wegener–Bergeron–Findeisen process** in recognition of everyone's contribution. **Alfred Wegener** (1880–1930), best known for *The Origins of Continents and Oceans* where he described his theory of continental drift, laying the foundation for the current theory of plate tectonics, was a polar meteorologist.

Storelvmo and Tan (2015) provide a summary of the origin of the Bergeron process. When studying the formation of hoar frost, Wegener realized that the coexistence of liquid water and ice was thermodynamically unstable. **Tor Bergeron** (1891–1977), a Swedish meteorologist born in England, was familiar with Wegener's meteorological research. While vacationing at a resort near Oslo, Norway, Bergeron observed that when it was cold and humid, areas near frost-covered trees were clear of fog while nearby areas were shrouded in fog. When conditions were warmer and humid, fog formed throughout the area. A few years later (1928), Bergeron described this phenomenon in his Ph.D. thesis. A decade later, German meteorologist **Walter Findeisen** (1909–1945) continued this work by building a cloud chamber in his laboratory to further develop and test Wegener and Bergeron's theoretical concepts, forming the first complete description of cloud microphysics (Storelvmo and Tan, 2015).

7.4.3 The Curvature or Kelvin Effect

The shape of water droplets and ice crystals affects precipitation formation. As a consequence of the flexible hydrogen bonds between water molecules, liquid water has a high surface tension, allowing the formation of small droplets. This inward force divided by the distance along the surface gives the units of surface tension as N m^{-1} or the equivalent J m^{-2}. When water vapor condenses on a CCN, the small droplets have a high peripheral surface tension, making it difficult for the droplet to increase in size by the addition of more liquid water. Regardless, small water droplets will often collide and coalesce to increase size and mass, furthering the chance of precipitation. This **collision-coalescence** precipitation formation process dominates in a **warm cloud** where liquid water prevails at temperatures above the freezing point. Large drops of rainfall associated with tropical thunderstorms are a result of the collision-coalescence process.

The droplet's size and the vapor pressure in the surrounding air are connected through the competing processes of condensation and evaporation. Peripheral surface tension increases as the droplet's radius decreases. And, as the surface tension increases, so does the exterior force required to keep water molecules from escaping the droplet through evaporation. This external force is the vapor pressure surrounding the droplet. As the water droplet's size increases, the surrounding vapor pressure required to keep the droplet from evaporating away decreases. Conversely, as the water droplet's size decreases, the surrounding vapor pressure required for a further reduction in size due to evaporation increases. Sometimes the term **equilibrium vapor pressure** is used for the vapor pressure required to maintain a constant droplet size, when the rates of evaporation and condensation are equal, synonymous with the term saturation vapor pressure as previously defined for a pure, flat, water surface. For a curved pure water surface, the equation to calculate the equilibrium vapor pressure was developed by **William Thomson**.

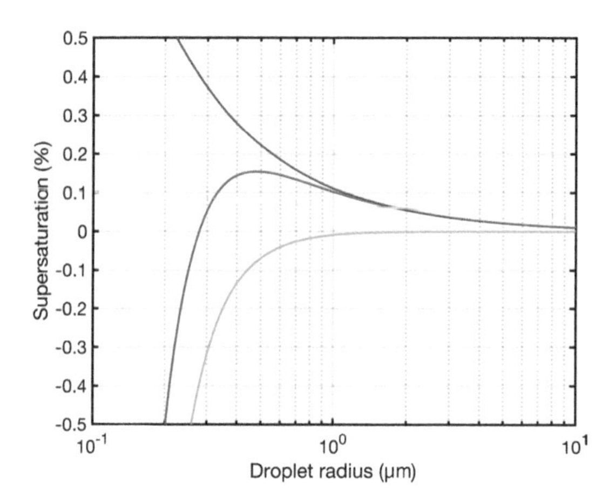

Figure 7.11 The curvature or Kelvin effect (red line), the solute or Raoult effect (orange line) and the combined effect based on Köhler's theory (blue line). The conditions used were $\gamma = 0.756$ J m^{-2}, $n_w = 5.55 \times 10^4$ mol m^{-3}, $R = 8.314$ J mol^{-1} K^{-1}, $T = 20$ °C (293.15 K), and $N_s = 1 \times 10^{-17}$ moles of NaCl ($i = 2$).

Since water droplets are far from flat, there was a need to calculate their equilibrium (saturation) vapor pressures. William Thomson (1824–1907) was given the title of **Baron (Lord) Kelvin** in 1892 in recognition of research resulting in the determination that absolute zero was approximately −273.15 °C, forming the basis of the Kelvin temperature scale. In addition, Thomson's prolific research on various aspects of thermodynamics, some in collaboration with **James Prescott Joule** (1818–1889), resulted in the **Kelvin equation** that describes the **Kelvin** or **curvature effect**, the relationship between the saturation vapor pressure over a pure flat water surface (e_s) and that over a curved pure water surface (e_{sc}) (Eq. 7.11):

$$e_{sc} = e_s \exp\left(\frac{2\gamma}{n_w R T r}\right) \tag{7.11}$$

where γ is the surface tension (approximately 0.0756 N m^{-1} or J m^{-2} for water), n_w is the molar volume of the water (5.55×10^4 mol m^{-3}), R is the universal gas constant (8.314 J K^{-1} mol^{-1}), T is temperature (K), and r is the droplet's radius (m). At a constant temperature (20 °C is used in this example), the ratio e_{sc}/e_s can be plotted against r to illustrate how large e_{sc} is relative to e_s as the droplet's radius, hence curvature and surface tension, increases (Figure 7.11). Instead of plotting the e_{sc}/e_s indicating the degree of saturation, the **degree of supersaturation**, $e_{sc}/e_s - 1$, is shown in Figure 7.11. This allows both the solute effect and the combined curvature and solute effects to be shown. The curvature effect shows that as water first condenses on a very small condensation nucleus, a vapor pressure much greater than that over flat water is required. A slightly **supersaturated** air (meaning the relative humidity is greater than 1) is required because the surface tension of the small droplet is so large. As droplets grow, surface tension decreases, and so does the required degree of supersaturation.

7.4.4 The Solute or Raoult Effect

The solution created when water condenses also influences precipitation formation. The relative humidity in the air adjacent to droplets containing solutes can be less than 100 percent without having the droplet evaporate. This is because once a solution is made, the periphery of the droplet is no longer completely occupied by water molecules. Since some of the periphery now contains other elements, a lower vapor pressure surrounding the droplet can be maintained.

As the size (volume) of the droplet decreases owing to a decrease in relative humidity resulting in increased evaporation, or the droplet's radius increases through collision-coalescence, the solute concentration will also change. The solute concentration will increase as the droplet's volume decreases (i.e., the radius decreases) and the solute concentration will decrease as the droplet's volume increases (i.e., the radius increases).

To understand how the saturation (equilibrium) vapor pressure changes above a solution compared to pure water (both flat surfaces), **Raoult's Law** can be used. **François-Marie Raoult** (1830–1901) was a French chemist known for his work with chemical solutions, including how solutes lower the freezing point. The law for which he is known states that the partial pressure of each component of an ideal mixture is equal to the sum of the mole fraction multiplied by the vapor pressure of each component. Raoult's Law can be written as (Eq. 7.12):

$$e_{sol} = e_s(1 - \chi_s) \tag{7.12}$$

where e_{sol} is the saturation vapor pressure of the solution, e_s is the saturation vapor pressure for pure water, and χ_s is the mole fraction of the solution, equal to the ratio of the volume mole fraction of the solute (η_s; mol m^{-3}) to the sum of the mole fraction of water (η_w; mol m^{-3}) and the solute (Eq. 7.13):

$$\chi_s = \frac{\eta_s}{\eta_w + \eta_s}. \tag{7.13}$$

When the solution is dilute, η_s is small relative to η_w so the sum of the two terms can be approximated by the volume mole fraction of the water, η_L (Eq. 7.14):

$$\chi_s = \frac{\eta_s}{\eta_w + \eta_s} \approx \frac{\eta_s}{\eta_L}. \tag{7.14}$$

Raoult's Law can then be approximated by Eq. (7.15):

$$\chi_s = \frac{iN_s}{n_L V_{sphere}} = \frac{BiN_s}{r^3} \tag{7.15}$$

where $B = \frac{3}{4\pi n_L}$ m^3 mol^{-1}, i is the **van't Hoff factor** determined based on whether the compound splits into ions when dissolved (e.g., NaCl dissolved in water forms Na^+ and Cl^- so $i = 2$), N_s is the number of moles of the solute, and r is the droplet's radius (m). The numbers 3, 4 and π appear in the term B because the volume of the droplet is required ($V_{sphere} = \frac{4}{3}\pi r^3$) in the denominator.

Now that the mole fraction of the dilute solution can be calculated, the ratio of the saturation vapor pressure for the solution to the saturation vapor pressure for pure water, e_{sol}/e_s, can be plotted against the droplet's radius since the radius is used to calculate the mole fraction. To plot solute and curvature effects on the same graph, the solute effect is shown with supersaturation on the y-axis rather than the ratio e_{sol}/e_s (Figure 7.11).

7.4.5 Combined Curvature and Solute Effects: Köhler Theory

A suspended water droplet is subject to several coupled physical and chemical forces that ultimately determine whether that droplet forms precipitation. When water condenses upon a condensation nucleus, the shape is not flat but spherical, not pure but a solution, and any changes in the droplet's size change both the surface tension and the solute concentration. These complicated, connected processes were described and quantified by Köhler (1936) and summarized by Brune (2020).

The saturation vapor pressure over pure flat water (e_s) is modified by both the curvature (e_{sc}) and solute effects (e_{sol}), as stated by Eq. (7.16) from Köhler (1936):

$$e_{sc+sol} = e_s \times e_{sc} \times e_{sol} = e_s \times \exp\left(\frac{2\gamma}{n_w RTr}\right) \times \left(1 - \chi_s\right). \tag{7.16}$$

Taking the ratio of the saturation vapor pressure for a curved solution (the droplet) relative to that for pure flat water, and after expanding the exponential term, gives Eq. (7.17):

$$\frac{e_{sc+sol}}{e_s} = \exp\left(\frac{2\gamma}{n_w RTr}\right) \times \left(1 - \chi_s\right) = 1 + \frac{A}{r} - \frac{BiN_s}{r^3} \tag{7.17}$$

where the second term is the curvature term with $A = \dfrac{2\gamma}{n_w RT}$ and the third term is the solute term with $B = \dfrac{3}{4\pi n_w}$. Expressing the combined equation not as a ratio e_{sc+sol}/e_s but as the supersaturation S equal to ($e_{sc+sol}/e_s - 1$) gives Eq. (7.18):

$$S = \frac{A}{r} - \frac{BiN_s}{r^3}. \tag{7.18}$$

Köhler's theory and resulting equation (Eq. 7.18) provide the means to calculate the combined curvature and solute effects as shown in Figure 7.11.

To summarize what is shown in Figure 7.11, the processes involved for water to condense in the atmosphere and create precipitation are complex. There are two competing effects. The high surface tension of small radii droplets requires the air to be supersaturated, yet the solution that forms allows for the air to be below saturation. Since both the curvature and solute effects occur simultaneously, their combined effect reveals that the solute effect dominates for small droplet radii. This means that tiny droplets of water can condense at low humidity. As the droplet size increases in size to greater than approximately 0.28 μm for the example conditions shown in Figure 7.11, the curvature effect dominates. Humidity increases, reaching a maximum at a radius of approximately 0.48 μm. Humidity must exceed this supersaturation at the maximum droplet radius (the critical radius) before the particle will serve as a cloud condensation nucleus and continue to grow. With droplet growth, the humidity then decreases, approaching that for pure flat water (supersaturation equals zero). As the droplet's growth continues, its mass increases, and it may eventually fall to the Earth's surface as precipitation. The Bergeron process may be simultaneously acting to reduce droplet size if there is ice present.

7.5 SUMMARY

Evaporation is the largest mass flux in the water cycle. In 1802, Dalton described how evaporation was predictable, varying with the difference between the saturation and atmospheric vapor pressure and air flow. A century and a half later, Dalton's equations were refined by

Penman to specifically include wind speed to predict the evaporation rate from open water and moist, well-watered surfaces in natural settings. The quantification of a source of water vapor, and a sink (or demand) for water vapor in the atmosphere by Dalton and Penman resulted in the framework still used today to measure and model evaporation from wet surfaces.

Limitations on the direct measurement of evaporation (Chapter 12), and its spatial and temporal variation, have to some extent been overcome by models and remote sensing observations. Studies have shown that, globally, 58% of the precipitation falling on land evaporates, with 80% of that from transpiration, indicating the importance of vegetation in the global water cycle (Chapters 11 and 12). Evaporation from the oceans is nearly six times that from the land, with 86% of global evaporation occurring from oceans. The highest evaporation rates occur near the Equator where the availability of liquid water and energy are greatest.

Condensation returns water vapor to liquid form, releasing the latent heat of vaporization in the process. As with evaporation, condensation plays a major role in the water cycle, especially in the formation of precipitation that resupplies surface water for evaporation. For condensation to occur, a surface for the water vapor to condense onto (the ground for dew formation or CCN for cloud formation), and a high humidity and low temperature are required. The combined humidity and temperature requirement can be quantified through the calculation of the dew-point temperature. Air temperature decreases rapidly with altitude, so getting air to rise is an effective way for air to cool to the dew-point temperature. Convection, convergence, frontal systems, and topography (orographic) are all effective means of getting air to rise and cool, thus promoting cloud and precipitation formation.

Suspended water droplets in a cloud are subject to several processes that ultimately determine the evaporation and the condensation rates. When evaporation dominates, the droplets shrink, and the cloud vanishes. When condensation dominates, the droplet grows, and precipitation is likely. When ice is present in a cloud along with liquid water droplets, the Bergeron process describes how the higher saturation vapor pressure near the liquid droplets compared to near the ice crystals results in ice crystal growth and water droplet shrinkage. The combination of small water droplets (as described by the curvature or Kelvin effect) and the solute concentration (as described by the solute or Raoult effect) results in a dynamic interplay between the droplet's radius (volume) and humidity as described by Köhler's theory. Without the high surface tension and high solvent properties of water, precipitation formation would be unlikely.

Evaporation, condensation, and the associated formation of precipitation are critical components of the liquid–gas forms of water in the water cycle, having a profound influence on the abiotic (Chapters 9 and 10) and biotic (Chapters 11 through 14) environments. Water cycling involving water in its solid form, ice, is the topic of Chapter 8.

7.6 QUESTIONS

7.1 Calculate the saturation vapor pressure and latent heat of vaporization for pure liquid water at 10, 20, and 30 °C. Explain at the molecular level why both the saturation vapor pressure and the latent heat of vaporization vary with temperature.

7.2 The observations of Dalton and Penman were based on experiments conducted over open water and well-watered bare soil or grass. Evaporation from dry surfaces and the effects of vegetation were not yet explored. In the context of the processes and controls of evaporation described in this chapter, how and why would evaporation from a dry forest differ from the evaporation observed by Dalton or Penman?

7.3 Using Penman's equation (Eq. 7.5), calculate the expected daily total millimeters of evaporation when the average daily wind speed at 2 m above the surface was 2 m s^{-1}, the air temperature was 18 °C, the relative humidity was 40 percent, and the water surface temperature was 20 °C. Be careful with the units. Convert your answer in units of millimeters per day to the equivalent watts per square meter.

7.4 What is the equivalent water vapor density in grams per cubic meter for a volume of air at 10 °C with a vapor pressure of 0.4 kPa?

7.5 In a closed system, the term "saturated vapor pressure" is used to describe the vapor pressure when the rates of evaporation and condensation are equal. Why is the word "saturated" used, and how does temperature influence the saturation vapor pressure?

7.6 Describe the conditions in terms of temperature and vapor pressure vertical gradients that are conducive to the formation of dew.

7.7 If the air temperature is 24 °C and vapor pressure is 3 kPa, is condensation likely or not? Calculate the dew-point temperature to justify your response.

7.8 Humans are notoriously poor in estimating the altitude of clouds since there are no objects in the sky to use as a reference. Based on the dry adiabatic lapse rate and surface air temperature of 16 °C and a relative humidity of 25 percent, estimate the altitude of a cloud base.

7.9 Which would be the dominant mechanism resulting in air to ascend and cool under each of the following scenarios or situations? (i) A large thunderstorm. (ii) Cold air from the coast moves onshore displacing the warmer air. (iii) Air flows up and over a mountain range.

7.10 The adjectives "hot, humid, and heavy" are sometimes used to describe oppressive conditions associated with high heat and humidity. Is the use of the adjective "heavy" accurate for such a comparison between humid and dry air?

7.11 Why is sodium chloride such a common and effective cloud condensation nucleus?

7.12 In a cloud that contains a mixture of liquid water and ice, how is it possible to have liquid water, and what are the dynamics that develop between the liquid and ice?

7.13 As the radius of a droplet of water changes, the humidity in the surrounding air required to maintain equal evaporation and condensation rates (hence maintain the droplet's radius) also changes (see Figure 7.11). Describe the properties of water that account for this complex relationship.

7.14 What are some implications of the properties of water for evaporation and/or condensation? For example, if water were not a bipolar molecule, would the formation and occurrence of precipitation (hence the water cycle) be affected? If the hydrogen bonding between water molecules did not permit a high surface tension, would either the rates of evaporation or condensation and precipitation formation change? Explore these or other hypothetical scenarios.

REFERENCES

Brune, W. (2020) *Fundamentals of Atmospheric Science*. Available at: https://chem.libretexts.org/@go/page/3347.

Buck, A. L. (1981) 'New equations for computing vapor pressure and enhancement factor', *Journal of Applied Meteorology and Climatology*, 20, pp. 1527–1532.

Croft, B., Martin, R. V., Moore, R. H. et al. (2021) 'Factors controlling marine aerosol size distributions and their climate effects over the northwest Atlantic Ocean region', *Atmospheric Chemistry and Physics*, 21, pp. 1889–1916. doi: 10.5194/acp-21-1889-2021.

Dalton, J. (1802a) 'Experimental Essays, on the Constitution of mixed Gases; on the Force of Steam of Vapour from Water and other Liquids in different temperature, both in a Torricellian Vacuum', in *Memoirs of the Literary and Philosophical Society of Manchester*. R. & W. Dean & Co., pp. 535–602.

Dalton, J. (1802b) 'Experiments and Observations to determine whether the Quantity of Rain and Dew is equal to the Quantity of Water carried off by the Rivers and raised by Evaporation; with an Enquiry into the Origin of Springs', in *Memoirs of the Literary and Philosophical Society of Manchester*. R. & W. Dean & Co., pp. 346–372.

Hao, X. M., Li, C., Guo, B. et al. (2012) 'Dew formation and its long-term trend in a desert riparian forest ecosystem on the eastern edge of the Taklimakan desert in China', *Journal of Hydrology*, 472–473, pp. 90–98. doi: 10.1016/j.jhydrol.2012.09.015.

Henderson-Sellers, B. (1984) 'A new formula for latent heat of vaporization of water as a function of temperature', *Quarterly Journal of the Royal Meteorological Society*, 110, pp. 1186–1190. doi: 10.1002/qj.49711046626.

Huang, J. (2018) 'A simple accurate formula for calculating saturation vapor pressure of water and ice', *Journal of Applied Meteorology and Climatology*, 57, pp. 1265–1272. doi: 10.1175/JAMC-D-17-0334.1.

Köhler, H. (1936) 'The nucleus in and the growth of hygroscopic droplets', *Transactions of the Faraday Society*, 32, pp. 1152–1161. doi: 10.1039/TF9363201152.

Miralles, D. G., De Jeu, R. A. M., Gash, J. H., Holmes, T. R. H. and Dolman, A. J. (2011) 'Magnitude and variability of land evaporation and its components at the global scale', *Hydrology and Earth System Sciences*, 15, pp. 967–981. doi: 10.5194/hess-15-967-2011.

Monteith, J. L. (1957) 'Dew', *Quarterly Journal of the Royal Meteorological Society*, 83, pp. 322–341.

Moore, R. D., Spittlehouse, D. L., Whitefield, P. H. and Stahl, K. (2010) Chapter 3 – Weather and Climate, in *Compendium of Forest Hydrology and Geomorphology in British Columbia*. Land Management Handbook 66. Forest Science Program/FORREX, pp. 47–84.

Penman, H. L. (1948) 'Natural evaporation from open water, bare soil and grass', *Proceedings of the Royal Society of London, Series A*, 193, pp. 120–145.

Storelvmo, T. and Tan, I. (2015) 'The Wegener-Bergeron-Findeisen process – Its discovery and vital importance for weather and climate', *Meteorologische Zeitschrift*, 24, pp. 455–461. doi: 10.1127/metz/2015/0626.

Sun, J. and Ariya, P. A. (2006) 'Atmospheric organic and bio-aerosols as cloud condensation nuclei (CCN): A review', *Atmospheric Environment*, 40, pp. 795–820. doi: 10.1016/j.atmosenv.2005.05.052.

Vuollekoski, H., Vogt, M., Sinclair, V. A. et al. (2015) 'Estimates of global dew collection potential on artificial surfaces', *Hydrology and Earth System Sciences*, 19, pp. 601–613. doi: 10.5194/hess-19-601-2015.

Yu, L. (2007) 'Global variations in oceanic evaporation (1958–2005): The role of the changing wind speed', *Journal of Climate*, 20, pp. 5376–5390. doi: 10.1175/2007JCLI1714.1.

8 Melting and Freezing, Sublimation and Deposition

Key Learning Objectives

After reading this chapter you will be able to:

1. Describe the processes and energy requirements as water changes state to or from its solid form.
2. Calculate the energy required to melt a volume of ice or snow and use the energy balance concept to explain the sources of energy available for melting.
3. Explain how ice formation affects and alters both the abiotic and biotic environments.
4. Provide examples of the importance of sublimation in the winter water balance of forests and explain the factors that influence sublimation.

8.1 Introduction

Water could cycle between the liquid and vapor states through the processes of evaporation and condensation alone. However, Earth's surface temperature in the past and present are low enough for water to freeze, so an understanding of processes involving ice is required. Ice plays an important role in the temperature–albedo feedback and precipitation formation, and has a major influence on the landscape, in addition to affecting the distribution and life cycles of plants and animals. Ice can deform under its own weight, slowly flow to scour and erode the landscape, and directly change state to gas without becoming a liquid. When water freezes within confined spaces, the volume expansion can shatter rocks, mix soils, and damage or kill plant or animal tissues. When ice melts, liquid water can enter soils, becoming available for transpiration, dissolving nutrients, entering groundwater, and eventually joining rivers and then oceans.

This chapter begins by explaining the process of melting ice based on the amount of energy required for water to change state from solid to liquid. The response of the temperature of melting ice through time as energy is applied is described, and sample calculations are provided to show the energy required to increase the temperature of a volume of ice (or snow) to the melting-point temperature. The energy balance of a volume of ice is used to illustrate where the energy required to warm and melt ice originates. The effect of lowering the melting-point temperature of ice through the addition of pressure is explained based on classic experiments to explain glacial ice motion.

The impacts of water's volume expansion upon freezing to both the abiotic and biotic environments are reviewed. Freezing tolerance mechanisms in vascular plants, as well as the effects of frostbite in humans, are discussed. Changes of state directly between solid and liquid are reviewed in the context of the rates and controls of sublimation from various types of forests, and deposition (frost formation) on an ice or snow surface.

8.2 Melting

Melting is the change of state from solid to liquid and is affected by both temperature and pressure. Melting water involves the breaking of hydrogen bonds between water molecules, requiring energy supplied by the surrounding environment (an **endothermic process**; Chapter 4). The amount of energy required is determined not only by the characteristics of hydrogen bonds, but also the temperature and pressure imposed on the ice. When the ice temperature is below the melting point, energy must be applied to raise the temperature to the melting point as specified by the cold content before melting begins. Alternatively, pressure could be applied to melt ice. The high pressure at the base of ice sheets or large glaciers can be sufficient to melt ice and form a thin layer of liquid water, assisting in ice movement, or provide liquid water to subglacial lakes (e.g., Lake Vostok under the Antarctic Ice Sheet).

Energy supplied by incident solar radiation is usually sufficient to melt ice. The **melting-point temperature** is the temperature at which a substance changes state from solid to liquid. For pure water at sea level, the melting point is 0 °C, which may or may not be the same as the **freezing-point temperature**, the temperature at which a substance changes state from liquid to solid. As ice forms, solutes dissolved or dissociated in the water, such as Na^+ and Cl^-, are expelled. Ocean ice that has survived at least one melt season (**multiyear ice**) thus has a low salinity with a melting-point temperature near 0 °C. The temperature of the ice will remain at the melting-point temperature as the available energy is used to break the bonds between water molecules. Only after all the ice has melted will the temperature increase with continued energy input. The hidden or "latent" heats keep the temperature of water constant despite continued energy input as ice melts and water boils (Figure 8.1).

The melting-point temperature for water varies with pressure, as was first observed in the nineteenth century. **Michael Faraday** (1791–1867), known for his research on electromagnetic fields, demonstrated this effect. In 1850, Faraday performed experiments showing that two blocks of ice at 0 °C with moist surfaces adhered together as the liquid water froze upon contact. The same effect was observed when ice was placed in much warmer locations (Faraday, 1860). The term **regelation** was coined in reference to this refreezing (re-gel) between blocks of ice. **James Thomson** (1822–1892) predicted that pressure lowered the freezing point of water (T_f) by 0.0075 °C for every increase in pressure above one atmosphere (Thomson, 1849), which was experimentally verified by his brother William (Thomson, 1850). The idea that pressure must be applied for sections of ice to come in contact, and that under such conditions regelation could occur since the increased pressure lowered the melting point, was proposed by **John Tyndall** (1820–1893) as a mechanism to explain glacier ice motion. Tyndall and Huxley (1857) proposed that fragments of ice crushed under great

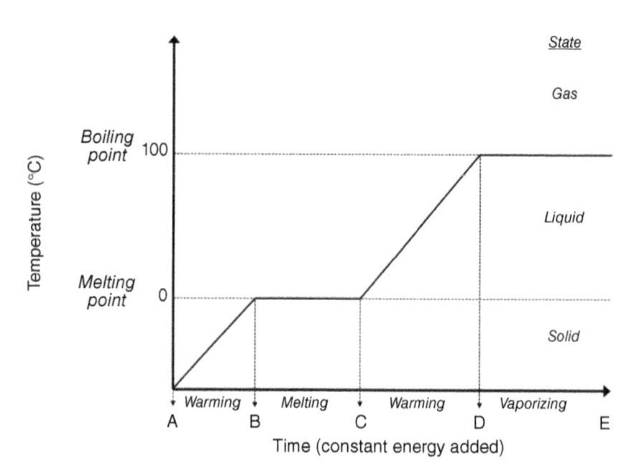

Figure 8.1 Changes in temperature and state of pure water at sea level over time with a constant supply of energy applied to ice at time A. Warming of the ice occurs until it reaches its melting-point temperature at B. Thereafter, no increase in temperature occurs until all the ice is melted at time C. The now-liquid water warms until time D when the boiling-point temperature is reached. The water continues to boil until all of it is in vapor form at time E.

pressure could reconnect though regelation to produce the continuous mass of ice required for viscous glacial ice flow.

This relationship between pressure, melting, and refreezing was shown in a classic demonstration first performed by **James Bottomley** (Bottomley, 1872) in a lecture on heat at Glasgow University in 1871. At the beginning of the lecture, Bottomley placed a lump of ice the size of an apple on top of a wire gauze. A flat board was placed over the ice, and a 12-pound (5.55 kg) weight was placed on the board. At the end of the lecture, the block of ice had slowly passed through the wire mesh and reformed underneath. Next, Bottomley placed a wire over a block of ice with a 2 lb weight (0.91 kg) attached to each end. The wire passed through the ice but left it undivided afterwards. He repeated the experiment several times using different weights and wire thicknesses, with the same result: the wire cut the ice in half, but the ice reformed or "regelled" immediately afterwards. Bottomley's experiments were the first experimental confirmations of the theories of Faraday and the Thomson brothers (Drake and Shreve, 1973).

These theories, experiments, and observations provided a foundation for understanding how pressure influences melting, and therefore an explanation for how glaciers and ice sheets could flow. Doubling atmospheric pressure would only lower the freezing-point temperature by 0.015 °C (0.0075 °C × 2), so only under glaciers or ice sheets does the influence of pressure on temperature need consideration. Only when pressures equal 100 times sea-level pressure does an appreciable decrease in the melting-point temperature occur. Without the assistance of pressure, the temperature of ice must reach the melting-point temperature of 0 °C before interconnecting water molecule bonds begin to break and melting commences.

The energy required to raise the ice volume's temperature (T_i; °C or K) over a given depth (z_i in m) to the melting-point temperature (T_m; °C or K) is known as the **cold content** (Q_{cc}; J per m^2

of surface area), which also varies with ice density (ρ_i; kg m^{-3}) and the specific heat capacity of ice (C_i; typically 2,108 J kg^{-1} °C^{-1} or K^{-1}) (Eq. 8.1):

$$Q_{cc} = -\rho_i C_p (T_i - T_m) z_i. \tag{8.1}$$

The Q_{cc} indicates the energy required to raise the temperature before melting can commence. Once the ice volume's temperature is **isothermal** at the melting-point temperature (the temperature is the same over a given depth), melting begins, yet the temperature remains at the melting point since the energy is being used as the latent heat of fusion (334 kJ kg^{-1}) to change the state of water from solid to liquid.

Melting cannot commence until the cold content energy requirement has been satisfied. If the average ice temperature over a 10-cm depth (0.10 m) was −5 °C (or K) below the melting-point temperature, and the average ice density was 917 kg m^{-3}, the Q_{cc} per unit area would be (Eq. 8.1):

$$Q_{cc} = -917 \frac{\text{kg}}{\text{m}^3} \times 2{,}108 \frac{\text{J}}{\text{kg °C}} \times (-5°\text{C} - 0°\text{C}) \times 0.10 \text{ m} = 966{,}518 \frac{\text{J}}{\text{m}^2}.$$

So 966,518 joules are required to raise the ice temperature to the melting point for each square meter of ice at the surface over a depth of 10 cm. If the energy available at the ice surface was supplied at a rate of 200 W m^{-2} (J m^{-2} s^{-1}), it would take roughly 80 minutes to raise the temperature to the melting point (966,518 J m^{-2} / 200 J m^{-2} s^{-1} = 4,833 s).

After ice reaches the melting-point temperature, any available energy can be used to melt ice. The **latent heat of melting** (usually referred to as the **latent heat of fusion**; λ_f) is 334,000 joules per kilogram of ice (334 kJ kg^{-1}), which is equal to 334 J g^{-1}, the energy required to break the bonds between water molecules in their solid state thus forming a liquid. This latent heat is the least of all the latent heats because of the relatively weak hydrogen bonds between water molecules. Since energy is required to melt ice, this is an **endothermic process**, and despite the requirement of energy for melting to occur, the temperature will not rise above the melting point until the change from solid to liquid is complete (Figure 8.1). The water molecules cannot move freely until all the bonds are broken.

The mass of ice multiplied by the latent heat of fusion gives the energy in joules required to break the hydrogen bonds between water molecules (Q_m). To express this energy in units of J m^{-2}, for comparison to Q_{cc}, Q_m can be calculated over a known depth of ice, z_i (m) (Eq. 8.2):

$$Q_m = \lambda_f \rho_i z_i. \tag{8.2}$$

Following the example provided above for Q_{cc}, the energy required to melt the layer of ice 10 cm deep with an average density of 917 kg m^{-3} is:

$$Q_m = 334{,}000 \frac{\text{J}}{\text{kg}} \times 917 \frac{\text{kg}}{\text{m}^3} \times 0.1 \text{ m} = 3.06 \times 10^7 \frac{\text{J}}{\text{m}^2}.$$

If the energy available at the ice surface was 200 W m^{-2} (200 J m^{-2} s^{-1}) as used previously, it would take roughly 42.5 hours to melt each square meter of 10-cm-deep ice (3.06 × 10^7 J m^{-2} / 200 J m^{-2} s^{-1} = 153,000 s).

8.3 Sources of Energy to Warm and Melt Ice

The total energy required to melt ice (Q_{mt}) is equal to the sum of the energy required to raise the temperature to the melting-point temperature (Q_{cc}) and the energy required to then melt the ice (Q_m), all in units of J m^{-2} (Eq. 8.3):

$$Q_{mt} = Q_{cc} + Q_m. \tag{8.3}$$

Presumably, for the ice to form and persist, the temperature was below the melting point for some time, but if not then Q_{cc} is equal to zero. Following the example of the 10 cm of ice initially at 5 °C below the melting point, the total melt energy is:

$$Q_{mt} = Q_{cc} + Q_m = 9.67 \times 10^5 \frac{J}{m^2} + 3.06 \times 10^7 \frac{J}{m^2} = 3.16 \times 10^7 \frac{J}{m^2}.$$

With energy available at the ice surface of 200 W m^{-2}, it would take roughly 43 hours and 50 minutes for the ice to melt (Q_{cc} required 80 minutes and Q_m required 42.5 hours).

The balance between inputs and outputs of energy can be used to determine how much energy is available to warm and melt ice or snow (Eq. 8.4) (Figure 8.2):

$$Q_{mt} = S^* + L^* + \lambda E + H + G + R - Q_s \tag{8.4}$$

where S^* is the net solar (shortwave) radiation ($S^* = S\downarrow - S\uparrow$), L^* is the net terrestrial (longwave) radiation ($L^* = L\downarrow - L\uparrow$), λE is the latent heat flux (sublimation, melting, freezing, or other changes of state, expressed as latent heat of vaporization multiplied by the evaporation rate E), H is the sensible heat flux (convection), G is the ground (soil) heat flux, R is the transfer of energy due to rainfall refreezing, and Q_s is the internal snowpack energy content, all in units of W m^{-2}. In this example, positive values indicate a gain of energy into the ice volume and negative values indicate a loss of energy. Melting can only occur when Q_{mt} is positive. The net shortwave and longwave radiation terms represent the difference between downwelling ($\downarrow$) and upwelling ($\uparrow$) radiation.

Equation (8.4) and Figure 8.2 together show how changes in the energy balance terms can explain the current widespread melting of snow and ice. **Solar** or **shortwave radiation** has wavelengths between roughly 0.1 and 4.0 μm and is emitted by the Sun at a surface temperature near 6,000 K. Snow and ice have a high **albedo**, the ratio of reflected to incident solar radiation ($\alpha = S\uparrow/S\downarrow$), so although solar radiation plays a major role in snowmelt, most (70–90 percent) of the incident solar radiation is reflected from the snow surface, thus decreasing S^*.

Longwave radiation or **terrestrial radiation** has wavelengths between 4.0 and 100 μm and is emitted by terrestrial objects much cooler (roughly 300 K) than the Sun. The difference between the incident longwave radiation received from the sky and the longwave radiation emitted from the ice or snow surface tends to be small and often negative owing to the low emissivity and temperature of the sky compared with the high emissivity and nearly constant 0 °C surface temperature of a melting snow surface. **Emissivity** is a dimensionless number between 0 and 1 indicating how well the object emits radiation. Objects with an emissivity of 1 are referred to as **blackbodies** that radiate all the radiation they absorb. Objects with an emissivity less than 1 are referred to as graybodies that radiate only a portion of the radiation they absorb. Objects

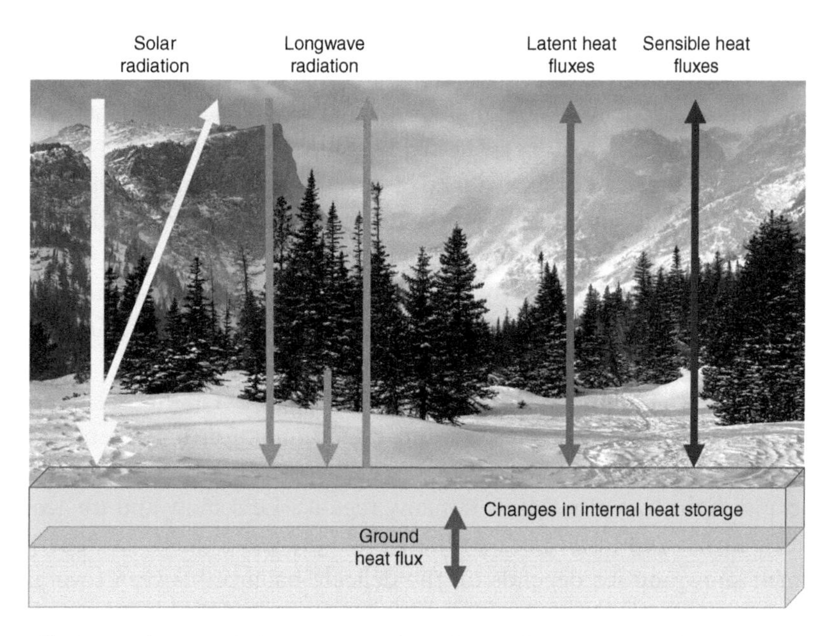

Figure 8.2 Radiation and energy balance terms of a volume of ice or snow, with energy inputs or outputs indicated by downwards or upward arrows, respectively. A portion of the solar (shortwave) radiation received at the snow surface is reflected. Longwave radiation is received from the atmosphere or trees and other surfaces and is emitted from the surface back to the atmosphere. Energy can be released or taken from the snowpack due to changes of state (latent heat fluxes) from rain-on-snow events, or sublimation, for example. The snowpack may gain or lose energy by convection exchanges with the atmosphere (sensible heat flux), or by conduction at the base of the snowpack (ground heat flux). Whether the snowpack warms or cools (the snowpack internal heat storage) is determined by the energy balance equation (warming if energy inputs exceed outputs; cooling if energy outputs exceed inputs). Background image: Rocky Mountain National Park, Colorado, USA. Photo credit: P. D. Blanken.

with a high liquid water content tend to have high emissivity (e.g., 0.98) since liquid water is such a strong absorber of solar radiation. The **Stefan–Boltzmann Law** (Eq. 8.5) allows the calculation of the total energy emitted by an object over all wavelengths (L; W m^{-2}) based on its emissivity (ε) and surface (or radiative skin temperature) temperature (T_0; K) when using the Stefan–Boltzmann constant $\sigma = 5.67 \times 10^{-8}$ W m^{-2} K^{-4}:

$$L = \varepsilon \sigma {T_0}^4. \tag{8.5}$$

Using Eq. (8.5), the energy emitted by ice with a surface temperature of 0 °C (273.15 K) and emissivity of 0.98 is 309 W m^{-2},

$$L = 0.98 \times 5.67 \times 10^{-8}\ \mathrm{W\,m^{-2}\,K^{-4}} \times 273.15\,\mathrm{K}^4 = 309.32\,\mathrm{W\,m^{-2}},$$

the maximum possible since the surface temperature of melting ice cannot exceed 0 °C.

Most studies show that the sum of the radiation terms S^* and L^* provides most of the energy to melt snow. The transfer of energy into the snow or ice surface from condensation (surface hoar frost) tends to be small since cold air has a low saturation vapor pressure (Chapter 3). Air above 0 °C can transfer **sensible heat** (energy transfer through the process of **convection**) to the surface but this too tends to be small except for events such as **Foehn winds** (warm, downsloping winds in mountain regions). The **soil heat flux** (energy transfer through conduction) measured at the base of snow or ice tends to be close to zero owing to their high insulative property, resulting in a small temperature gradient across the ground/ice interface. Rain falling on ice or snow can transfer a significant amount of energy to it since heat is released when the liquid water freezes. This is particularly relevant in regions that have a rain/snow mix in the spring, and in high latitudes and altitudes where the shift from snow to rain-dominated precipitation events is expected to increase with climate change.

Why are glaciers and ice sheets melting so rapidly? The mass balance of glaciers and ice sheets is the difference between snow and ice accumulation and ablation (melting, sublimation, and/or calving in the case of ice shelves; see Chapter 5). A decrease in snow and ice accumulation explains the loss of ice mass in many regions. Less snow and ice accumulation means ice will melt earlier and faster because there is simply less to melt. The energy available to warm and melt snow and ice depends on the delicate balance between several terms, all of which tend to be small in cold regions, making for variations in local sensitivity to what controls melting. Glacial melting near the Tibetan Plateau was found to be controlled primarily by S^* through changes in the surface albedo. In contrast, glacier melting in warmer and humid maritime continental locations was dominated by λE and H (Che et al., 2019). Plateau Glacier on Kilimanjaro in Tanzania lost ice mass primarily not from melting but sublimation (Cullen et al., 2007). At the summit of Greenland, the surface energy balance was largely controlled by changes in the downwelling radiation, indicating the importance of clouds, since periods of cloud cover decrease the surface incident shortwave radiation and increase the surface incident longwave radiation (Miller et al., 2017). In the eastern Antarctica plateau far from the coast, the summertime surface energy balance was also dominated by the radiation terms with small turbulent fluxes (van As et al., 2005). Outside of changes in the surface energy balance, increasing surface temperature not only results in melting through the advection of warm air over the glaciers' surface, but can also change the structure of ice crystals, making the surface more porous, to accelerate further melting (McGrath et al., 2013).

8.4 Freezing

Freezing occurs when water changes state from liquid to solid. The latent heat of fusion is released as bonds form and water molecules align, and the surrounding environment warms through this exothermic process. Once all the water has frozen, the release of the latent heat of fusion energy ends, and the ice temperature is determined by the energy balance of its surroundings. The change of state from liquid to solid has several important consequences for abiotic and biotic processes. The volume expansion upon freezing (roughly 9 percent; Chapter 3) results in ice having a lower density than liquid water, thus ice floats. Floating ice shelves at ice sheet margins have been linked to the acceleration of ice loss from the ice sheets themselves (Chapter 5). In the Polar Seas, sea ice increases the albedo and limits the latent and sensible heat loss to the atmosphere. The expulsion of salt with repeated freezing cycles results in low-density freshwater inputs into the ocean once the sea ice melts, influencing thermohaline ocean circulation. The presence and duration of ice on lakes influences thermal structure, hence vertical mixing as indicated by the lake classification (Chapter 5). Lakes and oceans with extensive ice cover have a higher albedo and decreased latent and sensible heat exchange with air above compared to when they are ice-free.

When water freezes in a confined space such as in soil or rock, volume expansion plays a major role in weathering, erosion, and soil formation (Chapter 10). Repeated freeze–thaw cycles, especially when occurring frequently in the spring and fall, can shatter and fragment rocks: the first step in soil formation. Within existing soils, volume expansion can effectively mix soils and increase soil porosity. The powerful force exerted by the expanding volume of water when freezing plays a major role in shaping the Earth's abiotic landscape on both small and large scales (Chapter 10).

The expansion of freezing water also significantly affects the biotic environment. Within plant cells, volume expansion can easily rupture vacuoles and cell walls, causing significant damage. Plants with internal water-conducting tissue (**vascular plants**; Chapters 11 and 12) either avoid seasonal freezing by completing their life cycle within a limited period (**annual plants**) or have evolved means to tolerate freezing conditions and survive more than two years (**perennials**). Perennials include **deciduous** species which shed their leaves for several months to avoid the long period of winter drought when liquid water is not available in the soil, yet is still present within the tree. Many freeze-tolerant species expel water from within cells to extracellular spaces where the formation of any ice is less harmful. An increase in the cellular fluid's solute concentration serves as a natural antifreeze by lowering the freezing-point temperature. As a last resort, some species produce specialized proteins that inhibit ice nucleation within cells. These examples of **freezing tolerance mechanisms** are especially well developed in Arctic and alpine vegetation, and likely arose through the adoption of drought-tolerance mechanisms, since in both freezing and drought conditions water is not available (Guy, 2003; Neuner, 2014).

There is one beneficial aspect of water freezing for vegetation. During periods when sub-zero air temperatures are expected or occurring, farmers sometimes spray their crops (typically fruit or citrus trees) with water. Pictures of oranges covered in ice often translate to a spike in the price of orange juice. The reason the crops are sprayed is to take advantage of the exothermic freezing process. Each gram of water that freezes releases 334 J (the latent heat of fusion), offering temporary protection from frost damage as the temperature where water is freezing will not fall below the freezing point until all the water is frozen, assuming that ample liquid water is available to protect the crop during such times (Hamer, 1980).

For animals and people, the formation of ice on or within the skin causes significant tissue damage. For humans, there are four stages of **frostbite**, the general term used to denote freezing injury to tissues that occurs when the skin's temperature decreases below its freezing-point temperature, typically −0.55 °C (Hallam et al., 2010). The first stage, **frostnip**, results in exposed skin feeling numb but does not cause permanent damage. The second stage, **superficial frostbite**, results in pain, often a burning sensation, and sometimes blistering on the affected area after rewarming. The second stage progresses to the third when layers of tissue beneath the skin also freeze and are damaged. The fourth stage, **deep frostbite**, damages not only the surface skin and deeper underlying tissue but also muscles, tendons, and bones. The tissue dies, turning black and hard, with the victim feeling no sensation of pain or discomfort.

Since there is no standardized reporting system or database for frostbite injuries, broad-scale and long-term studies are rare. Studies that do exist report that in the United States nearly 500 patients annually are hospitalized with frostbite, at an annual rate of 0.83 per 100,000 population. For comparison, the annual rate per 100,000 is 6.0 in Siberia, 2.5 in Finland, and 0.07 in the Australian Alps (Lipatov et al., 2022). Circumstances that result in frostbite injury vary. Over a 12-year study period of frostbite injuries in Saskatchewan, Canada, alcohol consumption was a predisposing factor in 46% of the injuries, followed by psychiatric illness (17%), vehicle trauma (19%), vehicle failure (15%), and drug use (4%) (Valnicek, Chasmar and Clapson, 1993). A study in Finland reported that men had a much higher rate of hospitalization from frostbite injuries (86%) than women, explained, according to the authors, by men working in the cold more often (Juopperi et al., 2002).

8.5 Sublimation

Sublimation is the change of state directly from solid to gas, and requires sufficient energy to break the bonds connecting water in its solid form directly to vapor. Therefore, the **latent heat of sublimation** ($\lambda_s \sim 2.834$ million J kg^{-1}) is the largest of any latent heats, equal to the sum of the **latent heat of vaporization** ($\lambda_v \sim 2.5$ million J kg^{-1}) and the **latent heat of fusion** ($\lambda_f \sim 334$ thousand J kg^{-1}). Sublimation requires energy and is therefore an endothermic process that results in a temperature decrease in the surrounding environment.

To sublimate ice requires a lot of energy, and where ice exists energy is typically not available. Under clear-sky, windy conditions, solar radiation can supply much of the latent heat of sublimation, and high wind speeds can entrain snow into the air and maintain high sublimation rates. Such conditions explain the high rate of sublimation observed from tropical high-elevation glaciers such as those found on Mount Kilimanjaro in Tanzania, the equatorial region of the Andes, or the Himalayan region, where 21 percent of the annual snowfall is lost by sublimation (Stigter et al., 2018). Even under optimal conditions of high solar radiation, wind, and the low humidity found in high-elevation equatorial regions, sublimation rates seldom exceed 1 mm per day, so this is likely sublimation's upper limit (Stigter et al., 2018).

In continental climates with long, cold, dry winters, sublimation rates are not high but can be a significant fraction of the total precipitation, especially in coniferous forests. The high surface area of snow intercepted by the needles and branches means that exposure to solar radiation and the dry, windy air increases the likelihood of sublimation. The resistance to sublimation of canopy-intercepted snow can be an order of magnitude less than from the snow on the ground, resulting in large sublimation losses from forests (Frank et al., 2018). Although rare, sublimation measurements have shown that 40% of the total winter snowfall can be sublimated from snow intercepted by the forest canopy (Table 8.1). The amount of snow sublimated increases with the forest's **Leaf Area Index** (LAI; area of leaves per area of ground) since more snow is intercepted as the surface area of leaves increases. In high-elevation regions, clear skies and dry,

Table 8.1 Measured sublimation from various surfaces and locations

Site	Climate	Location	Location of sublimation	Sublimation as a percent of the total winter snowfall (total mm per winter)	Reference
Mature black spruce southern boreal forest	Cool continental	Prince Albert National Park, Saskatchewan, Canada	Canopy	40% (39 mm)	(Pomeroy et al., 1998)
Jack pine southern boreal forest	Cool continental	Prince Albert National Park, Saskatchewan, Canada	Canopy	31% (30 mm)	(Pomeroy et al., 1998)
Subalpine fir	Continental Subarctic	Fraser Experimental Forest, Colorado, USA	Canopy	20% (98 mm)	(Montesi et al., 2004)
High elevation wind-scoured alpine treeless tundra	Continental Subarctic	Niwot Ridge, Colorado	Ground	19% (101 mm)	(Knowles et al., 2012)
Alpine spruce and pine	Continental warm-summer humid	Berchtesgaden National Park, Germany	Canopy	13% (85 mm)	(Strasser et al., 2008)
Mature mixed white spruce and aspen, southern boreal forest	Cool continental	Prince Albert National Park, Saskatchewan, Canada	Canopy	13% (13 mm)	(Pomeroy et al., 1998)
Douglas fir	Maritime	Umpqua National Forest, Oregon, USA	Canopy	10% (100 mm)	(Storck, Lettenmaier and Bolton, 2002)
Alpinc spruce and pine	Continental warm-summer humid	Berchtesgaden National Park, Germany	Ground	7% (45 mm)	(Strasser et al., 2008)

windy conditions are common, resulting in measured canopy-intercepted snow sublimation rates of 20% of the total winter snowfall (Montesi et al., 2004). In coastal regions with a mild and humid winter, the sublimation of canopy-intercepted snow is much less than in continental climate regions (e.g., 10% of the total winter snowfall). However, the amount of sublimation expressed as the liquid water equivalent is much higher than in continental regions, since the total winter snowfall tends to be much higher in coastal regions.

Sublimated snow can affect local water resources, since it is no longer available for melting and spring runoff. As discussed in Chapter 7, air rising over mountain regions is an effective way for air to cool and generate **orographic precipitation**. Especially in arid and semi-arid continental regions, mountains are often referred to as **towers of water**, as the winter snowfall can supply most of the water to the surrounding regions for the entire year. Therefore, the winter sublimation losses shown in Table 8.1 can significantly impact water resources as sublimated water in vapor form may leave the region and not enter local reservoirs as springtime liquid water runoff. Since most of the sublimation occurs from snow intercepted by the forest canopy, one could question if reducing the forest cover would result in a net increase in spring runoff. Indeed, Frank et al. (2018) found that the loss of canopy resulting from spruce bark beetle in a high-elevation spruce-fir forest increased snowpack sublimation by 3% but decreased canopy sublimation by 24%, equivalent to 6.1% of the snowfall (4.4% of the total annual precipitation). Would similar results be found elsewhere? Lundquist et al. (2013) found that in mild and wet regions with an early spring snow melt and an average winter (December–January–February) air temperature greater than −1 °C, snow duration was reduced by 1–2 weeks in forested areas compared with adjacent open areas. In these mild regions, forests increased snowmelt because the longwave radiation emitted from their warm foliage outweighed the protection provided to the ground by tree shading and decreased wind speeds. In cold regions with average winter air temperatures around −6 °C, the opposite was found; snow lasted longer in forests than in open areas. Lundquist et al. (2013) hypothesized that the dominant term of energy available for snowmelt (Q_{mt}, Eq. 8.4) shifts from shortwave radiation in regions where snow lasts longer in open areas to longwave radiation in regions where snow lasts longer in the forest. The forest management implication based on maximization of snow retention alone is that forests should be preserved in regions with cold winters and reduced through thinning or gap-cutting in regions with warm winters (Lundquist et al., 2013).

8.6 Deposition

The change of state from gas to solid is called **deposition**, the reverse process to sublimation. Deposition releases energy and is therefore an **exothermic process** resulting in a temperature increase in the surrounding environment. The energy released with deposition is equal to the **latent heat of sublimation** $\lambda_s \sim 2.834$ million J kg^{-1}. Since ice can only exist at temperatures below the freezing point, temperatures must be at or below the freezing point on the surface where deposition occurs. **Frost** is an example of deposition, whereas **dew** is an example of **condensation** (Chapter 7). Even though both changes of state begin with water vapor, the surface temperature determines whether the gas changes state to a solid (deposition) or liquid (condensation).

For deposition to occur, the vapor pressure in the air must be greater than that at the adjacent surface, and that surface must also be below the freezing point. As described in Chapter 7 (Eq. 7.2), the saturation vapor pressure over ice is always less than that over pure liquid water, meaning that when surface ice is present the vapor pressure gradient is directed towards the surface, thus aiding in surface deposition. A common example of deposition is the formation of **surface hoar frost** on a snow surface, often making the snowpack prone to avalanches after subsequent snowfalls. If the snow surface temperature is −10 °C, the saturation vapor pressure at the surface would be 0.26 kPa (using Eq. 7.2 with ice coefficients $a = 0.61115$ kPa, $b = 22.452$, $c = 272.55$ °C; Buck, 1981)

$$e_s = 0.61115\ \text{kPa} \times \exp\left(\frac{22.452 \times -10\ ^\circ\text{C}}{272.55^\circ\text{C} - 10\ ^\circ\text{C}}\right) = 0.26\ \text{kPa}$$

and equal to the surface vapor pressure assuming a relative humidity of 100%. The vapor pressure in the air immediately above the cold snow surface needs to exceed 0.26 kPa for deposition to occur. Based on the dew-point temperature equation (Chapter 7, Eq. 7.7), a saturation vapor pressure of 0.26 kPa can be achieved at air temperatures of –11.2 °C or higher, indicating that surface hoar frost is likely with air temperatures above –11.2 °C and 100% relative humidity (RH):

$$T_d = \frac{240.97\ ^\circ\text{C} \times \ln\left(0.26\ \text{kPa} / 0.61121\ \text{kPa}\right)}{17.502 - \ln\left(0.26\ \text{kPa} / 0.61121\ \text{kPa}\right)} = -11.2\ ^\circ\text{C}.$$

If the atmosphere's relative humidity decreased to 50%, an air temperature of only −2.2 °C would be required for the vapor pressure in the air to exceed that at the −10 °C snow surface:

$$e_a = \text{RH} \times e_s = (0.50) \times 0.61121\ \text{kPa} \times \exp\left(\frac{17.502 \times -2.2\ ^\circ\text{C}}{240.97\ ^\circ\text{C} - 2.2\ ^\circ\text{C}}\right) = 0.26\ \text{kPa}.$$

Following the same calculation, for very dry air with a relative humidity of 10%, an air temperature of 21.7 °C would be required for deposition to occur on a surface at −10 °C, a very unrealistic scenario. As these calculations illustrate, a low air temperature and high relative humidity are favorable conditions for deposition. These conditions occur in clouds that contain a mixture of supercooled liquid droplets and ice (mixed cloud) where, through the Bergeron process (Chapter 7), water vapor is deposited onto the ice crystals.

8.7 SUMMARY

Changes of state involving ice influence not only water and energy cycling but also shape the landscape and life upon it. Melting is an endothermic process requiring energy as expressed by the latent heat of fusion. Before ice melts, the temperature must be raised to the melting point. Faraday and others performed experiments showing that pressure decreases the melting-point temperature, helping to explain why glaciers and ice sheets flow. In the absence of high pressure, the energy required to raise the temperature of ice to the melting point can be calculated using the cold content equation. Once the temperature is raised to the melting point, further energy input results in the melting process with no further increase in temperature until the process is complete. The radiation and energy balances can be used as a framework to examine the sources of energy available to warm and subsequently melt snow and ice. Despite the high albedo of snow and ice, studies have shown that most of the melt energy is supplied from the net solar and longwave radiation terms.

When water freezes, it returns to a lower-energy state, so the latent heat of fusion is released and the surrounding environment warms. The volume expansion upon freezing plays a major role in weathering and erosion. Snow and ice also affect temperature through the temperature–albedo feedback, influence ocean circulation through changes in salinity and air–sea heat exchange, and influence the vertical mixing of water in lakes and oceans. Some species of vegetation have developed various mechanisms to tolerate freezing conditions, many of which involve the relocation of water and increases in the cellular solute concentration. The exposure of human skin tissue to freezing conditions results in various stages of progressively worse frostbite damage.

Water can change states directly between solid and gas, if sufficient energy is available. Sublimation rates tend to be small, less than 1 mm per day. High values of solar radiation and a dry, windy atmosphere offer the best conditions to sublimate snow, especially when the snow is intercepted by coniferous forest and exposed to solar radiation and wind. Measurements have shown that up to 40 percent of the total winter snowfall can be lost to sublimation, decreasing the liquid water supplied by snowmelt to soil, streams, and reservoirs. The effect of forests on the radiation and energy balance of the underlying snowpack is complex; in mild regions, forests tend to increase snowmelt, whereas in cold regions, snow lasts longer in forests than in open areas.

Deposition, the change of state directly from gas to solid, releases the latent heat of sublimation as bonds form. The surface must be below the freezing point temperature, and the vapor pressure must be higher in the air than at the surface for deposition to occur. This is a relatively rare occurrence most likely achieved with low temperatures and high relative humidity. The formation of surface hoar frost on snow, and the growth of ice crystals in a mixed cloud through the Bergeron process (Chapter 7), are examples of deposition.

8.8 QUESTIONS

8.1 Sketch a graph of how the average temperature of a volume of snow or ice changes over several days (or weeks) starting from when the ice is initially below 0 °C, warms to 0 °C, starts to melt, and then finally melts completely. Explain if, when, and how changes of the state of water influence how temperature changes with time.

8.2 Lake Vostok in the Antarctic is the largest known subglacial lake. Located roughly 4 km beneath the Antarctic Ice sheet's surface (500 m below sea level), Lake Vostok's 5,400 km^3 of fresh water contains cavities of liquid water at an estimated water temperature of −3 °C. Calculate the pressure (in units of atmospheres) that would be required to lower the freezing point of water by 3 °C.

8.3 Calculate the energy required to raise the temperature of a 10-cm-deep snowpack with a density of 250 kg m^{-3} and a specific heat capacity of 2,108 J kg^{-1} $°C^{-1}$ from an average temperature of −2 °C to 0 °C.

8.4 Using the values given in Question 8.3, calculate the total amount of energy required to completely melt all the snow, starting from an initial average temperature of −2 °C. At your location, use Eq. (8.4) and Figure 8.2 to explain where this energy required to melt the snow could originate. Select your favorite location that has snow if you live in a location without snow.

8.5 Explain why snow or ice that has reached the melting-point temperature cannot emit more than roughly 316 W m^{-2} of longwave radiation.
8.6 Sometimes citrus orchard farmers will spray water on their crops when there is a risk of frost damage. Why would they take such action that appears counterintuitive?
8.7 The increase in the volume of water upon freezing has a strong influence on both landscape and life. Provide examples of this influence on the landscape, vegetation, and animals. Describe any means vegetation has developed to tolerate freezing conditions.
8.8 Some studies (e.g., Stares and Kosatsky, 2015) have found that most cases of hypothermia occurred not at very low air temperatures below the freezing point, but at a mean air temperature of 0.56 °C. Provide discussion and some possible explanations for this observation.
8.9 The sublimation of snow that has been intercepted by forest can vary from 7 percent to 40 percent of the total winter snowfall (Table 8.1). Imagine you are an advocate for protecting forests for the purpose of retaining and protecting water resources, yet the point is raised that removing forests could provide for more water by reducing sublimation by canopy-intercepted snow. Provide an argument either for or against forest thinning with regards to minimizing sublimation water loss.
8.10 The formation of hoar frost on the surface of a snowpack in alpine regions is a major factor resulting in deadly avalanches worldwide. If the temperature at the surface of a snowpack is −1 °C, calculate the minimum relative humidity required for surface hoar frost formation if the air temperature is 0 °C.

REFERENCES

van As, D., van den Broeke, M., Reijmer, C. and van de Wal, R. (2005) 'The summer surface energy balance of the high Antarctic plateau', *Boundary-Layer Meteorology*, 115, pp. 289–317. doi: 10.1007/s10546-004-4631-1.

Bottomley, J. T. (1872) 'Melting and regelation of ice', *Nature*, 5, p. 185.

Buck, A. L. (1981) 'New equations for computing vapor pressure and enhancement factor', *Journal of Applied Meteorology and Climatology*, 20, pp. 1527–1532.

Che, Y., Zhang, M., Li, Z. et al. (2019) 'Energy balance model of mass balance and its sensitivity to meteorological variability on Urumqi River Glacier No.1 in the Chinese Tien Shan', *Scientific Reports*, 9, art. 13958. doi: 10.1038/s41598-019-50398-4.

Cullen, N. J., Mölg, T., Kaser, G., Steffen, K. and Hardy, D. R. (2007) 'Energy-balance model validation on the top of Kilimanjaro, Tanzania, using eddy covariance data', *Annals of Glaciology*, 46, pp. 227–233. doi: 10.3189/172756407782871224.

Drake, L. D. and Shreve, R. L. (1973) 'Pressure melting and regelation of ice by round wires', *Proceedings of the Royal Society of London A. Mathematical and Physical Sciences*, 332, pp. 51–83. doi: 10.1098/rspa.1973.0013.

Faraday, M. (1860) 'Note on regelation', *Proceedings of the Royal Society of London*, 10, pp. 440–450. doi: 10.1098/rspl.1859.0082.

Frank, J. M., Massman, W. J., Ewers, B. E. and Williams, D. G. (2018) 'Bayesian analyses of 17 winters of water vapor fluxes show Bark Beetles reduce sublimation', *Water Resources Research*, 55, pp. 1598–1623. doi: 10.1029/2018WR023054.

Guy, C. L. (2003) 'Freezing tolerance of plants: Current understanding and selected emerging concepts', *Canadian Journal of Botany*, 81, pp. 1216–1223. doi: 10.1139/b03-130.

Hallam, M. J., Cubison, T., Dheansa, B. and Imray, C. (2010) 'Managing frostbite', *BMJ*, 341, pp. 1151–1156. doi: 10.1136/bmj.c5864.

Hamer, P. J. C. (1980) 'An automatic sprinkler system giving variable irrigation rates matched to measured frost protection needs', *Agricultural Meteorology*, 21, pp. 281–293. doi: 10.1016/0002-1571(80)90072-2.

Juopperi, K., Hassi, J., Ervasti, O., Drebs, A. and Näyhä, S. (2002) 'Incidence of frostbite and ambient temperature in Finland, 1986–1995. A national study based on hospital admissions.', *International Journal of Circumpolar Health*, 61, pp. 352–362. doi: 10.3402/ijch.v61i4.17493.

Knowles, J. F., Blanken, P. D., Williams, M. W. and Chowanski, K. M. (2012) 'Energy and surface moisture seasonally limit evaporation and sublimation from snow-free alpine tundra', *Agricultural and Forest Meteorology*, 157. doi: 10.1016/j.agrformet.2012.01.017.

Lipatov, K., Komarova, E., Asatryan, A. et al. (2022) 'Frostbite of the upper extremities: Hot issues in diagnosis and surgical treatment (review)', *Burns*, 48, pp. 1279–1286. doi: 10.1016/j.burns.2022.03.006.

Lundquist, J. D., Dickerson-Lange, S. E., Lutz, J. A. and Cristea, N. C. (2013) 'Lower forest density enhances snow retention in regions with warmer winters: A global framework developed from plot-scale observations and modeling', *Water Resources Research*, 49, pp. 6356–6370. doi: 10.1002/wrcr.20504.

McGrath, D., Colgan, W., Bayou, N., Muto, A. and Steffen, K. (2013) 'Recent warming at Summit, Greenland: Global context and implications', *Geophysical Research Letters*, 40, pp. 2091–2096. doi: 10.1002/grl.50456.

Miller, N. B., Shupe, M. D., Cox, C. J. et al. (2017) 'Surface energy budget responses to radiative forcing at Summit, Greenland', *Cryosphere*, 11, pp. 497–516. doi: 10.5194/tc-11-497-2017.

Montesi, J., Elder, K., Schmidt, R. A. and Davis, R. E. (2004) 'Sublimation of intercepted snow within a subalpine forest canopy at two elevations', *Journal of Hydrometeorology*, 5, pp. 763–773. doi: 10.1175/1525-7541(2004)005<0763:SOISWA>2.0.CO;2.

Neuner, G. (2014) 'Frost resistance in alpine woody plants', *Frontiers in Plant Science*, 5. doi: 10.3389/fpls.2014.00654.

Pomeroy, J. W., Parviainen, J., Hedstrom, N. and Gray, D. M. (1998) 'Coupled modelling of forest snow interception and sublimation', *Hydrological Processes*, 12, pp. 2317–2337. doi: 10.1002/(SICI)1099-1085(199812)12:15<2317::AID-HYP799>3.0.CO;2-X.

Stares, J. and Kosatsky, T. (2015) 'Hypothermia as a cause of death in British Columbia, 1998–2012: a descriptive assessment', *CMAJ Open*, 3, pp. E352–E358. doi: 10.9778/cmajo.20150013.

Stigter, E. E., Litt, M., Steiner, J. F. et al. (2018) 'The importance of snow sublimation on a Himalayan glacier', *Frontiers in Earth Science*, 6, pp. 1–16. doi: 10.3389/feart.2018.00108.

Storck, P., Lettenmaier, D. P. and Bolton, S. M. (2002) 'Measurement of snow interception and canopy effects on snow accumulation and melt in a mountainous maritime climate, Oregon, United States', *Water Resources Research*, 38, pp. 5-1-5–16. doi: 10.1029/2002wr001281.

Strasser, U., Bernhardt, M., Weber, M., Liston, G. E. and Mauser, W. (2008) 'Is snow sublimation important in the alpine water balance?', *The Cryosphere*, 2, pp. 53–66. doi: 10.5194/tc-2-53-2008.

Thomson, J. (1849) 'Theoretical considerations on the effect of pressure in lowering the freezing point of water', *Transactions of the Royal Society of Edinburgh*, 16, pp. 575–580. doi: 10.1017/S0080456800022493.

Thomson, W. (1850) 'The effect of pressure in lowering the freezing-point of water experimentally demonstrated.', *Proceedings of the Royal Society of Edinburgh*, 2, pp. 267–271. doi: 10.1017/S0370164000036695.

Tyndall, J. and Huxley, T. H. (1857) 'On the structure and motion of glaciers', *Philosophical Transactions of the Royal Society of London*, 147, pp. 327–346. doi: 10.1098/rstl.1857.0016.

Valnicek, S. M., Chasmar, L. R. and Clapson, J. B. (1993) 'Frostbite in the prairies: A 12-year review', *Plastic and Reconstructive Surgery*, 92, pp. 663–641.

9 Warm-Region Abiotic Processes and Landforms

Key Learning Objectives

After reading this chapter you will be able to:

1. Summarize how surface runoff is generated to form rivers and how rivers are classified.
2. Compare calculations of riverbed to sediment shear stress to determine whether sediment transport may occur and use this information to reconstruct historic river flows.
3. Describe the formation of landforms created by fluvial processes.
4. Explain mechanical and chemical denudation and how denudation rates vary geographically.
5. Identify how water participates in weathering through hydration, hydrolysis, and dissolution.

9.1 Introduction

Flowing liquid water shapes and alters the landscape across a range of temporal and spatial scales. This chapter describes the processes that erode, transport, and deposit material in regions warm enough for liquid water to flow across the landscape most of the year. The generation of surface runoff needed for river formation requires that inputs of surface water exceed outputs, so both precipitation characteristics and surface properties pertaining to water infiltration are examined. With the cumulative addition of water from connected sloped surfaces, streams and rivers ultimately form and are classified by connective channels that define the surface area they drain.

Rivers connect terrestrial and aquatic environments in the landscape through the accumulation of surface runoff. In addition to the river's course, the rate of water discharge is significant since the rate and volume of water flow determine the river's ability to erode and transport material. The pressure, or shear stress, the river exerts on the riverbed depends on the weight of the water above (i.e., depth) and the slope of the surface (i.e., velocity). To transport sediment, the shear stress needs to exceed the force holding the sediment in place. In addition to examples of calculations for the transport of material along the river's bed, the mechanisms for transporting suspended sediment are provided. Knowledge of the forces required to erode and subsequently deposit material can be used to reconstruct the characteristics of former rivers. These paleohydraulic discharge reconstruction methods help illuminate conditions that once existed.

Landforms created by flowing water, fluvial landforms, not only indicate the locations of former rivers but also provide an indication of their characteristics. Where rivers form, rapids, canyons, and waterfalls clearly illustrate the power of flowing water to erode and transport material. Where rivers end, the meandering, sinuous course, oxbow lakes, and deltas easily recognizable in many large rivers provide unmistakable evidence of the rivers' erosion and deposition ability. Whereas many of these fluvial landforms form slowly over thousands or millions of years, some form within minutes or days with flooding events, debris flows, or landslides.

On large spatial and long temporal scales, the river erosion rate of solid and dissolved material can be estimated to provide insight of the role that flowing water plays in shaping the environment. Studies reveal that although the rate at which rivers remove solid material far exceeds the rate at which dissolved material is removed, the role water plays in chemical weathering is still important both for the formation of new materials and the breakdown of others. Therefore, the three main chemical processes involving water in the chemical weathering of the landscape, hydration, hydrolysis, and dissolution, are described.

9.2 Generating Surface Runoff

Imagine a simple, flat, permeable soil surface that receives water from rain or melting snow that enters a volume of soil beneath until the soil is **saturated**, meaning that all pore spaces are filled with water. Whether saturation occurs depends on the soil's **porosity** (the ratio of pore space to total soil volume), the **degree of saturation** (the ratio of volume of water to volume of pore space) prior to the rain or melt event, and the amount and duration of rainfall, which is the **rainfall intensity** (mm of rain per hour) or the rate of snow melt. Using the water balance concept, whether saturation occurs also depends on any losses of water from our simple flat surface. Water could drain vertically and move into deeper, unsaturated layers. Assuming the rate of water input exceeds the rate of water output, the volume of water stored in the soil will increase until the soil substrate is saturated. Then, water will appear on the surface and ponding will occur. The volume of surface water will increase, decrease, or remain the same over time as a function of the water balance equation for that system; inputs equal outputs plus any changes in water storage in the volume (Table 9.1). If the surface is sloped or if the water-holding capacity of any surface depression is reached, overland surface flow will begin – the first step in the formation of streams and rivers.

Table 9.1 **Variables that affect surface water formation and therefore surface flow**

Water inputs	Soil properties that affect water storage	Water outputs
Rainfall intensity	Porosity	Vertical drainage
Rate of water provided by snowmelt	Hydraulic conductivity	Slope of the surface
	Antecedent water content	Evaporation rate
		Transpiration rate (if vegetated)
		Canopy intercepted evaporation rate (if vegetated)

In addition to precipitation, any other water inputs (and outputs) must be considered. On a sloped surface, water from upslope locations could flow into and out of each location being considered. The hydrologic properties of the substrate including its porosity, the antecedent water content (degree of saturation), and the **hydraulic conductivity** (rate of water flow through the substate) all determine the **infiltration rate**, the rate at which water enters the substrate, and therefore how quickly the soil saturates. Water outputs include surface runoff and/or vertical drainage that depend on the slope, and any evaporative water losses. A summary of the variables to be considered is given in Table 9.1.

As the water's depth increases, so does the pressure on the material beneath. If the water's horizontal flow is negligible, the **static pressure** (P; N m^{-2} or pascals) can be calculated using the **hydrostatic equation** $P = \rho g D$ where ρ is the liquid water density (kg m^{-3}), g is the acceleration due to gravity (9.81 m s^{-2}), and D is the depth of water (m). For example, a column of stationary water 10 m deep with a density of 1,000 kg m^{-3} exerts a pressure of 98.1 kPa (Eq. 9.1), close to air pressure at sea level:

$$P = \rho g D = 1{,}000\frac{\text{kg}}{\text{m}^3} \times 9.81\frac{\text{m}}{\text{s}^2} \times 10\text{ m} = 98{,}100\frac{\text{kg}}{\text{m s}^2} \text{ or } \frac{\text{N}}{\text{m}^2} \text{ or Pa.} \tag{9.1}$$

For a flat surface, the force moving the water vertically (perpendicular to the flat surface) F_g is equal to the mass of water (m; kg) multiplied by the acceleration due to gravity (g); $F_g = mg$, which is simply the water's weight. This **gravitational force** is always directed downward towards the Earth's center. The **normal force** (F_n) always acts perpendicular to and away from the surface, so on a flat surface, F_n is directly opposite to F_g. When the water is not moving, the opposing forces are equal in magnitude, opposite in direction; $F_g = F_n$ (Figure 9.1).

On an inclined surface such as a riverbed, F_g is still directed toward Earth's center, and F_n is still directed away from and perpendicular to the surface (Figure 9.1). The gravitational force can be resolved into two components or **vectors** having both magnitude and direction; one parallel (F_s), and one perpendicular (F_{gp}) to the sloped surface (when the vectors F_s and F_{gp} are added "head to tail", the resultant vector will equal F_g). The normal force and the perpendicular component of the gravitational force are balanced. However, the parallel component of the gravitational force is not balanced by an opposing force so the water will accelerate downslope. Assuming friction is negligible, the net force acting parallel to the sloped surface is (Eq. 9.2):

$$F_s = m \times g \times \sin(\theta) \tag{9.2}$$

where θ is the angle between the inclined surface and a flat surface. From Eq. (9.2), the acceleration of the surface water down the incline is equal to (Eq. 9.3):

$$a = g\sin(\theta) \tag{9.3}$$

($a = F/m$) neglecting any friction. Since θ can vary from >0° to <90° for a sloped surface, $\sin(\theta)$ is always greater than zero but less than 1, and the object will increase its acceleration as the slope angle increases. This additional force vector imparted by the slope is required to initiate the flow of surface water, hence the formation of what may become a **stream** or **river**. Streams or **creeks** are generally much smaller and shorter than rivers, although name usage varies with culture and location.

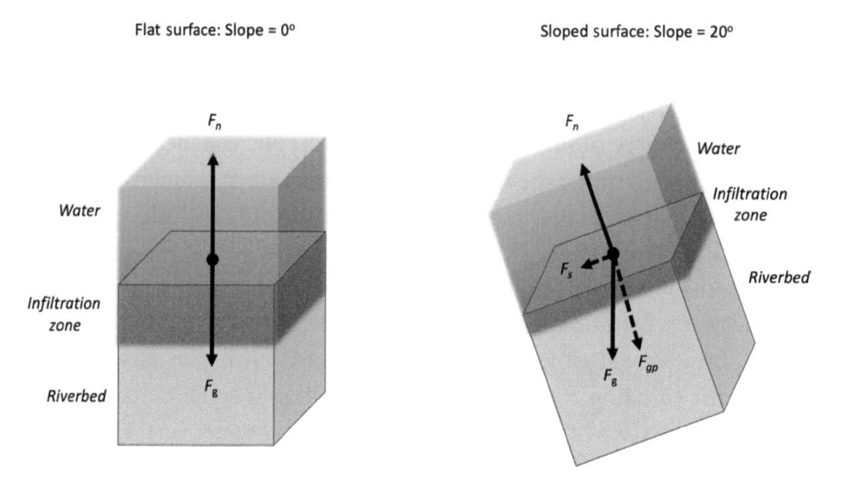

Figure 9.1 Forces acting on a column of water on a flat (left) and sloped (right) surface. The gravitational force (F_g) is always directed towards Earth's center. The force opposing F_g is the normal force (F_n) acting perpendicular to the surface. On a sloped surface, F_g can be resolved into parallel (F_s) and perpendicular (F_{gp}) vector components, with the water flowing downslope due to the unopposed F_s component (disregarding any surface friction).

9.3 Stream or River Classification

Once surface runoff begins, runoff from adjacent sections of land could merge and combine their **discharge**, the rate at which a volume of water passes a location (Q; $m^3\ s^{-1}$). As additional sections of streams merge or connect to existing channels, discharge increases until eventually rivers form. When viewed from above, this hierarchical network of streams resembles an animal's circulatory system or a tree-like branch dendritic pattern (Figure 9.2). **Robert Horton** (1875–1945) described a **stream order** system to quantify the contribution of streams to river flow. Horton's stream order consists of whole numbers starting with 1st-order streams with no tributaries (Horton, 1945). Streams that receive only 1st-order tributaries are 2nd order. Streams that receive 2nd- or 1st- and 2nd-order tributaries are 3rd-order streams, and so on, with the highest order characterizing the **drainage basin** defined as the area based on surface topography that contributes the stream discharge. **Arthur Strahler** (1918–2002) derived a modified approach to quantify streams and their drainage basins. In Strahler's stream order system, 1st-order streams are the outermost without any tributaries. When two 1st-order streams merge, a 2nd-order forms. A higher whole number is used when two streams of the same order merge, and when two streams of different orders merge, the higher number is given. Strahler's system avoids what he referred to as "...the necessity of subjective decisions, inherent in Horton's methods, and assures that there will be only one stream bearing the highest order number" (Strahler, 1952). This common stream ordering system is referred to as the **Horton–Strahler stream order**. With this system, the Amazon River has a stream order of 12, and the Mississippi River 10. The majority of the world's rivers have a stream order less than 3.

Figure 9.2 Dendritic drainage patterns (foreground) and pivot irrigation fields (circles and semicircles) near the Platte River in southeastern Nebraska. Viewed from above, both the stream river patterns resemble the roots or branches of trees, hence the use of the prefix "dendron". Photo credit: P. D. Blanken.

A concept that expands on the stream order system is the **River Continuum Concept** which describes rivers as open, flowing ecosystems in constant change from source to mouth and in constant interaction with the landscape (Vannote et al., 1980). Rather than focusing on small, discrete sections, this concept stresses that rivers should be studied as a continuum, since they are changing constantly as they move downstream. In the River Continuum Concept, all streams and rivers are classified as having headwater sections (or reaches) with a stream order of 1 to 3, mid-reaches (stream order 4 to 6), and lower reaches (stream order greater than 6). The River Continuum Concept postulates that these sections do not act as independent systems but as one continuous system, with outputs from one section providing inputs to the next. Such a functionally linked river structure can be used to explain patterns in river biological communities.

What is the value of systems of river classification? For decades, the study of the physical features of the Earth's surface, **geomorphology**, centered upon qualitative descriptions. Many early geomorphology studies and subsequent theories focused on how the so-called **cycle of erosion** could explain landscape features. The cycle of erosion or **geographic cycle**, proposed in 1889 by **William Davis** (1850–1934), stated that following tectonic uplifting, rivers erode the landscape back to some base level. Davis proposed that rivers have three main development stages; youthful, mature, and old age, with distinct landforms associated with each. The theory assumes a **uniformitarian principle**, where processes at work in the past are assumed to be present today and apply across time. Prior to the 1950s, geomorphology studies were largely descriptive, lacking any quantification or predictive capacity. From the 1950s, studies such as those by Horton and Strahler focused on measurements and scaling methods. This resulted in an understanding of the processes behind landscape formation, enabling us to make predictive models of landscape formation and change.

9.4 River Discharge

The rate at which water flows past a given location, the **discharge**, is calculated by multiplying the cross-sectional area of a section of river (depth multiplied by width) by the water's velocity: Q (m^3 s^{-1}) = A (m^2) × V (m s^{-1}). A river is typically divided into smaller subsections, with the area and flow velocity measured in each subsection and added to give the total discharge. The flow velocity can be measured with a **current meter** consisting of several metal cups that rotate with the current and calibrated so that the rate of rotation is proportional to stream velocity. Since using a current meter poses a risk for the user who must stand in or be suspended above the river, newer technology can be used to measure the water's velocity and depth. An **acoustic Doppler current profiler** (ADCP) traveling across the water surface by boat or float sends a sound pulse downwards through the water column. The depth is determined by measuring how long it takes for the sound pulse to return to the sensor after reaching the river's bottom. Sediment or other particles being transported by the water change the frequency of the sound wave (the **Doppler effect**) and from this change in frequency the water's velocity can be determined, since the particles are moving at the same velocity as the water.

As a river's depth changes, the cross-sectional area will also change. Assuming there is no change in width (i.e., the river is confined within its banks and there is no overbank flow and flooding), a change in area or velocity will result in a change in discharge. Repeated

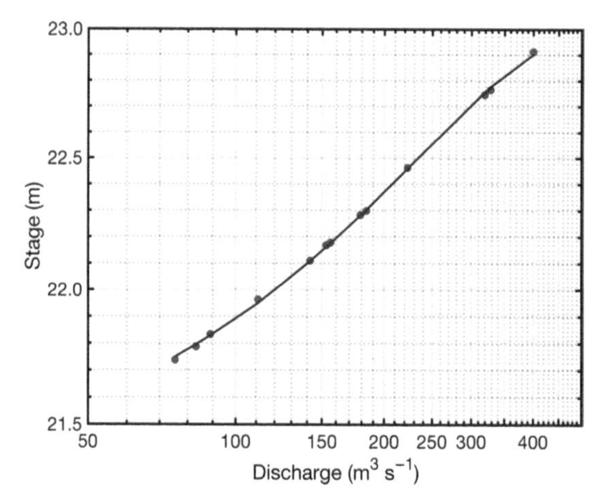

Figure 9.3 A rating curve showing the relationship between river stage and discharge from the Churchill River in northern Manitoba, Canada (Station 06FD001). Data are monthly summertime means (July through October) during the ice-free season from 2002 through 2004. Curve fit (blue line) is a quadratic equation fitted to the monthly means (red circles). Data were extracted from the Environment and Climate Change Canada Historical Hydrometric Data web site (https://wateroffice.ec.gc.ca/mainmenu/historical_data_index_e.html) on March 3, 2023.

Figure 9.4 River depth being measured by a float located in calm water inside a stilling well (vertical cylinders) with the recorder enclosed on top (left). Alternatively, river stage can be measured using an acoustic sensor mounted on a bridge as shown on the Left Hand Creek in Colorado (right). Photo credit: P. D. Blanken.

measurements of discharge across a range of river depths (often referred to as **river stage**) show a nonlinear relationship between Q and depth known as a **stage–discharge relationship** or a **rating curve** (Figure 9.3). Although rating curves are unique for each river and location, they provide a valuable means to estimate Q by only measuring river depth. River depth can readily be measured by a stilling well consisting of a float attached to a recording device, or acoustically by measuring the length of time for a sound wave to travel from a bridge-mounted sensor to the water surface (and back) (Figure 9.4).

9.5 Shear Stress

The force that flowing water exerts on the material below is referred to as **shear stress** (τ_b; N m^{-2} or pascals). Since shear stress has units of pressure (force divided by area), the net force acting on the sloped surface (F_s, see Figure 9.1) is divided by the surface area of the bed (A; m^2) (Eq. 9.4):

$$\tau_b = \frac{F_s}{A} = \frac{mg \times \sin(\theta)}{A} = \rho g D \times \sin(\theta). \tag{9.4}$$

Expressing the shear stress in units of pressure is convenient since it allows τ_b to be calculated from the river's depth (D) instead of mass, since mass divided by area is equal to density multiplied by depth:

$$\frac{M(\text{kg})}{A(\text{m}^2)} = \rho\left(\frac{\text{kg}}{\text{m}^3}\right) \times D(\text{m}).$$

For river segments that do not have a steep slope (e.g., $\theta < 15°$), the sine of the angle can be approximated by its tangent: $\sin(\theta) \cong \tan(\theta)$. Since the definition of a tangent in a right-handed triangle (opposite length divided by adjacent length) is equal to the slope (S = rise divided by run), $\sin(\theta) \cong \tan(\theta) = S$ and therefore the shear stress can be approximated knowing the river's depth and slope (Eq. 9.5):

$$\tau_b = \rho g D S. \tag{9.5}$$

Equation (9.5) shows that shear stress increases in direct proportion to the river's depth and slope. This is significant because when the magnitude of τ_b exceeds the pressure required to dislodge material on the riverbed, erosion will occur, material will be transported, and geomorphic features will develop on the landscape. The deeper the river, the greater the water's mass, hence pressure exerted on the river's bed. Similarly, the steeper the river, the greater the water's acceleration, hence pressure exerted on the river's bed. Together, a river's depth and slope increase its ability to erode and transport material.

9.6 Sediment Transport

What is required to transport river sediment? The processes and factors involved are many. In fact, it is believed that when Albert Einstein's son Hans was studying sediment transport, Albert said he preferred to study the simpler problem of General Relativity since it has fewer variables (Kleinhans, 2005). Consider the **Colorado River**, the river responsible for the creation of the **Grand Canyon**. This example uses average conditions and omits factors including whether the flow at the riverbed is laminar or turbulent, and friction of the sediment against the bed and with other sediment. First, the shear stress on the riverbed sediment needs to be calculated, so the river's depth and slope are required. The Colorado River changes elevation by an average of 4 m for every one horizontal kilometer (Cohen et al., 2018). This is equivalent to a slope gradient of 0.004 m (vertical) for every 1 m (horizontal), 0.004 m m^{-1}, or 0.23° (arctan(0.004) = 0.23°). With this shallow slope angle, $\tan(\theta) = S$ and τ_b can be approximated, knowing the river's mean depth of approximately 6 m. Using Eq. (9.5), 235 N m^{-2} of shear stress would be generated:

$$\tau_b = \rho g D S = 1{,}000 \frac{\text{kg}}{\text{m}^3} \times 9.81 \frac{\text{m}}{\text{s}^2} \times 6\text{ m} \times 0.004 \frac{\text{m}}{\text{m}} = 235 \frac{\text{kg}}{\text{ms}^2} \left(\text{N m}^{-2}\right).$$

Following Andrews (1983), a particle such as a piece of gravel on the riverbed with a diameter d_s (m) is subject to the downward force of gravity (F_g; N or kg m s^{-2}) (Eq. 9.6) and the opposing entrainment force (F_e; N or kg m s^{-2}) (Eq. 9.7):

$$F_g = c_1 (\rho_s - \rho_w) g d_s^3 \tag{9.6}$$

$$F_e = c_2 \tau_s d_s^2 \tag{9.7}$$

where c_1 and c_2 are constants, ρ_s and ρ_w are the densities of the sediment and liquid water (kg m^{-3}) respectively, g is the acceleration due to gravity (9.81 m s^{-2}), and τ_s is the sediment shear stress. When the sediment particle is at rest, the upward (F_e) and downward (F_g) forces are equal (Eq. 9.8):

$$c_1(\rho_s - \rho_w)gd_s^3 = c_2\tau_s d_s^2. \tag{9.8}$$

Solving Eq. (9.8) for τ_s gives the **critical shear stress** τ_{cs}, the particle shear stress that must be exceeded by the bed shear stress τ_b in order for the river's flow to initiate movement of the particle (Eq. 9.9):

$$\tau_{cs} = \tau_s = \frac{c_1}{c_2}(\rho_s - \rho_w)gd_s. \tag{9.9}$$

Now the task is to compare the magnitude of τ_b to τ_{cs}. If $\tau_b > \tau_{cs}$, sediment transport will occur; if $\tau_b < \tau_{cs}$, the river is unable to transport the sediment. The shear stress exerted on the sediment particle by the river (τ_b) can be calculated as shown above ($\tau_b = \rho gDS$). To calculate the particle's critical shear stress (τ_{cs}) requires knowing the value of c_1/c_2, the **critical dimensionless shear stress** τ^*_{ci} that depends on a combination of several sediment properties such as shape, size, and arrangement. Classic experiments performed in a laboratory **flume** (long rectangular channels filled with water and sediment) by Shields (1936), using homogeneous gravel beds, determined that c_1/c_2 equaled 0.06, a value that became known as **Shields' critical dimensionless shear stress**. While there remains much discussion as to which value to use for τ^*_{ci} for various river flow conditions, natural rivers seldom have a uniform particle diameter size, so the use of 0.06 for τ^*_{ci} is questionable. Based on bedload measurements taken in three self-formed natural rivers (the East Fork River in Wyoming, and the Snake and Clearwater Rivers in Idaho), Andrews (1983) found that τ^*_{ci} could be estimated using Eq. (9.10):

$$\tau^*_{ci} = 0.0834\left(d_s/\overline{d_s}\right)^{-0.872} \tag{9.10}$$

where $\overline{d_s}$ is the mean particle diameter. The potential effects of a variable particle size distribution could be incorporated in the estimation of the critical shear stress by replacing $c_1/c_2 = 0.06$ by Andrews' (1983) equation (Eq. 9.10).

Returning to the Colorado River example, $\tau_b = \rho gDS = 235\,\mathrm{N\,m^{-2}}$ based on the average slope and depth. Using Shields' critical dimensionless shear stress of $c_1/c_2 = 0.06$, a particle density of 2,650 kg m^{-3}, and a particle diameter of 2 mm (0.002 m) for a very fine gravel, the particle's critical shear stress is 1.94 N m^{-2} (Eq. 9.9):

$$\tau_{cs} = \frac{c_1}{c_2}(\rho_s - \rho_w)gd_s = 0.06 \times \left(2{,}650\frac{\mathrm{kg}}{\mathrm{m}^3} - 1{,}000\frac{\mathrm{kg}}{\mathrm{m}^3}\right) \times 9.81\frac{\mathrm{m}}{\mathrm{s}^2} \times 0.002\,\mathrm{m}$$
$$= 1.94\frac{\mathrm{kg}}{\mathrm{m}^2\mathrm{s}}\left(\mathrm{N\,m^{-2}}\right).$$

Since τ_b is much greater than τ_{cs}, from these calculations the Colorado River could easily transport very fine gravel. Repeating the calculations for very coarse gravel ($d_s = 60$ mm) gives $\tau_{cs} = 58.3$ N m^{-2}, indicating that the river can transport material of this size also. Substituting increasing values of d_s in the calculation of τ_{cs} shows that when $d_s = 243$ mm (a **cobble**, larger than a pebble but smaller than a boulder), $\tau_b \sim \tau_{cs}$, indicating the expected maximum diameter

Table 9.2 Sediment transport mechanism based on the Rouse number

Modified from Yuill and Nichols (2011).

Rouse number (P)	Sediment transport category
$P \geq 2.6$	Bed load
$0.8 < P < 2.6$	Suspended load
$P \leq 0.8$	Wash load

size of material that the river could transport. The actual maximum diameter of material that could be transported by the Colorado River is likely much smaller, owing to friction and interlocking of sediment of different sizes.

Once erosion occurs and particles are liberated from the bed, the material can be transported downstream. Sediment can be rolled or dragged along the bed (**traction**) or move in a jumping action consisting of sporadic motion known as **saltation**. The sediment transported by any of these mechanisms is referred to as **bed load**. Sediment can be suspended in the flow without any contact with the riverbed, as **suspended sediment load**. Suspended sediment typically gives rivers a lasting brown, cloudy appearance since the smaller particles remain in suspension for long periods. As described by Hans Einstein, **wash load** refers to fine sediment entirely in suspension with a grain size smaller than those of the bed, so washed into the river from another source (Einstein, 1950). Once suspended, the relationship between the flow velocity and the particle's diameter (hence mass) determines if the sediment can remain suspended.

The **Rouse number** (Rouse, 1937) can be used to determine a river's likely sediment transport mechanism: bed, suspended, or wash. The Rouse number (P; dimensionless) is the ratio of the downward settling velocity of the sediment (ω_S; m s^{-1}) to the upward velocity of the sediment generated by the shear velocity (u_{*S}; m s^{-1}) multiplied by the dimensionless von Karman constant ($k \sim 0.41$) (Eq. 9.11):

$$P = \frac{\omega_S}{k u_{*S}}. \tag{9.11}$$

When P exceeds 1, the downward velocity exceeds the upward, so bed load transport is more likely to occur than suspended transport when P is less than 1. Refinements based on observations have resulted in a better-defined relationship between P and the sediment transport mechanism (Table 9.2).

Shields or **Hjulström plots** can be used to estimate the mode of sediment transport based on shear stress or particle diameter and river velocity, respectively (e.g., see Sundborg, 1956).

9.7 Estimates of Historic River Flows

Calculations of the bed and the critical shear stress can be used to estimate the river size required to transport material. Buried fluvial material consisting of well-rounded rocks deposited in dendritic patterns indicates that there was once a river present. Using the equations presented above, it is possible to back-calculate the depth and slope of the river required to transport the discovered fluvial deposits. This technique of **paleohydraulic discharge reconstruction** has been used with reasonable accuracy to reconstruct the size of flash floods based on the size of the deposited boulders in steep watersheds along Colorado's Front Range (Costa, 1983). The technique has also been used to help reconstruct the discharge from glacial Lake Missoula (Northwestern United States) that formed at the front of the Cordilleran Ice Sheet during the late Pleistocene. Based on evidence provided by the surrounding landforms, sudden catastrophic draining of Lake Missoula's estimated 2,600 km^3 of water due to the failure of an ice dam resulted in the Earth's largest known peak flood or **megaflood**. Calculations show that the flows resulting from the drainage of the lake could suspend gravel 76 mm in diameter at mean flow velocities of at least 20 m s^{-1}, or 72 kilometers per hour (Alho, Baker and Smith, 2010). For comparison, this velocity is double that of the Niagara River at the lower Whirlpool Rapids where Class 5–6 rapids exist.

Paleohydraulic discharge reconstruction could be used on other planets to estimate the volume of liquid water required to erode the surface and transport material deposited in fluvial forms. Based on satellite orbital and surface observations of Mars, there is evidence of extensive fluvial erosion activity, indicating that there was once an active hydrologic cycle (Howard, Moore and Irwin, 2005). Using flow and sediment transport models tested on 190 rivers on Earth, Kleinhans (2005) summarized Martian flow discharge and velocity reconstruction based on channel depth, slope, and bed sediment characteristics. Similar to Earth, the channels and deltas on Mars likely formed quickly over multiple events with several reloadings of water, hence providing insights into the planet's past water cycle and climate (Kleinhans, 2005).

9.8 Fluvial Landforms

A river's ability to erode and transport material during short-duration floods or over decades and centuries of persistent flow can leave signature features on the landscape, broadly referred to as **fluvial landforms**. Areas where rivers cut channels into the landscape result in an **inverted relief** or **landscape inversion** where previously low-standing landscape features become high-standing as constant erosion over the same location lowers the river channel's elevation below that of the surrounding area, even lower than the previously lowest-lying features. A sinuous or **meandering** river channel with an incised valley or canyon is an unmistakable indicator of river erosion and one of the most easily recognized fluvial landforms, but why does a river meander instead of flowing in a more efficient straight line?

Imagine some imperfection along water's downslope path, such as a root or rock, that resists erosion. The water will turn slightly at this location and preferentially erode the softer material. The water's velocity will be higher at the outside than at the inside of the curve in its path around the erosion-resistant material. This higher velocity will result in a higher bed shear stress, and therefore greater erosion on the outside of the curve. Meanwhile, the slower velocity on the inside of the curve will reduce the bed shear stress, decreasing further erosion, and reducing the ability to transport material; thus the water will deposit any suspended sediment

Figure 9.5 The meandering Mississippi River with several oxbow lakes. Note the deposition of light-colored material on the inside curves of the Mississippi River (main channel), and the connecting Arkansas River (middle, right side). Photo: M. Blanken.

(Figure 9.5). Deeper in the water, **secondary flow** (flow in a different direction to the primary downslope flow) is directed laterally toward the inside of the curve, further adding to sediment deposition. As this process continues, the river's curvature increases as the radius decreases, increasing the river's outer bank's velocity and further erosion, while decreasing the inner bank's velocity and further deposition, and so on. As a result, the river's meandering shape increases over time through this positive feedback. Eventually an **oxbow** or **cutoff lake** may form as the meandering persists and the river cuts through, leaving the meander behind (Figure 9.5). Meandering rivers and oxbow lakes are common in the lower reaches of large rivers in gently sloping erodible terrain.

Changes in river velocity can result in the formation of several other fluvial landforms. For material to be deposited, the bed shear stress must decrease below the particle's critical shear stress. At the terminus of some rivers, bed shear stress decreases both from shallower water and decreased velocity from the shallow slope (Eq. 9.5). Consequently, the deposition of sediment forming a **river delta** is common. Sediment deposition is also common in reservoirs after the creation of dams slows or stops the current. In regions where river flow varies seasonally owing to snowmelt or monsoon rains, periods of erosion followed by periods of deposition are common. Solid material that has been transported by liquid water is distinctly recognizable from material that has been affected by other processes such as freeze–thaw cycles, thus making it relatively easy to identify the erosion process even long after the process has ended (Figure 9.6).

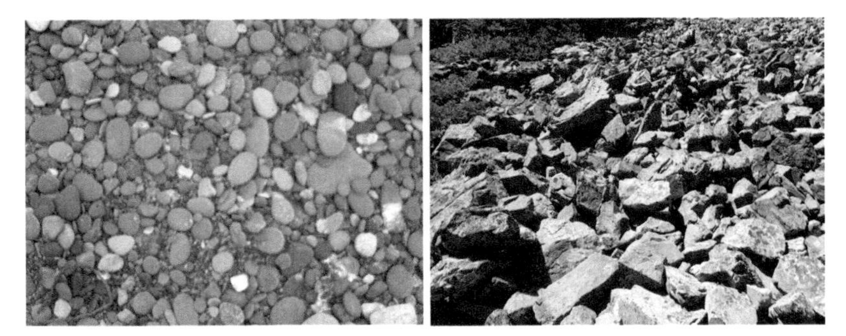

Figure 9.6 Material transported by flowing water has distinct rounded features and is often sorted by size (left; small stones a few centimeters in length). Frost-shattered rocks have characteristic sharp edges and are often much larger (e.g., a meter in length) than material capable of being transported by liquid water (right). The role of liquid or freezing water results in vastly different features. Photo credit: P. D. Blanken.

Figure 9.7 Example of a deeply incised canyon, the Grand Canyon of the Yellowstone in Yellowstone National Park, Wyoming, United States, created by the erosion of the Yellowstone River. Photo credit: P. D. Blanken.

In general, sediment deposition dominates over erosion in the lower reaches of a river with a stream order greater than 6 where the slope is small and the river's volume of water is large. Upstream near the river's origin with a stream order of 1 or 2, erosion dominates. Steep slopes and fast-moving water create sufficient erosion to carve steep canyons, such as the **Grand Canyon of the Yellowstone** in Wyoming, United States, created by erosion by the Yellowstone River (Figure 9.7). Since the headwater regions of rivers are often in tectonically active mountain regions, the uplift rate needs to be considered in relation to the erosion rate. The erosion rate depends not only on the river's properties, but also the erodibility of the underlying material. **Waterfalls** may form when a change in erodibility is encountered, especially when a harder rock overlies softer material. Such is the case along the Niagara River and Niagara Falls along the Canada–United States border, where the erosion of softer shale and clay material continually undermines the harder limestone above (Figure 9.8).

Figure 9.8 Niagara Falls, as the Niagara River flows over erosion-resistant limestone capping erodible shale beneath. Photo credit: P. D. Blanken.

Depending on the substrate materials, flat river benches or terraces may form along the river valley as alternate layers of erodible material are successively eroded.

9.9 Intermittent-Flow Fluvial Landscapes

Fluvial erosion and deposition may occur slowly and steadily over thousands or millions of years or within seconds or minutes as a result of flash flooding, earthquakes, or volcanic eruptions. In either case, some of the most dramatic and impressive features on Earth can form. The Grand Canyon in the southwestern United States, currently 446 km long, 29 km wide, with a maximum depth of 1,857 m, was created by fluvial erosion over millions of years. Although sections of the canyon were carved by paleorivers that existed 15 to 70 million years ago, today's Colorado River has widened and deepened the canyon over the past 4 million years at an incision rate of 100–200 m every million years (Karlstrom et al., 2014).

Floods are usually defined as the condition during which the river's flow increases to the extent that it can no longer be contained within the river's banks. More broadly, a flood occurs when water is temporarily in a location where it is not normally. An extreme example of a flood was the aforementioned drainage of Glacial Lake Missoula, considered the world's largest flood event. The sudden draining due to the collapse of an ice dam roughly 12,000 years ago dramatically altered the landscape at a scale similar to the Grand Canyon in only a few days. On a much smaller scale, on July 15, 1982, a 79-year-old earthen dam located in **Rocky Mountain National Park**, Colorado, suddenly failed (likely because of the failure of a seal around an outlet pipe), releasing 831,000 m^3 of water from **Lawn Lake**. Water flowing down a steep slope at an estimated peak discharge of 510 $m^3\ s^{-1}$ killed three people and caused nearly $30 million in damages (Jarrett and Costa, 1993). The high bed shear stress generated by the rapid flows severely eroded the valley to depths up to 152 m and deposited sediment at the valley floor 3 m deep. A large **alluvial fan** (fluvial deposits spread in a fan-shaped pattern) with an estimated 278,000 m^3 of sediment (753,000 tonnes) with a maximum thickness of 13.4 m formed where the steep canyon reached the valley floor, with sediment deposits extending over

Figure 9.9 A large alluvial fan (triangle shape in the foreground) in the Alps near Wassen, Switzerland, formed as material eroded from the mountains is transported by the stream as it enters the glaciated valley below. Photo credit: P. D. Blanken.

a continuous 2-km-long sedimentation zone (Pitlick, 1993). Under natural conditions, alluvial fans are common in environments where ample precipitation drives high surface runoff over erodible material. When the streams decrease their velocity upon reaching a valley floor, the material is deposited in symmetrical fan-shaped patterns (Figure 9.9).

Flooding and sudden erosion and depositing may occur as a result of intense precipitation events. During September 9–16, 2013, a record-setting 450 mm of rain (close to the mean annual precipitation) fell across much of Colorado's Front Range, creating catastrophic flooding with eight lives lost and damages exceeding $2 billion (Gochis et al., 2015). In this flood, known as the **2013 Colorado Front Range flood**, Boulder Creek increased discharge from 2.6 to 210 $m^3 s^{-1}$ (Langston and Temme, 2019). Calculations of the shear stress in Boulder Creek showed that 1-m diameter boulders were transported during the flood. The bed shear stress was so large that it was capable of "plucking" shale bedrock from the river bed and banks, in addition to removing large vegetation (cottonwood trees) from the river banks (Langston and Temme, 2019). Across the area where flooding occurred, **braided streams** were created, containing several channels separated by large boulder-sized material, common near glaciers where seasonal melting can suddenly result in large discharge capable of transporting large volumes of material.

Periods of intense precipitation resulting in saturated soil conditions can also cause sudden erosion and deposition events in locations far removed from any stream or river. The 2013 Colorado Front Range flood created 11 large **debris flows**, characterized by extremely fast-moving saturated debris often containing entire trees in addition to large boulders and mud, flowing down steep channels. The **Twin Sisters West debris flow** transported 3,780 m^3 of material a distance of 1.36 km (Patton et al., 2018). **Landslides**, characterized by the transport of larger volumes of material than debris flows, can be much more devastating. The massive **Frank Slide** on Turtle Mountain in southern Alberta, Canada, in 1903 removed an estimated 30 million cubic meters of rock, destroying the small mining town of Frank and taking 70 lives. Excessive rain over a 4-year period prior to the landslide, ice wedging, mining activity, and geological factors all played a role (Benko and Stead, 1998).

The frequency of and damage caused by landslides and debris flows are expected to increase with climate and land-use change. Landslides and debris flows triggered by excessive rainfall affected an estimated 4.8 million people, and resulted in 18,414 fatalities and an economic loss of $8 billion (US dollars; landslides, volcanic activity, and mass movement combined) worldwide between 1998 and 2017 (Wallemacq and House, 2017). In Japan, more than 10,000 rainfall-triggered landslides are reported every year (Polemio and Petrucci, 2000). The majority (80 percent) of studies find a causal relationship between landslides and climate change, and the frequency of landslides is likely to increase with climate and land-use change, increasing the number of people exposed to landslide risk (Gariano and Guzzetti, 2016). Antecedent rainfall and degree of soil saturation also contribute to the occurrence of landslides. Increases in both the antecedent and short-term precipitation resulted in a predicted 28 percent increase in debris flows along the south coast of British Columbia, Canada (Jakob and Lambert, 2009). If the intensity of precipitation associated with hurricanes increases, and/or their velocity decreases resulting in an increase in accumulated rainfall, the impact of landslides could also increase. When **Hurricane Otto** produced 300 mm of rain in Costa Rica in an area that had recently experienced an earthquake, 942 landslides over a 27-km^2 area resulted in eight casualties and $103 million in damages (Quesada-Román et al., 2019). Changes in land use and land cover that replace vegetation with surfaces less pervious to water infiltration, and even periods of drought, affect the frequency of debris flows and landslides. In Eastern Uganda, expansion of deforestation for agriculture from lower to higher elevations coincided with the occurrence of landslides (Mugagga, Kakembo and Buyinza, 2012). Prolonged drought events followed by shifts to extreme rainfall events have been found to decrease slope stability (Handwerger et al., 2019). With drought often comes an increase in fire frequency and severity, and numerous studies have shown an increased occurrence of debris flows following wildfire (Cannon and Gartner, 2007).

9.10 Denudation Rates

Whereas the occurrence of a recent landslide is obvious, erosion also occurs slowly and imperceptibly over hundreds of thousands of years. To quantify the erosion rate over such long durations, a combination of river discharge and total sediment transport can be used to estimate the **denudation rate**, the rate at which material is removed, or denuded, from a drainage basin. This provides an estimate of how erosion from both mechanical and chemical weather processes lowers the elevation of the landscape (mm per thousands of years), which can then be compared to the uplift rate to provide a holistic representation of Earth processes and landscape evolution.

Mechanical denudation refers to the export rate of material from a drainage basin from the processes previously described: river bedload, river suspended sediment, debris flows,

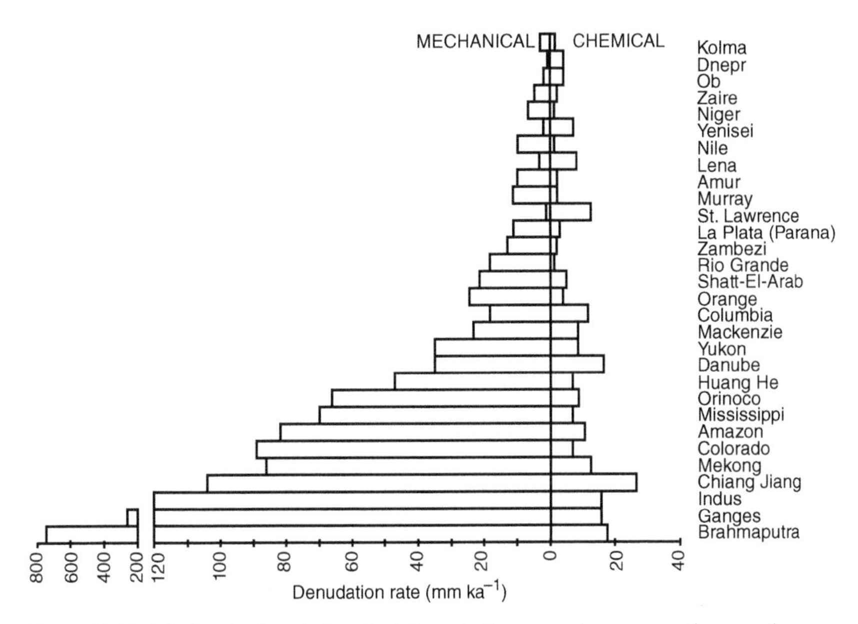

Figure 9.10 Mechanical and chemical denudation rates in mm per thousand years from several major river basins worldwide. Reproduced with permission from Summerfield and Hulton (1994).

landslides, without any change in the chemical composition of the material. Whereas paleohydraulic studies can be used to infer a former river's flow characteristics, sediment cores and samples taken from fluvial deposits can be used to infer the river's potential to transport solid sediment. Combined, this information can be used to estimate the mechanical denudation rate of the region (typically the drainage basin) over time.

Mechanical denudation rate varies widely through time and across space. A compilation of studies of the rate and controls of fluvial denudation rates in the world's major drainage basins by Summerfield and Hulton (1994) found that mechanical denudation rates ranged from 1 mm every thousand years (1 mm kyr^{-1}) to around 750 mm kyr^{-1} (Figure 9.10). The Amazon and Colorado River basins had similar mechanical denudation rates of 82 and 89 mm kyr^{-1}, respectively, surprising given the large differences in basin area, elevation, climate, and runoff. There was a strong statistical correlation between the mechanical denudation rates and basin relief (maximum–minimum basin elevation), relief ratio (basin relief/basin length), the mean local relief (mean maximum–minimum elevation within grid cells), and the mean modal elevation (mean of the modal grid-cell elevation). Correlations with these measures of basin topography rather than climate controls were attributed to the idea that topographic factors such as the availability of highly erodible fracture rock in high-elevation steep terrain have a larger influence than climate in producing high sediment yields.

Chemical denudation refers to the export rate of material that has undergone a chemical reaction involving water (see Section 9.11). Chemical denudation rates are much less than the mechanical rates, but also show considerable temporal and spatial variation. Summerfield and Hulton (1994) reported chemical denudation rates ranging from 1 to 27 mm kyr^{-1}

(Figure 9.10). In the Amazon and Colorado River basins, the chemical denudation rates were again similar at 11 and 7 mm kyr^{-1}, respectively. Similar to mechanical denudation rates, the chemical denudation rates were statistically correlated more with measures of basin topography than climate controls. When the mechanical and chemical denudation rates were added for the total rate, basin relief and runoff alone explained over 62 percent of the total denudation rate.

9.11 Chemical Weathering

Although Summerfield and Hulton (1994) showed that the chemical denudation rate is small at rates typically 10 mm or less over 1,000 years, chemical weathering is still an important process in shaping the landscape and providing ecologically important nutrients. There are three main processes describing how water participates in chemical weathering that are primarily based on the degree of solvency of the mineral. **Hydration** refers to the bonding of water molecules or hydrogen and oxygen atoms to a mineral without dissolving any of the mineral. **Hydrolysis** refers to the chemical bonding of water to the mineral resulting in some of the mineral dissolving into solution. Lastly, **dissolution** refers to the bonding of water with the mineral to dissolve it completely without any new solid structures forming.

9.11.1 Hydration

The hydration reaction does not result in the creation of a dissolved solution but does create new solid compounds, and is often the first step in the chemical weathering process. After a rock's surface has been exposed to water, the bipolar charge of the water molecule means that water molecules are attracted to either a positive or negative charge in the exposed mineral surface. The addition of liquid water to the solid mineral surface can then result in the formation of new compounds. The rusting of iron (Fe) to produce the new compound iron hydroxide ($Fe(OH)_x$) (s: solid) occurs when iron reacts with oxygen gas and liquid water (Eq. 9.12):

$$4Fe(s) + 3O_2(g) + 6H_2O(l) \rightarrow 4Fe(OH)_3(s). \quad (9.12)$$

In another example of hydration, the new solid compound **gypsum** ($CaSO_4 \cdot 2H_2O$) forms when calcium sulfate ($CaSO_4$) containing water (known as **Plaster of Paris**) reacts with liquid water (Eq. 9.13):

$$CaSO_4 \cdot \tfrac{1}{2}H_2O(s) + \tfrac{3}{2}H_2O(l) \rightarrow CaSO_4 \cdot 2H_2O(s). \quad (9.13)$$

Large gypsum deposits can be found in regions where water containing dissolved gypsum has evaporated away. In the **White Sands National Park**, New Mexico, the gypsum sand dunes are the largest on Earth. After lakes created at the end of the last Ice Age rich in dissolved gypsum eventually evaporated away as the climate warmed and dried, wind collected the gypsum to form the unique gypsum dune landscape (Figure 9.11).

Figure 9.11 Gypsum dunes in White Sands National Park, New Mexico, USA. Photo credit: Raul Touzon/ Photodisc via Getty Images.

Hydration does not create a solution, but rather new solid products such as iron hydroxides or gypsum. Hydration, however, does aid in further exposing a mineral surface for chemical weathering such as hydrolysis.

9.11.2 Hydrolysis

The **hydrolysis** reaction produces a solution containing part of the reactant mineral, with the remainder of the reactant forming a new product in solid form. The word translates to "water" (Greek *hydor*) "unbind" (Greek *lysis*). The addition of liquid water breaks the chemical bonds between molecules, forming a new molecule in the solid form. If the water molecule is split during the reaction, hydrogen can bond to the newly formed molecule. For example, rainwater is naturally slightly acidic (pH 5.6), and hydrogen ions associated with the slightly acidic rainwater react with the minerals. Acid rain, created through the chemical reaction of rainwater with sulfur dioxide or nitrous oxides, increases the acidity further, accelerating the hydrolysis reaction.

A common example of hydrolysis chemical weathering is the formation of the new solid white clay mineral **kaolinite** through the reaction of feldspar, found in granite, with water. As described in Chapter 3, rainwater (H_2O) can react with carbon dioxide (CO_2) in the atmosphere or soil to produce carbonic acid (H_2CO_3) (Eq. 9.14):

$$H_2O(l) + CO_2(g) \rightarrow H_2CO_3(aq). \tag{9.14}$$

The dissociation of the carbonic acid into a hydrogen ion (H+) and carbonate ion (HCO_3^-) further increases the acidity by supplying additional hydrogen ions. The carbonate acid can then react with feldspar ($CaAl_2Si_2O_8$, calcium plagioclase feldspar) to produce the solid clay kaolinite ($Al_2Si_2O_5(OH)_4$), and a solution containing dissolved calcium (Ca^{2+}) and carbonate ions (CO_3^{2-}) (Eq. 9.15):

$$CaAl_2Si_2O_8(s) + H_2CO_3(l) + \tfrac{1}{2}O_2(g) \rightarrow Al_2Si_2O_5(OH)_4(s) + Ca^{2+}(l) + CO_3^{2-}(l). \tag{9.15}$$

Kaolinite is mined for commercial applications including use as a blood-clotting agent, in ceramics, and in cosmetics. **Kaolin**, a clay rich in kaolinite, is mostly mined for the paper

Figure 9.12 Limestone pavement along the Swiss/French Alps created by the dissolution of limestone by carbonate acid (left). Large cave formations formed by the dissolution of limestone deposits at the Cave of the Winds National Park in South Dakota (right). Photo credits: P. D. Blanken (left) and M. Blanken (right).

industry, to produce a glossy finish on packaging. Most coating-grade kaolin is produced from mines in the United States (Georgia and South Carolina), England (Cornwall), and Brazil (Lower Amazon region) (Murray, Alves and Bastos, 2007).

9.11.3 Dissolution

As the term implies, a **dissolution** reaction completely dissolves the original mineral by water breaking the bonds between atoms. No new solid compounds are formed. A common example of chemical weathering through dissolution is the weathering of **limestone** (calcium carbonate, $CaCO_3$). The carbonic acid (H_2CO_3) produced from the reaction between rainwater and carbon dioxide can readily dissolve limestone to produce calcium bicarbonate, $Ca(HCO_3)_2$ (Eq. 9.16):

$$H_2CO_3(l) + CaCO_3(s) \rightarrow Ca(HCO_3)_2(aq). \tag{9.16}$$

This chemical weathering process is often referred to as **carbonation** since CO_2 is required along with H_2O to produce the required carbonic acid. Over short periods, water collection in exposed joints can dissolve the limestone, creating **limestone pavement** (Figure 9.12). Over longer periods, fluctuating water tables can create, then expose, vast cave networks created by the dissolution of limestone (Figure 9.12). The term **karst** is used to described landscapes formed by the dissolution of limestone.

9.12 SUMMARY

Surface runoff forms when the input of water from precipitation or from adjacent surfaces exceeds the substrate's infiltration capacity. When the substrate is saturated and the water cannot infiltrate into deeper levels at a rate faster than the input rate, water pools at the surface. If the surface is sloped, small individual streams of water merge and connect, increasing the discharge (volume of water per time) and drainage area contributing to discharge at each junction, as quantified by the stream order classification system or the River Continuum Concept. A river's discharge at a given location can be estimated by the river's depth using a rating curve, the relationship between discharge and depth.

Liquid water's large density exerts a large static pressure, proportional to the water's depth, that can be calculated using the hydrostatic equation. Surface water will begin to flow on an inclined surface and create a dynamic pressure as the net force parallel to the slope and resulting acceleration increases with the sine of the angle. The dynamic pressure, or shear stress on the riverbed, is therefore calculated by the sine of the slope's angle multiplied by the static pressure. This shear stress must exceed the force required to move the sediment, the critical shear stress, before bed material can be transported. The sediment's critical shear stress depends on the particle's diameter and is complicated by the size distribution and arrangement of sediment. The transport of solid material along the riverbed can be supplemented by the transport of suspended sediment that originated from locations other than the riverbed. The Rouse number, the ratio of the downward to upward particle settling velocity, can be used to estimate the mode of sediment transport.

When river flow is intermittent, such as in dry regions or during flooding, erosion and depositional landforms can appear quickly. The sudden removal of vegetation, soil, and material can result in hillslope scars, and the deposition of this material in the forms of alluvial fans can be visible for decades. Disturbances such as hurricanes or earthquakes, often coupled with human activity, can result in large debris flows and landslides, responsible for thousands of fatalities and billions of dollars in economic loss. The theory and calculations used to understand and quantify these processes can also be used to estimate historic river flows at locations where fluvial landforms are present but the river no longer exists. Such paleohydraulic discharge reconstructions can help us understand and quantify past events such as megafloods or provide evidence of liquid water flows on other planets.

The calculation of the denudation rate, the long-term rate of the removal of material from river bedload, suspected sediment transport, debris flows, and landslides combined, is useful to understand broad-scale fluvial erosion processes and for comparisons to geologic terrain uplift rates. Mechanical denudation rates generally correspond better with measures of basin topography than climate, and, with a global average range up to 750 mm kyr^{-1}, are much larger than the chemical denudation rates that average up to 27 mm kyr^{-1}.

The chemical weathering responsible for the chemical denudation results from the ability of liquid water to dissolve most substances. Chemical weathering by liquid water can be classified into three categories depending on the solution formed. Hydration occurs when new solid compounds are formed by the addition of water, but a solution containing dissolved molecules does not form. A hydrolysis reaction produces a solution containing only part of the reactant material, in addition to a solid product containing the remainder of the reactant. A dissolution reaction results in the complete transformation of a solid to a liquid containing all of the reactants in an aqueous solution.

This chapter has shown the power and processes behind water's ability to shape and transform the landscape across a range of spatial and temporal scales in warm regions where liquid water is available. In cold regions where ice dominates, processes that result in unique landforms also occur, as described in Chapter 10.

9.13 QUESTIONS

9.1 Describe the relevant variables that determine when surface runoff will be generated. How could human activity influence these variables and therefore alter surface runoff?

9.2 What is the purpose or value of a river classification system such as that devised by Robert Horton or Arthur Strahler? Would a stream with a large stream order number be more, or less, susceptible to non-point sources of pollution than a stream with 1st-order stream? How does the Horton–Strahler stream order system differ from the classification used in the River Continuum Concept?

9.3 Why does the stage–discharge relationship shown in a rating curve become unreliable when a river floods? After a flood event changes the shape of the riverbed, would the rating curve derived prior to the flood be reliable at that location? Why or why not?

9.4 A river 2 m deep is flowing over a surface with a slope of 2 degrees. What is the static pressure exerted on the riverbed by the water above? Calculate the shear stress exerted on the riverbed created by the dynamic force of the flowing water.

9.5 At times, water managers release large volumes of water from dammed river systems to simulate spring meltwater runoff events. This helps to recreate the river's ecology and aquatic and riparian habitats by allowing the fast-flowing water to remove any accumulated bed sediment downstream of the dam. If managers require that sediment with a density of 2,650 kg m^{-3} and a mean particle diameter of less than 25 mm be removed, what depth of water is required if the river's slope is 0.006 m m^{-1}?

9.6 Describe how a river can transport sediment that is in suspension and was not liberated from the riverbed. Which ratio can be used to determine whether sediment was likely transported by bed, suspended, or wash load?

9.7 An esker is a fluvial landform consisting of a riverbed created by a river that once flowed within or beneath a glacier (Chapter 10). Estimate the depth of the river that once existed if the slope of the region is 0.003 m m^{-1} and the esker's sediment has a density of 2.650 kg m^{-3} and a mean particle density of 0.5 m. If the esker is 6 m wide, what was the volume of river that created this esker?

9.8 Describe the characteristics (e.g., geology, soils, slope, land use, land cover) and geographic location (e.g., climate) of a river basin where you might expect a high mechanical denudation rate and/or a high chemical denudation rate. Provide reason(s) that you would expect these high rates.

9.9 Why are landslides and debris flows expected to increase in frequency and cause more damage? Is there anything that could be done to help mitigate against the damages caused by these events?

9.10 Water can participate in chemical weathering in three ways. Provide a summary and example of how water participates in each of these reactions.

REFERENCES

Alho, P., Baker, V. R. and Smith, L. N. (2010) 'Paleohydraulic reconstruction of the largest Glacial Lake Missoula draining(s)', *Quaternary Science Reviews*, 29, pp. 3067–3078. doi: 10.1016/j.quascirev.2010.07.015.

Andrews, E. D. (1983) 'Entrainment of gravel from naturally sorted riverbed material.', *Geological Society of America Bulletin*, 94, pp. 1225–1231. doi: 10.1130/0016-7606(1983)94<1225:EOGFNS>2.0.CO;2.

Benko, B. and Stead, D. (1998) 'The Frank slide: A reexamination of the failure mechanism', *Canadian Geotechnical Journal*, 35, pp. 299–311. doi: 10.1139/t98-005.

Cannon, S. H. and Gartner, J. E. (2007) 'Wildfire-related debris flow from a hazards perspective', in Jakob, M. and Hungr, O. (eds.) *Debris-Flow Hazards and Related Phenomena*. Springer, pp. 363–385. doi: 10.1007/3-540-27129-5_15.

Cohen, S., Wan, T., Islam, M. T. and Syvitski, J. P. M. (2018) 'Global river slope: A new geospatial dataset and global-scale analysis', *Journal of Hydrology*, 563, pp. 1057–1067. doi: 10.1016/j.jhydrol.2018.06.066.

Costa, J. E. (1983) 'Paleohydraulic reconstruction of flash-flood peaks from boulder deposits in the Colorado Front Range', *Geological Society of America Bulletin*, 94(8), pp. 986–1004. doi: 10.1130/0016-7606(1983)94<986:PROFPF>2.0.CO;2.

Einstein, A. H. (1950) *The Bed-Load Function for Sediment Transportation in Open Channel Flows*. Technical Bulletin No. 1026, US Department of Agriculture. Available at: https://naldc.nal.usda.gov/download/CAT86201017/PDF.

Gariano, S. L. and Guzzetti, F. (2016) 'Landslides in a changing climate', *Earth-Science Reviews*, 162, pp. 227–252. doi: 10.1016/j.earscirev.2016.08.011.

Gochis, D., Schumacher, R., Friedrich, K. et al. (2015) 'The great Colorado flood of September 2013', *Bulletin of the American Meteorological Society*, 96, pp. 1461–1487. doi: 10.1175/BAMS-D-13-00241.1.

Handwerger, A. L., Huang, M.-H., Fielding, E. J. et al. (2019) 'A shift from drought to extreme rainfall drives a stable landslide to catastrophic failure', *Scientific Reports*, 9, article no. 1569. doi: 10.1038/s41598-018-38300-0.

Horton, R. E. (1945) 'Erosional development of streams and their drainage basins; Hydrophysical approach to quantitative morphology', *Bulletin of the Geological Society of America*, 56, pp. 275–370.

Howard, A. D., Moore, J. M. and Irwin, R. P. (2005) 'An intense terminal epoch of widespread fluvial activity on early Mars: 1. Valley network incision and associated deposits', *Journal of Geophysical Research E: Planets*, 110, pp. 1–20. doi: 10.1029/2005JE002459.

Jakob, M. and Lambert, S. (2009) 'Climate change effects on landslides along the southwest coast of British Columbia', *Geomorphology*, 107, pp. 275–284. doi: 10.1016/j.geomorph.2008.12.009.

Jarrett, R. D. and Costa, J. E. (1993) 'Hydrology and geomorphology of the 1982 Lawn Lake Dam Failure, Colorado', in McCutchen, H. E., Herrmann, R., and Stevens, D. R. (eds.) *Ecological Effects of the Lawn Lake Flood of 1982, Rocky Mountain National Park*. Scientific Monograph. US Department of the Interior, National Park Service, pp. 1–17.

Karlstrom, K. E., Lee, J. P., Kelley, S. A. et al. (2014) 'Formation of the Grand Canyon 5 to 6 million years ago through integration of older palaeocanyons', *Nature Geoscience*, 7, pp. 239–244. doi: 10.1038/ngeo2065.

Kleinhans, M. G. (2005) 'Flow discharge and sediment transport models for estimating a minimum timescale of hydrological activity and channel and delta formation on Mars', *Journal of Geophysical Research E*, 110, pp. 1–23. doi: 10.1029/2005JE002521.

Langston, A. L. and Temme, A. J. A. M. (2019) 'Bedrock erosion and changes in bed sediment lithology in response to an extreme flood event: The 2013 Colorado Front Range flood', *Geomorphology*, 328, pp. 1–14. doi: 10.1016/j.geomorph.2018.11.015.

Mugagga, F., Kakembo, V. and Buyinza, M. (2012) 'Land use changes on the slopes of Mount Elgon and the implications for the occurrence of landslides', *Catena*, 90, pp. 39–46. doi: 10.1016/j.catena.2011.11.004.

Murray, H. H., Alves, C. A. and Bastos, C. H. (2007) 'Mining, processing and applications of the Capim Basin kaolin, Brazil', *Clay Minerals*, 42, pp. 145–151. doi: 10.1180/claymin.2007.042.2.01.

Patton, A. I., Rathburn, S. L., Bilderback, E. L. and Lukens, C. E. (2018) 'Patterns of debris flow initiation and periglacial sediment sourcing in the Colorado Front Range', *Earth Surface Processes and Landforms*, 43, pp. 2998–3008. doi: 10.1002/esp.4463.

Pitlick, J. (1993) 'Geomorphic response of the Fall River, Rocky Mountain National Park, Colorado', in McCutchen, H. E., Herrmann, R., and Stevens, D. R. (eds.) *Hydrology and Geomorphology of the 1982 Lawn Lake Dam Failure, Colorado*. Scientific Monograph. US Department of the Interior, National Park Service, pp. 18–32.

Polemio, M. and Petrucci, O. (2000) 'Rainfall as a landslide triggering factor: An overview of recent international research', in Bromhead, E., Dixon, N., and Ibsen, M.-L. (eds.) *Landslides in Research, Theory, and Practice*. Volume 3. Thomas Telford, pp. 1219–1226.

Quesada-Román, A., Fallas-López, B., Hernández-Espinoza, K., Stoffel, M. and Ballesteros-Cánovas, J. A. (2019) 'Relationships between earthquakes, hurricanes, and landslides in Costa Rica', *Landslides*, 16, pp. 1539–1550. doi: 10.1007/s10346-019-01209-4.

Rouse, H. (1937) 'Modern conceptions of the mechanics of fluid turbulence', *Transactions of the American Society of Civil Engineering*, 102, pp. 463–554.

Shields, A. (1936) *Application of Similarity Principles and Turbulence Research to Bed-Load Movement*. California Institute of Technology; translated from Ott, W. P. and van Uchelen, J. C. (1936) 'Anwendungen der Aehnlichkeitsmechanik und der Turbulenzforschung auf die Geschiebebewegung,' *Mitteilungen der Preussischen Versuchsanstalt für Wasserbau Schiffbau.*

Strahler, A. N. (1952) 'Hypsometric (area-altitude) analysis of erosional topography', *Bulletin of the Geological Society of America*, 63, pp. 1117–1142.

Summerfield, M. A. and Hulton, N. J. (1994) 'Natural controls of fluvial denudation rates in major world drainage basins', *Journal of Geophysical Research: Solid Earth*, 99, pp. 13871–13883. doi: 10.1029/94jb00715.

Sundborg, A. (1956) 'The River Klarälven: A study of fluvial processes', *Geografiska Annler*, 38, pp. 238–316.

Vannote, R. L., Minshall, G. W., Cummins, K. W., Sedell, J. R. and Cushing, C. E. (1980) 'The river continuum concept', *Canadian Journal of Fisheries and Aquatic Sciences*, 37, pp. 130–137.

Wallemacq, P. and House, R. (2017) *Economic Losses, Poverty and Disasters 1998–2017*. UNISDR.

Yuill, B. T. and Nichols, M. H. (2011) 'Patterns of grain-size dependent sediment transport in low-ordered, ephemeral channels', *Earth Surface Processes and Landforms*, 36, pp. 334–346. doi: 10.1002/esp.2041.

10 Cold-Region Abiotic Processes and Landforms

Key Learning Objectives

After reading this chapter you will be able to:

1. Describe the ice-related processes that result in landforms at the small, medium, and large scales.
2. Explain why circular features and patterns in the landscape are so common in cold regions.
3. State how the thermal properties of snow play an important role in snowpack metamorphism and avalanche formation.
4. Discuss how ice sheets can alter the landscape across vast spatial scales, and how the landforms that appear today provide evidence of their past characteristics.

10.1 Introduction

Ice has a powerful impact on the landscape in **cold regions**. Here, cold regions are defined as locations that currently experience, or have in the past experienced, conditions resulting in ice formation. At small spatial scales (e.g., meters) such as on the surface of or within small fissures in rocks, the pressure exerted by the volume expansion can fragment and weather rock, facilitating the process of soil formation. When freeze–thaw cycles are frequent, whether daily or seasonal, ice can effectively mix surface materials to further aid in soil formation and improve soil fertility. In addition, latent heat exchanges driven by vapor pressure gradients and capillary rise are active at small spatial scales where freeze–thaw cycles are prevalent, and these energy exchanges also influence soil formation and processes.

Over medium spatial scales spanning hundreds of meters, the effects of freeze–thaw cycles are evident from the landforms that develop. Ice formation within soils can move large volumes of soil downhill on gently sloped surfaces. In flat regions, ice formation can sort material based on particle size, resulting in unique, symmetrical surface patterns, and where liquid water is available in the warmer months, circular mounds or depressions form as liquid water is drawn into the freezing location. In alpine regions prone to frequent freeze–thaw cycles, a different suite of landforms can develop from frost action on steeply sloped surfaces. These sloped alpine surfaces, when combined with the process of water changing state with the snowpack, can result in avalanche formation.

At large spatial scales (e.g., kilometers) and generally over much longer time periods (e.g., centuries), glaciers slowly but continuously erode and transform the landscape. The fluid flow behavior of ice, made possible by the immense pressure at the base of the ice sheet (Chapter 8), allows bedrock to be scoured and material to be removed and deposited far from its source, creating several distinct erosion and deposition landforms. Owing to the immense size and volume of ice contained within ice sheets, the lithosphere is depressed and then slowly rebounds after the ice has melted. Depressions in the landscape created by ice are often filled with meltwater. The spectrum of processes related to the freezing of water has created varied and dynamic landscapes in regions around us that are or were once cold.

10.2 Small-Scale Processes

Even the large-scale landforms created as a consequence of ice formation result from processes that occur at small spatial scales (i.e., on the scale of millimeters). When liquid water collects in small depressions on exposed rock and then freezes, the volume expansion can **weather** the rock's surface (break apart the material without any change in chemical composition) in a process called **granulation** where small fragments of the rock detach and remain nearby, giving the rock a distinct "crumbly" appearance (Figure 10.1). Even a few drops of water repeatedly freezing and thawing can weather the rock's surface, making it appear softly rounded with a loose, flakey textured surface. In addition to water, dust and organic material may collect in small depressions or at the base of rocks, and, together with the minerals provided through the process of chemical weathering (Chapter 9), can cause soil formation. Moss, lichens, and small plants may establish in these depressions, further weathering the rock through the addition of organic matter or the application of root pressure. As vegetation growth continues, the added organic matter increases water retention, further promoting freeze–thaw cycle weathering, and soil formation.

Vapor pressure gradients within soils when the surface temperature is near the freezing point can result in the vertical movement and mixing of material. When the soil is not saturated, the flux of water vapor is often directed toward the surface. This occurs because, especially in the spring and fall in temperate regions, nighttime soil temperatures just beneath the surface are

Figure 10.1 Granulated rock surfaces on the summit of Longs Peak, Colorado, elevation 4,346 m above sea level, on August 23, 2015. Photo credit: P. D. Blanken.

Figure 10.2 Ice needles in frozen soil. Photo credit: Michele D'Amico supersky77/ Getty Images.

often higher than in the air right at the surface. The warmer, moist soil conditions translate to a soil pore-space vapor pressure close to the saturation vapor pressure, higher than the vapor pressure near the soil surface. This is especially common at night during cool, clear-sky conditions with ample surface radiative cooling in the spring and fall seasons. Water can condense on the underside of cool surfaces such as small pebbles or rocks, or freeze if the temperature is below the freezing-point temperature. Thus, all three states of water are realized over the space of a few millimeters and over the course of minutes to hours. The volume expansion that occurs with ice formation on the underside of a pebble can exert enough pressure to dislodge it, relocating numerous small pebbles and stones to the surface. Ice forms in small columns (needles) as water moves by capillary rise (Chapter 11) between narrow spaces in the soil and freezes upon contact with cold air. These small columns of ice, typically under 5 cm in length, are referred to as **ice needles** and are the reason why, to the dismay of gardeners, small pebbles continue to appear in gardens each year (Figure 10.2).

The mixing of surface soil layers due to the formation of ice needles can be important in the weathering process and soil development, especially in alpine regions where freeze–thaw cycles occur frequently and chemical weathering rates are slow (Ponti, Cannone and Guglielmin, 2018). The formation of ice also dries the soil, and the additional heat released when water freezes (the **latent heat of fusion**) can supply more energy to dry the soil further (Soons and Greenland, 1970).

In polar and alpine regions often underlain by **permafrost**, ground that remains at or below 0 °C for two or more consecutive years, a dominant mechanism resulting in the mixing of the often poorly developed soils occurs owing to the seasonal formation and thawing of ice in the layers above the permafrost, the **active layer**. The mixing of organic matter from the surface down to deeper layers is often hindered in these regions by a short growing season with limited soil biological activity such as microbial activity and root growth. The vertical mixing (**turbation**) of surface nutrients to deeper soil layers can be achieved through ice formation and thawing: **cryoturbation** (Figure 10.3). The expansion of water when it freezes in soils can result in dramatic vertical movement (generally referred to as **frost heave**) when the soils are thin and saturated, and the expansion volume is confined by the permafrost below. In regions such as the Rocky Mountains of Colorado, the annual formation and thawing of ice in the soil has resulted in vertical soil displacements of up to 30 cm, indicating how vigorous the frost heaving process may be (Fahey, 1974).

Figure 10.3 Cryoturbation, the mixing of soils due to the freezing of water and thawing of ice, results in dramatic deformation of what would otherwise be horizontal sediment layers. Photo credit: Andrew Mattox, Bretwood Higman, Ground Truth Trekking under Creative Commons Attribution license CC BY-SA 3.0.

10.3 Medium-Scale Processes and Landforms

At medium spatial scales (hundreds of meters), ice formation can result in weathering and erosion creating several recognizable landforms. Fairly common in gentle to moderately sloped surfaces with the same aspect, needle ice growth can result in **soil creep**, the slow downward movement of the upper soil layers. Soil creep, and the **solifluction sheets** or **solifluction lobes** that indicate the occurrence of soil creep over large areas, play a major role in slope evolution, carbon and nutrient cycling, and vegetation colonization (Ponti, Cannone and Guglielmin, 2018). Entire sides of valleys and hillslopes can be influenced by solifluction. In the southern Canadian Rocky Mountains, for example, surface (the upper 30 cm of soil) movement rates averaged 0.7 cm per year, transporting 10.9 cm^3 downslope 1 cm per year (Smith, 1988). In the eastern Austrian Alps, surface solifluction rates from 5 to 50 cm per year were measured, with frost heaves ranging from 5 to 8 cm per year (Jaesche, Veit and Huwe, 2003). The solifluction rates and frost heaves were strongly influenced by daily fluctuations in soil water content derived from the lateral influx of snowmelt-derived meltwater (Jaesche, Veit and Huwe, 2003).

Over flat terrain typically found in boreal regions, freeze–thaw action including needle ice growth may result in the formation of several types of **patterned ground** created by the volume expansion of water and the attraction of liquid water to the location of ice formation. These patterns of polygons, circles, or stripes are not only a consequence of the ~9 percent volume expansion upon freezing, but are also due to the addition of liquid water at the **freezing front**. As discussed in Chapter 3, the saturation vapor pressure over ice is always less than that over liquid water. Therefore, when liquid water is available in the soil below, it will be drawn towards the freezing front due to **capillary rise** and the decrease in the soil water potential (Chapter 11) that develops at the freezing front as the liquid soil moisture decreases. This important process is known as **cryostatic suction** and is dominant in **periglacial** (near glaciated areas) and other environments (such as alpine, subalpine, and boreal) with ample water availability that experience repeated freeze–thaw cycles (Peterson and Krantz, 2003).

Figure 10.4
A hummock-and-hollow tundra landscape underlain by continuous permafrost, near Churchill, Manitoba, Canada. Photo credit: P. D. Blanken.

Landscape patterns that develop because of frost heaves and cryostatic suction that keeps fueling the ice formation process are truly unique. In the flat, poorly drained Subarctic and Arctic regions, thin soils underlain by permafrost together with short summers result in saturated soils and numerous small shallow lakes (Chapter 5). When freezing does occur, there is ample liquid water available to be drawn to the freezing front. More so than the volume expansion, the supply of liquid water towards the surface where temperatures are lower continuously adds to surface freezing until the liquid water is no longer available. As a result, **frozen mounds** or **hummocks** are common. The scale of these features can vary. Some are small, typically ~30 cm diameter and height, hummocks containing ice under a distinct relatively thick layer of mosses, lichens, and vegetation providing thermal insulation to the ice beneath. Between the raised hummocks lie the lower depressions or **hollows** with saturated soils serving as a strong source of methane from the anaerobic conditions (Roulet et al., 1994). This **hummock–hollow landscape** extends for thousands of kilometers in the Hudson Bay Lowland and other similar Subarctic and Arctic wetlands (Figure 10.4). In regions with **discontinuous permafrost**, areas underlain by ice can form larger areas that are elevated a few meters above the surrounding terrain because of the volume expansion of ice. These areas (similar to a hummock–hollow landscape but with hummocks at a much larger horizontal scale) are known as **palsas** when covered by **peat** (Chapter 5), or as **lithalsas** when not covered by organic peat soils, and can approach heights of 100 m.

Pingos have a cycle that illustrates the dynamics of ice formation and melting. The growth of a pingo results in a convex dome shape as cryostatic suction draws liquid water from all sides to the freezing front, sometimes draining surrounding shallow lakes in the process (Figure 10.5). Eventually, the upper ice lens cracks and sheds the peat layers towards the sides. With the insulative peat layer removed, melting eventually occurs, resulting in surface subsidence, collapse of the pingo, and the formation of a small lake (Figure 10.6). In such cases, a berm of organic soil may remain along the perimeter of the former pingo, aiding retention of water, especially if impermeable clays have accumulated at the bottom of the lake (Bouchard et al., 2017).

In polar regions with relatively flat ground and sparse vegetation that experience some seasonal thaw to provide liquid water, unique symmetrical patterns may form on the surface. This

Figure 10.5 A pingo covered with insulative vegetation and organic matter, and surface cracks visible at the top. Photo credit: Pierre Longnus/Getty Images.

Figure 10.6 A collapsed pingo. Photo credit: Pierre Longnus/ Getty Images.

patterned ground consists of large areas of polygon or circle patterns of sorted material with large stones on the edges and smaller ones in the center (Figure 10.7). To understand how these patterns form, imagine a small circular depression where water collects and freezing occurs. As ice forms, volume expansion occurs, and the surrounding liquid water is drawn to the freezing front through the cryostatic suction process. Since ice growth occurs perpendicular to the freezing front (Nicholson, 1976), the newly formed ice lens will grow along all sides and therefore expand in a circular manner. Coarse materials such as stones tend to migrate away from the freezing front, more so than finer materials (Nicholson, 1976). If there is an adjacent area experiencing this same process, the sides of the expanding circle or polygon will eventually meet, aggregating the larger stones. Patterned ground may even form in desert landscapes with some soil moisture that experience below-freezing nighttime temperatures. Linear, striped patterns may form in sloped areas from the same process (Nicholson, 1976).

In alpine regions where freeze–thaw cycles are common and large areas of rocks are exposed, several landscape features can form. With variations in slope and aspect that result in a myriad of **microclimates** over small spatial scales (Barry and Blanken, 2016), exposed rock is vulnerable to repeated freeze–thaw cycles that often occur daily (as temperatures reach above the freezing point during the daytime and below it at night). The frequent freeze–thaw cycles result in the fracturing of rock, termed **frost-shattered rock**. The force generated by the expansion is sufficient to fracture rocks of impressive size, such as the large boulder shown in Figure 10.8.

Figure 10.7 Patterned ground near Svalbard, Norway. Photo: Danita Delimont/Getty Images.

Figure 10.8 Frost-shattered boulder in Rocky Mountain National Park, Colorado. Photo credit: P. D. Blanken.

In areas where the angular frost-shattered rocks lock together, they can remain in place over time, forming collections called **seas of rock**, or **felsenmeer** (Figure 10.9). On sloped surfaces, the repeated freeze–thaw cycles are common, as exposure to daytime solar radiation melts ice that formed overnight. In addition to causing numerous rockslides on highways resulting in casualties and road damage, the collection of frost-shattered rocks along the sides of cliffs results in the formation of loose **talus cones** (Figure 10.10).

The inherent instability of snow or ice that has accumulated on sloped surfaces can result in **avalanches**, with impacts to the local landscape and those in their path. In Canada, for example, there is an average of at least 1.5 million potentially destructive avalanches every year that are capable of human injury or death, with an average 12.5 fatalities each year during the 1990s (Stethem et al., 2003). In Austria since 1950, more than 1,600 avalanche fatalities, roughly 30 each year, have occurred (Höller, 2007). In the mountainous western United States, 301 avalanche fatalities occurred between 1998 and 2009, an average of 27.4 annually, with snowmobiling and skiing activities representing most of the fatalities (121 and 70 deaths respectively; Spencer and Ashley, 2011).

The simultaneous presence of ice and liquid water, and the resulting gradients in temperature and vapor pressure that develop, play a role in avalanche formation. Since snow is a good

Figure 10.9 Felsenmeer near Longs Peak, Colorado, USA. Photo credit: P. D. Blanken.

Figure 10.10 Talus cone at Lake Agnus, Colorado. Note people in the foreground for scale. Photo credit: P. D. Blanken.

thermal insulator, the temperature is usually near 0 °C at the base of the snowpack, changing to several degrees below 0 °C near the top. Air within the snowpack has a relative humidity near 100 percent, so a calculation of the saturation vapor pressure at the snow's temperature can be used to closely approximate the vapor pressure. The base of the snowpack has a higher vapor pressure than the surface, so the vapor pressure gradient is directed upwards from the warmer ground to the cooler surface. The resulting flux of water vapor results in a smoothing, or **faceting**, of the snow crystals at the base called **depth hoar**. At the snowpack's surface, frail ice crystals called **surface hoar** can form when the vapor pressure gradient is directed downward from a warmer, more humid atmosphere to a cooler, drier snow surface (i.e., deposition).

Hoar layers are loose and unstable as the mass of snow on top increases with subsequent snowfalls. When a trigger motion is induced by humans or animals, or simply by increasing snow mass above, the snow structure can fail along a hoar layer, resulting in a **slab avalanche** or **slough avalanche** when a low-density surface layer of snow slides above another layer. Depending on the volume and velocity of the snow involved, an avalanche can easily snap and relocate trees and most anything else in their path (Feistl et al., 2015). The resulting **avalanche scar** can remain visible for years, given the slow growth rates of vegetation in alpine regions. In general, the greater the mass and density of snow, the greater the destructive force of the avalanche. The most destructive avalanche in history was an **ice avalanche**, originating from the north peak of Huascaran, Peru, on May 31, 1970. Triggered by a magnitude 7.75 earthquake, a debris avalanche containing glacier ice from an elevation of 3,658 m above sea level traveled 11.27 km at an average speed of 322 km per hour (Cluff, 1971). This **Great Peruvian Earthquake** destroyed the villages of Yungay and Ranrahirca and caused 70,000 casualties and the loss of 200,000 structures (Cluff, 1971).

10.4 Large-Scale Processes and Landforms

Large-scale (i.e., kilometers) landforms created by ice formation and melting match the scale of processes that created them. The mass of ice sheets and glaciers, coupled with their ability to deform and flow over terrain, results in several of the well-known and widely studied landforms described here.

During cooler periods in Earth's climate history when polar ice sheets and glaciers advanced, the impacts on the global water cycle were large. Sea levels decreased, ocean circulation patterns shifted as ocean temperature and salinity changed or the circulation was entirely blocked by ice, and planetary albedo increased. Pronounced changes in Earth's landscape, water cycle, and climate occurred during glacial periods with positive feedbacks such as the **temperature–albedo feedback** (Chapter 8) serving to further enhance cooling, hence ice sheet growth and glacial advance. The most recent **Laurentide Ice Sheet**, the largest ice sheet in the Northern Hemisphere during the last glacial period that ended roughly 15,000 years ago, is estimated to have reached a height of 3,200 m (Lacelle et al., 2018) and covered nearly all of Canada, extending just south of 40° N latitude (Clark, 1992). The pressure created at the base of the ice sheet as it slowly advanced southwards was easily capable of removing any soil and loose surface rock down to bedrock. At locations where the base of the ice sheet or glacier was frozen to the surface, **plucking** occurred when sections of bedrock were removed (Figure 10.11A). Once these sections of bedrock became entrained at the base of the ice, long scratches or **striations** were made on the bedrock beneath, indicating the direction and number of ice sheet advances (Figure 10.11B). In alpine regions, glaciers carved **u-shaped valleys**, forming deep **fjords** where coastal valleys filled with ocean water.

The Laurentide was so massive that the land surface was displaced downwards, allowing meltwater to accumulate and forming glacial **Lake Agassiz**, likely containing more fresh water than all of today's lakes combined (Perkins, 2002). After the melting of the Laurentide Ice Sheet, evidence of the **isostatic rebound**, or **glacio-isostatic adjustment**, the response of the lithosphere to the removal of the ice sheet's weight, can be found. Located near the former geographic center of the ice sheet near Hudson Bay, Canada, rates of isostatic rebound at roughly 10 mm per year have been recorded near Churchill, Manitoba (Lambert et al., 2001). Physical evidence of this glacio-isostatic adjustment can be seen by the many **raised** or **abandoned beaches** located along the shore of Hudson Bay (Figure 10.11C).

Eventually, as the atmosphere warmed and Earth entered the current interglacial period, deposition processes created by flowing meltwater water created several landforms. Sea levels rose, ocean circulation changed with changes in temperature and salinity, glacial and kettle lakes formed as depressions filled with meltwater, and massive rivers flowed from the receding ice. These rivers carried and deposited sediment, creating **eskers** (former riverbeds) marking their location on, within, or beneath the ice (Figure 10.12). From the size and distribution of the

Figure 10.11 Erosion features created by large-scale ice sheets. (A) Plucking; (B) striations; and (C) abandoned raised beaches. Photographs taken near Churchill, Manitoba, on the shore of Hudson Bay close to the center of the Keewatin Ice Dome. Photo credits: P. D. Blanken.

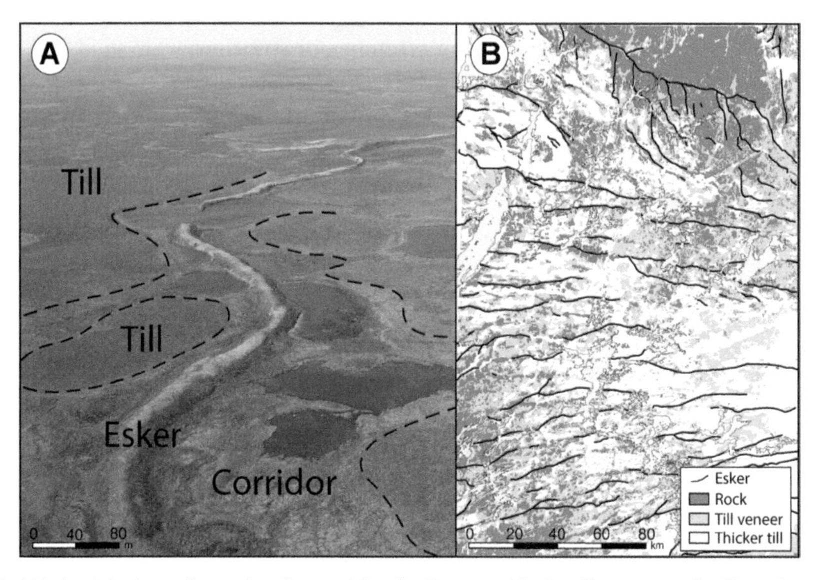

Figure 10.12 (A) Aerial view of an esker formed by the Laurentide Ice Sheet near the East Arm of Great Slave Lake in northwestern Canada. (B) Direction and composition of the eskers and glacial till indicates deposition of sediment in pressurized subglacial meltwater channels that occurred during a warm-period deglaciation. Reproduced with permission from Sharpe et al. (2017).

sediment, the discharge and flow conditions of these rivers, and hence the volume and extent of the ice, could be estimated (Chapter 9). The material scoured from the surface by the ice sheets or glaciers was also deposited along their sides and forward edge. **Terminal moraines** mark the maximum extent of the ice sheet or glaciers, and **lateral moraines** mark the lateral extent of glacial erosion. From these preserved terminal and lateral moraines, reconstruction of ice-sheet surface morphology is possible. At the terminus of the Laurentide Ice Sheet in the Wisconsin/Michigan region, ice lobe thicknesses based on the maximum height of the moraines are estimated from 250 to nearly 650 m, advancing at velocities of between 450 and 2,000 m per year (Clark, 1992).

In addition to moraines, large teardrop-shaped mounts of scoured debris were sometimes deposited in large mounds known as **drumlins** that appear as inverted spoons on the terrain (Figure 10.13). Drumlins, typically 100–300 m wide, 300–1,000 m long, and 30–50 m high,

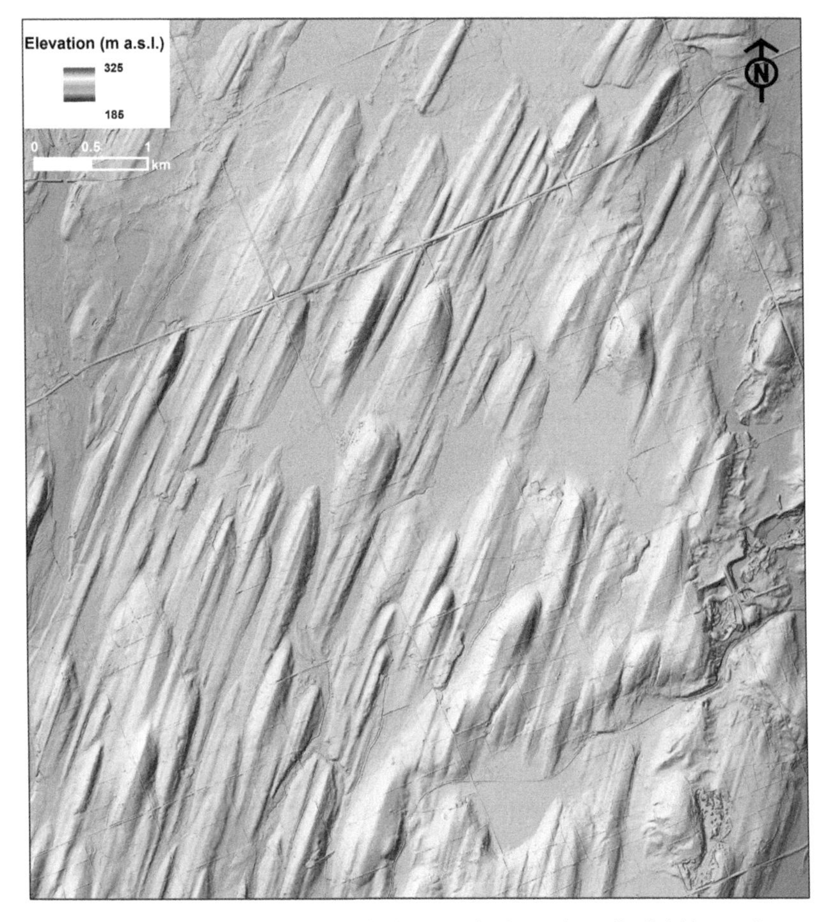

Figure 10.13 LiDAR-based digital elevation model image of a large drumlin field located near Peterborough, Ontario, Canada. The direction of ice flow movement from the Laurentide Ice Sheet was from the top right (north-northeast) to bottom left (south-southwest). Reproduced with permission from Sookhan, Eyles and Bukhari (2022).

usually occur in groups (fields) with the drumlin's steep end pointing towards the up-ice direction of ice flow and the tapered end towards the down-ice direction. The advancing ice sheet often not only removed all surface material but also eroded the hard bedrock beneath into teardrop shapes known as **roche moutonnees** (French meaning "sheep rock") that indicate the direction of icc movement, with plucking on the down-ice side. Large boulders were trapped and carried at the base of ice sheets and abruptly dropped as the ice melted. These **glacial erratics** mark the location of the ice sheet or glacier at thc time melting began (see Chapter 6, Figure 6.2).

10.5 SUMMARY

Ice formation, movement, and subsequent melting involve a suite of processes that alter the landscape resulting in a spectrum of landforms at scales spanning meters to kilometers. On the surface or within rocks, the increase in the volume of water upon freezing exerts sufficient pressure to break small or large fragments of rocks, a primary step towards the formation of soils. The greater the frequency of freeze–thaw cycles, the greater the weathering of the material. Within soils, ice formation and subsequent melting can effectively mix surface inorganic and organic material, improving soil fertility. The formation of ice within soils also serves to dry the soil as liquid water is drawn towards the location of freezing through capillary rise and differences in vapor pressure.

Small-scale processes involving freezing are reflected in various medium-scale (hundreds of meters) landforms. Needle ice growth near the surface can result in the formation of solifluction sheets or solifluction lobes in areas with gently sloped surfaces. The preferential sorting of surface material by size due to ice formation results in the formation of rings or patterned ground in poorly drained flat regions. Such circular patterns on the landscape, such as the hummock–hollow landscape, palsas, and pingos, and small shallow lakes are often common in poorly drained cold regions typically underlain by permafrost, owing to this cryostatic suction of liquid water to the initial location of freezing, volume expansion upon freezing, and the insulative properties of the organic peat layer above. Alpine regions are especially prone to frequent freeze–thaw cycles resulting in different landforms appearing on the landscape. Frost-shattered angular rocks can collect on flat surfaces (felsenmeer) or on steeply sloped surfaces forming talus cones. Avalanches can leave tell-tale signs on the landscape when they have sufficient force to remove trees and soil. Vapor and temperature gradients at the base or surface (after subsequent snowfalls) of a snowpack often result in a weak structure, making the snowpack prone to avalanches in sloping terrain.

At large spatial scales of kilometers, ice sheets and their associated glaciers have a profound influence on the landscape often lasting thousands of years after the disappearance of the ice. Landscape alteration by ice sheets is possible through the massive weight that can not only displace the lithosphere, but can also flow with the properties of a thick liquid and thus erode material beneath. Raised or abandoned beaches along the coast of Hudson Bay provide evidence and rates of the isostatic rebound resulting from the removal of the Laurentide Ice Sheet. Geological evidence of the erosion created by the advancement of the ice is shown on the hard bedrock surface that was once beneath the ice sheet. Striations, plucking, u-shaped valleys, fiords, and glacial erratics provide evidence of the pressure, location, and direction of ice movement. Deposits of eroded material by meltwater rivers (eskers), the ice sheet (drumlins), and glaciers (lateral and terminal moraines) provide key information as to the maximum extent of the ice sheet's advance.

The physical changes associated with the formation of ice, and the resulting impacts on the surface across scales ranging from meters to kilometers, have altered and continue to alter the abiotic landscape to a degree that life has become adjusted to ice and depends upon it. Earth's landscape without water cycling between states including ice would look very different. With the aid of water, the weathering, erosion, and deposition formed and continuously alters the landscape. This helped set the stage for the life on Earth to flourish, and the role of water in the biotic environment is the topic of the next chapters.

10.6 QUESTIONS

10.1 What are the fundamental processes related to the properties of water that result in so many of the unique landforms related to the presence of ice?

10.2 Describe the process(es) responsible for the widespread occurrence of granulated rocks on the summit of Longs Peak, Colorado (see Figure 10.1).

10.3 Gardeners and farmers are often frustrated by the prevalence of small stones that reappear at the surface of gardens and fields every spring. Explain to them why this occurs and whether there are any benefits of the process responsible.

10.4 Potholes, the small car-tire-sized holes that develop every spring in paved roads in regions that experience winter conditions, result in numerous car accidents and cost millions annually in car and road repairs. Why do these hazards form, and specifically what is the role that freeze–thaw cycles play in pothole formation?

10.5 In soils, liquid water will move over short distances to the location of freezing, the freezing front. What is (are) the physical reason(s) for this movement of water, and what are the implications for the availability of liquid water for plants or other organisms when this relocation and change of state of water is occurring?

10.6 Several landscape forms common in low-lying (flat) cold regions such as hummocks, patterned ground, palsas, lakes, and pingos tend to have a circular form. What is it about the formation and presence of ice that could explain these common circular patterns on the landscape in cold regions?

10.7 Compare and contrast the landforms created by freeze–thaw cycles found in alpine regions to those found in flatter Arctic regions. How do differences in the microclimate, geology (including soils), and topography influence the formation and appearance of these landforms?

10.8 What, if any, is the connection between vapor pressure and the risk of an avalanche occurring?

10.9 Along the shores of Hudson Bay near Churchill, Manitoba, Canada, one can find several examples of how the Laurentide Ice Sheet impacted the region during the last Ice Age. Describe at least three of these examples, how they formed, and why the Hudson Bay region is such a good location to find these examples.

10.10 Thousands of people walk through Central Park in New York City each day unaware that roughly 20,000 years ago they would have been standing under hundreds of meters of ice. If you were to visit Central Park today, what evidence could you provide to a tourist to support this claim?

REFERENCES

Barry, R. G. and Blanken, P. D. (2016) *Microclimate and Local Climate*. Cambridge University Press. doi: 10.1017/CBO9781316535981.

Bouchard, F., MacDonald, L. A., Turner, K. W. et al. (2017) 'Paleolimnology of thermokarst lakes: a window into permafrost landscape evolution', *Arctic Science*, 3, pp. 91–117. doi: 10.1139/as-2016-0022.

Clark, P. U. (1992) 'Surface form of the southern Laurentide Ice Sheet and its implications to ice-sheet dynamics', *Geological Society of America Bulletin*, 104, pp. 595–605. doi: 10.1130/0016-7606(1992)104<0595:SFOTSL>2.3.CO;2.

Cluff, L. S. (1971) 'Peru earthquake of May 31, 1970; Engineering geology observations', *Bulletin of the Seismological Society of America*, 61, pp. 511–533. doi: 10.1086/622062.

Fahey, B. D. (1974) 'Seasonal frost heave and frost penetration measurements in the Indian Peaks region of the Colorado Front Range', *Arctic and Alpine Research*, 6, pp. 63–70. doi: 10.2307/1550370.

Feistl, T., Bebi, P., Christen, M. et al. (2015) 'Forest damage and snow avalanche flow regime', *Natural Hazards and Earth System Sciences*, 15, pp. 1275–1288. doi: 10.5194/nhess-15-1275-2015.

Höller, P. (2007) 'Avalanche hazards and mitigation in Austria: A review', *Natural Hazards*, 43, pp. 81–101. doi: 10.1007/s11069-007-9109-2.

Jaesche, P., Veit, H. and Huwe, B. (2003) 'Snow cover and soil moisture controls on solifluction in an area of seasonal frost, Eastern Alps', *Permafrost and Periglacial Processes*, 14, pp. 399–410. doi: 10.1002/ppp.471.

Lacelle, D., Fisher, D. A., Coulombe, S., Fortier, D. and Frappier, R. (2018) 'Buried remnants of the Laurentide Ice Sheet and connections to its surface elevation', *Scientific Reports*, 8, pp. 1–10. doi: 10.1038/s41598-018-31166-2.

Lambert, A., Courtier, N., Sasagawa, G. S. et al. (2001) 'New constraints on Laurentide postglacial rebound from absolute gravity measurements', *Geophysical Research Letters*, 28, pp. 2109–2112. doi: 10.1029/2000GL012611.

Nicholson, F. H. (1976) 'Patterned ground formation and description as suggested by low Arctic and Subarctic examples', *Arctic and Alpine Research*, 8, p. 329. doi: 10.2307/1550437.

Perkins, S. (2002) 'Once upon a lake', *Science News*, 162, pp. 283–284. doi: 10.2307/4014064.

Peterson, R. A. and Krantz, W. B. (2003) 'A mechanism for differential frost heave and its implications for patterned-ground formation', *Journal of Glaciology*, 49, pp. 69–80. doi: 10.3189/172756503781830854.

Ponti, S., Cannone, N. and Guglielmin, M. (2018) 'Needle ice formation, induced frost heave, and frost creep: A case study through photogrammetry at Stelvio Pass (Italian Central Alps)', *Catena*, 164, pp. 62–70. doi: 10.1016/j.catena.2018.01.009.

Roulet, N. T., Jano, A., Kelly, C. A. et al. (1994) 'Role of the Hudson Bay lowland as a source of atmospheric methane', *Journal of Geophysical Research*, 99, p. 1439. doi: 10.1029/93jd00261.

Sharpe, D. R., Kjarsgaard, B. A., Knight, R. D., Russell, H. A. J. and Kerr, D. E. (2017) 'Glacial dispersal and flow history, East Arm area of Great Slave Lake', *Quaternary Science Reviews*, 165, pp. 49–72. doi: 10.1016/j.quascirev.2017.04.011.

Smith, D. J. (1988) 'Rates and controls of soil movement on a solifluction slope in the Mount Rae area, Canadian Rocky Mountains', *Zeitschrift für Geomorphologie, Supplementband*, 71, pp. 25–44.

Sookhan, S., Eyles, N. and Bukhari, S. (2022) 'Drumlins and mega-scale glacial lineations as a continuum of subglacial shear marks: A LiDAR based morphometric study of streamlined surfaces on the bed of a Canadian paleo-ice stream', *Quaternary Science Reviews*, 292, p. 107679. doi: 10.1016/j.quascirev.2022.107679.

Soons, J. M. and Greenland, D. E. (1970) 'Observations on the growth of needle ice', *Water Resources Research*, 6, pp. 579–593. doi: 10.1029/WR006i002p00579.

Spencer, J. M. and Ashley, W. S. (2011) 'Avalanche fatalities in the western United States: A comparison of three databases', *Natural Hazards*, 58, pp. 31–44. doi: 10.1007/s11069-010-9641-3.

Stethem, C., Jamieson, B., Schaerer, P. et al. (2003) 'Snow avalanche hazard in Canada – A review', *Natural Hazards*, 28, pp. 487–515. doi: 10.1023/A:1022998512227.

11 Water within Vegetation

Key Learning Objectives

After reading this chapter, you will be able to:

1. Describe the importance of water for terrestrial vegetation's function and form.
2. Analyze evidence for and against the Cohesion–Tension Theory.
3. Explain the concept of water potential in relation to water movement.
4. Define and calculate the solute (or osmotic), pressure (or turgor), matric (or soil), gravitational, and the liquid–air water potentials.
5. Explain how water can ascend in vegetation to heights well above 10 m.

11.1 Introduction

Water influences the abiotic environment through the processes of weather, erosion, and deposition as it flows over and through the landscape. Water also influences the form, structure, and phenology (life cycles) of plants and animals. To compete for sunlight for photosynthesis, plants are driven to grow tall, especially in forests where neighboring plants provide shade. Water must be available in the chloroplasts within leaves, where photosynthesis takes place, so to grow tall for access to sunlight also requires the availability of water in the highest leaves. This chapter describes the role of water in terrestrial vegetation, and describes how it is possible for water to move from the roots to the shoots against the force of gravity. To understand how vegetation has developed the means to extract water from within soils and raise it to heights sometimes exceeding 100 m requires an understanding of the chemical and physical properties of water described in previous chapters, and of how specialized tissues within vascular plants have evolved to harness these unique properties.

The ascent of water in vegetation is an impressive feat, resulting in a significant quantity of water being released into the atmosphere, with major consequences for the global water cycle and climate system. This chapter begins with an overview of several important roles water plays in terrestrial vegetation's function and form. The hydrostatic equation is used to show how atmospheric pressure alone can only raise a column of water to just over 10 m. However, measurements show the global forest canopy height is nearly double that height (17 m) and the tallest known tree is over 10 times that height (115.6 m).

To explain how water can ascend to such heights, the form and structure of specialized water-conducting structures found in vascular plants are described. A popular theory to explain the movement of water within plants is presented, along with measurements that either support or refute this theory. To explain the ascent of water in vegetation requires an understanding of water potential, a measure used to quantify the work required to move water from one location to another. Water potential has several variables including solute concentration, soil properties and water content, and relative humidity. How these variables contribute to the total water potential of a plant is discussed, along with measurement techniques and example calculations.

11.2 The Roles of Water in Terrestrial Vegetation

Liquid water serves several critical roles in terrestrial vegetation's function and form. First and foremost, water is required for **photosynthesis**, the conversion of solar radiation energy to chemical energy. Six water molecules combine with six CO_2 molecules to produce one glucose molecule, with the release of six oxygen gas molecules under the presence of **photosynthetic active radiation** with wavelengths between 400 and 700 nm (Eq. 11.1):

$$6H_2O + 6CO_2 \rightarrow C_6H_{12}O_6 + 6O_2 \,. \tag{11.1}$$

Water provides the hydrogen and oxygen that form the carbohydrates required for plant growth and maintenance. In turn, the plants provide a source of food and shelter for countless other organisms. Photosynthesis provides a direct coupling between vegetation and the climate system through the exchange of carbon, oxygen, and water vapor with the atmosphere, changes in surface albedo, and the cycling of numerous elements. The quest for exposure to solar radiation has resulted in canopy structures that provide habitats for numerous organisms, as well as many other ecosystem functions and services.

Water plays several roles in the chemical photosynthesis reaction. Water donates electrons in the **light-dependent photosynthesis reaction**, allowing the creation of new covalent bonds to form carbohydrates. When water is split into H_2 and O, the flow of electrons from electron donors to electron receptors creates **adenosine triphosphate** (ATP), a source of energy for the plant. Without the need for light (the **light-independent** or **dark reaction**), ATP is used to capture or "fix" the carbon from the CO_2 to produce the carbohydrates and reform water as a product. Without water, these new bonds could not form.

Water provides a means for the plant to transport materials. Nutrients required by plants such as nitrogen, phosphorus, and potassium dissolved in water can be transported into the plant through the roots when dissolved in water and relocated to different areas of the plant for processes such as seed development or leaf growth.

Water helps plant structure and shape. By relocating and moving water within and between cells, **vascular plants** (those plants that contain specialized tissues to conduct water) can regulate carbon uptake and water loss, and adjust leaf shape and angle to regulate solar radiation interception. Water pressure created within plant cells, called **turgor**, provides physical structure and support. Thus, even nonwoody species can grow tall in search of solar radiation.

Water helps plants regulate temperature. Evaporation requires energy, so evaporation on or within leaves results in leaf and canopy cooling as well as providing a significant quantity of water vapor to the atmosphere, which is important for the water cycle. With liquid water's high specific heat capacity, water provides the plant with a thermal buffer against periods of excessive heat or cold. The ability of water to dissolve substances creates a solute which lowers the freezing point, enabling the plant to tolerate below-freezing temperatures.

11.3 The Ascent of Water within Vegetation

The cohesive and adhesive forces that bind liquid water molecules to each other or to other surfaces (respectively) are largely responsible for how a continuous column of water ascends from the soil to the tops of trees. Vegetation has evolved an internal vascular system that effectively counteracts the relentless downward gravitational pull of water. As a result, vegetation can grow tall, gaining access to photosynthetic active radiation allowing growth and a competitive advantage compared with young and understory plants below (Figure 11.1).

Air pressure exerts a force that can raise a column of water, but how high? At sea level, the atmosphere exerts an average pressure of 101.3 kPa (1,013 mbar), equivalent to 101,300 N m^{-2} (1 Pa = 1 N m^{-2}). Pressure is a force per unit area, and force is mass multiplied by the acceleration due to

Figure 11.1 Kapok (*Ceiba pentrandra*) tree (center-background) reaching heights of 70 m protruding above the main canopy and gaining access to ample solar radiation in the Western Amazon rainforest in Peru. Note the darkness in the understory compared with above the canopy. Photo credit: P. D. Blanken.

gravity, so sea-level air pressure is equal to 10,326 kg m^{-2} (101,300 kg m^{-1} s^{-2} divided by 9.81 m s^{-2}). In other words, each square meter at sea level experiences a weight of 10,326 kg (14.7 psi, pounds per square inch) from the atmosphere above. Although this is an impressive value, even at sea level where air pressure is at its maximum this pressure can only produce enough force to raise a column of liquid water roughly 10 m. The **hydrostatic equation** (Eq. 11.2) can be used to calculate water pressure (P; Pa or N m^{-2}) as a function of gravity (g; m s^{-2}) and height (h; m):

$$P = \rho g h \,. \tag{11.2}$$

Rearranging Eq. (11.2) to find the height when P = 101,300 Pa gives:

$$h = \frac{P}{\rho g} = \frac{101{,}300\ \mathrm{N\,m^{-2}}}{1{,}000\ \mathrm{kg\,m^{-3}} \times 9.81\ \mathrm{m\,s^{-2}}} = 10.33\ \mathrm{m}. \tag{11.3}$$

A connection between changes in atmospheric pressure, the height of a water column, and weather conditions was first observed by a student of **Galileo de' Galilei** (1564–1642), **Evangelista Torricelli** (1608–1647). Galileo and Torricelli wondered why a suction pump could only raise water to 10 m. Torricelli proposed that the atmosphere exerted a force that balanced the force exerted by the weight of the column of water the pump extracted. Torricelli observed that variations in the height of the water column correlated with changes in weather, resulting in Torricelli's invention of the **barometer**. A simple barometer consists of fluid contained in an inverted tube with the air above the fluid at the top of the tube removed, and the fluid exposed to the atmosphere at the bottom of the tube. Changes in the water's height in the barometer precede changes in weather. Since 10 m is an impractical height for a barometer, the much denser liquid mercury (13,593 kg m^{-3}) was often used to measure atmospheric pressure in the popular units of 29.92 inches of Hg (at sea level). Mercury barometers have long since been replaced by digital barometers, but the use of the old units "inches of mercury" remains common in several countries.

In the example above (Eq. 11.3), the calculation is based on pressure, which is expressed as a force per unit surface area. Therefore, sea-level atmospheric pressure can support a column of water 10 m high regardless of any changes in surface area due to changes in the diameter of the water column, since both forces are expressed as pressure. The weight of the water column, however, does decrease as the diameter of a tube decreases, making it preferrable to have a narrow rather than a wide tube to raise a column of water. For water to ascend above 10.33 m, however, additional forces are required.

11.4 Xylem Tissue

How can trees be taller than 10 m? A 600–800-year-old tree named Hyperion, a California coastal redwood (*Sequoia sempervirens*), is currently, at 115.6 m in 2010, the world's tallest tree (Sillett et al., 2010). Using novel remote sensing methods, Simard et al. (2011) estimated a global mean canopy height of 16.9 m, with canopy height decreasing with latitude and elevation (Figure 11.2). Thousands of small-diameter tubes within these tall trees are used to

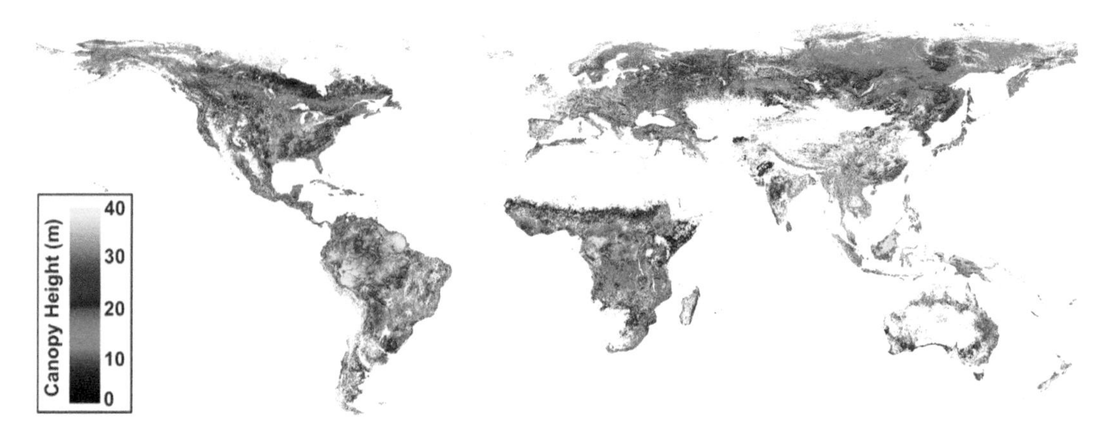

Figure 11.2 Canopy height at 1-km spatial resolution based on estimates from spaceborne light detection and ranging (LiDAR) from the Geoscience Laser Altimeter System (GLAS) in 2005. Reproduced with permission from Simard et al. (2011).

take advantage of water's cohesion and adhesion properties to raise water. The liquid water molecules electrically bond to the sides of a tube that itself has a weak positive or negative charge (adhesion). Similarly, water molecules can bond to each other (cohesion). Through adhesion–cohesion alone, water can ascend to heights greater than 10 m, with the additional gain in height increasing as the tube diameter decreases.

Vascular plants contain narrow tubes called **xylem** which transport water and water-soluble nutrients from roots to leaves or other locations, and **phloem** to transport organic molecules made during photosynthesis wherever needed throughout the plant (Figure 11.3). The xylem's narrow diameter (20–40 μm) allows a highly curved **meniscus** to form within each of the dozens of individual xylem tubes. This high surface tension is made possible by the cohesive property of water, and results in **capillary rise**. The height to which water can ascend by capillary rise is based on the difference in air pressure across the meniscus. This difference in air pressure, **Laplace pressure**, is calculated from the **Young–Laplace equation** $\Delta p = 2\gamma/r$ where γ is the fluid surface tension (N m^{-1}) and r is the radius of curvature of the spherical meniscus (m). The value for r can be calculated knowing the radius of the tube (r_0; m) and the contact angle of the liquid on the tube wall (meniscus angle, θ; degrees): $r = r_0/\cos(\theta)$. When the forces acting on the water column are in equilibrium, $\Delta p = \rho g h$ (the hydrostatic equation) combined with the Young–Laplace equation gives:

$$\rho g h = \frac{2\gamma}{r} = \frac{2\gamma\cos(\theta)}{r_0}. \tag{11.4}$$

Assuming r_0 is much less than the tube's length (a long, narrow tube), Eq. (11.4) can be solved for h to calculate the capillary rise, known as **Jurin's Law** (Eq. 11.5):

$$h = \frac{2\gamma\cos(\theta)}{\rho g r_0}. \tag{11.5}$$

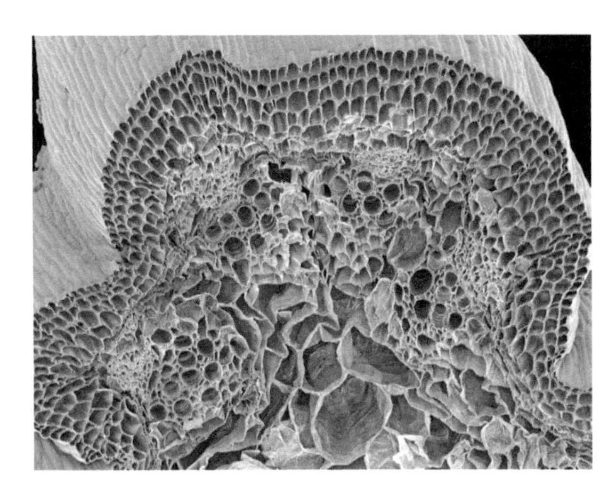

Figure 11.3 Scanning electron micrograph of a *Nasturtium* stem showing the inner xylem (pink) and outer phloem (yellow). Photo credit: Steve Gschmeissener/SPL/Getty Images.

From Jurin's Law (Eq. 11.5), the capillary rise in xylem with a diameter of 20 μm (r_0 = 1.0×10^{-5} m) using γ = 0.0728 N m^{-1}, ρ = 1,000 kg m^{-3}, g = 9.81 m s^{-2}, and a contact angle of 0° (since the water in the xylem is assumed to completely spread out over the internal hydrophilic surface) is 1.48 m:

$$h = \frac{2\gamma \cos(\theta)}{\rho g r_0} = \frac{2 \times 0.0728\,\mathrm{N\,m^{-1}} \times \cos(0°)}{1{,}000\,\mathrm{kg\,m^{-3}} \times 9.81\,\mathrm{m\,s^{-2}} \times 1.0 \times 10^{-5}\,\mathrm{m}} = 1.48\,\mathrm{m}. \quad (11.6)$$

The earliest evidence of xylem tissue, consisting of simple thin and smooth-walled microperforated tubes, dates to roughly 410 million years ago. Since then, xylem tissue has evolved adaptations to aid in water transfer (Sperry, 2003). These include thick xylem cell walls lined with lignin, allowing large surface tension to develop in the water column. Thick xylem cell walls are less prone to collapse as the water tension within increases (imagine taking a drink with a straw; a thin straw is much easier to collapse than a thick one). **Lignin**, made of complex organic polymers, lines the xylem to provide this physical support and is impermeable (hydrophobic), thus preventing xylem water loss (Liu, Luo and Zheng, 2018). **Tracheid cells**, or scalariform perforated plates located within the xylem, have evolved in some species to aid in water transport and reduce the risk of xylem embolism resulting from cavitation (Sperry, 2003). Even with these adaptations, xylem alone is not capable of transporting water to heights above 10 m unless the means for the vegetation to develop suction (negative pressure) can be identified.

11.5 The Cohesion–Tension Theory

As research into and measurements of barometric pressure progressed, it became clear that an explanation was needed for how tall trees can lift water to heights greater than any mechanical pump. The **Cohesion–Tension Theory**, or simply the **Cohesion Theory**, presented by Dixon and Joly (1894) was developed to explain the phenomenon. The theory states that at the liquid–air interface in the stomatal cavity, cohesion between water molecules, together with adhesion of water to the sides of the stomatal cavity, creates a highly curved meniscus. The surface tension

Figure 11.4 A pressure chamber used to measure xylem pressure potential. A small branch with the phloem tissue removed is placed inside the chamber (right side) with the end of the branch protruding through the small hole in the chamber's lid (held in hand). When pressured air is gradually applied to the chamber (black hose), the pressure when the water first appears at the branch's end is noted (large pressure gauge). The xylem pressure potential is equal to the negative value of this pressure. Photo credit: P. D. Blanken.

that develops in the stomatal cavities (see Section 11.6.5 and Chapter 12) results in the vertical movement of water down to the root level, aided by the cohesion between water molecules within the continuous column of water contained within the xylem. The surface tension at the liquid water–air interface can become excessive when the rate of water loss in the stomata exceeds the rate of water intake from the roots; thus the stomata close under such conditions of water stress. If the tension in the xylem's water column exceeds the cohesive force holding the water molecules together, the intermolecular bonds will break, resulting in the formation of an air pocket (cavitation), and the xylem will no longer be able to conduct water.

The Cohesion–Tension Theory is a well-known theory with widespread support (Brooks, 2004) despite some recent criticism (e.g., Zimmermann et al., 2004). The criticism concerns the measurement of the tension (negative pressure) in the xylem, known as the xylem pressure. The water tension in the xylem should decrease from the lowest root to the highest leaf for water to move upwards against the downward force of gravity. Xylem pressure can be measured directly using several methods, the most common being a **leaf pressure chamber** or **pressure bomb** (Figure 11.4). In a method developed by Scholander (Scholander et al., 1965), a small branch is cut from a tree and phloem tissue is carefully removed. The branch is then placed in a sealed chamber with the end of the branch protruding through a small, sealed hole. Compressed gas (usually nitrogen) is used to fill the sealed chamber. As the pressure inside the chamber increases, the branch is squeezed, resulting in the xylem water column returning to its pre-cut position. When the xylem water first appears, the required pressure is equal to the negative pressure (tension) that existed in the branch before it was cut.

Using a pressure chamber, Scholander et al. (1965) and others have shown that water within the xylem is indeed, under large tension, capable of raising water to 100-m heights such as in California redwood species. Measurements ranged from −5 to −80 atmospheres (−0.5 to −8.0 MPa) depending on species, location, and ambient conditions such as soil moisture and humidity. Low tensions (−5 atmospheres) were recorded in damp freshwater species, and high tensions (−80 atmospheres) were recorded in desert species. Samples from tall (100 m) Douglas fir and redwood species showed that the xylem tension did increase with height, as required for water to ascend (Scholander et al., 1965). Overall, the body of evidence based on xylem pressure measurements gave support to the Cohesion–Tension Theory.

The theory has, however, been questioned. Zimmermann et al. (2004) suggested that the cohesive and adhesive forces of water within xylem cannot withstand tensions exceeding 1 MPa, far less than the measurements made using a pressure chamber in tall or water-stressed vegetation. Some claim that pressure chambers provide xylem tensions that are too large for reasons related to the interference of air-filled spaces that develop during transpiration (Melcher et al., 1998). Therefore, they suggest that other measurements such as a cell pressure probe (Tomos, 2000) be used instead. Comparisons between pressure chamber and cell-probe xylem pressure measurements, however, show that **air seeding** limits the application of the cell pressure probe under high xylem tensions. Air seeding occurs when an air bubble is drawn under tension through a hole or crack in a cell wall, possibly resulting in embolism (Lewis, 1988). Some believe that overlooking this limitation of the cell pressure probe has led to claims that the Cohesion–Tension Theory is invalid (Wei, Steudle and Tyree, 1999).

Another criticism of the Cohesion–Tension Theory is based not on the method used to measure xylem tension but on the physics of the cohesive water bond itself. As the ambient pressure exerted on a fluid decreases, an air parcel will form when the cohesion between molecules can no longer keep them bonded together. The pressure when the intermolecular bonds break, forming air pockets (cavitation), or when boiling occurs, is the **saturation vapor pressure**. Based on the strength of the cohesive bond, liquid water should be capable of tensions beyond an impressive −120 MPa (−1,184 atmospheres) at room temperature before air pockets form and cavitation begins (Davitt, Arvengas and Caupin, 2010). Hundreds of experiments over the past two centuries have been conducted to determine water's actual cavitation threshold. Water is in a stable state of matter when in liquid, solid, or vapor form; these states can exist for long periods of time since they are in a state of chemical equilibrium. When liquid water is under tension, a **metastable state** develops, meaning that water remains in liquid form even though the system is not in equilibrium. Under tension, this metastable state should not last long, and cavitation should occur, yet xylem tension measurements indicate that this metastable state does exist for long periods.

To dissolve a gas in a fluid and achieve a metastable state, the fluid can be superheated above the boiling-point temperature, or tension below the saturation vapor pressure can be applied (around 2 kPa for water at 20 °C). Of course, only the latter applies to plants. An extensive review of dozens of laboratory cavitation experiments using different techniques concluded that the tension required to induce cavitation in water at room temperature is around −25 MPa (−247 atmospheres) (Caupin and Herbert, 2006). Thus, under laboratory conditions, it is possible for water to experience the −80 atmosphere (−8 MPa) tensions that were observed by Scholander et al. (1965). The roles of impurities in the water and sides of xylem tissue, air seeding and the entrance of air through the xylem wall, and choice of measurement technique mean that the lively debate regarding Cohesion–Tension Theory will likely continue (e.g., Brooks, 2004).

11.6 Water Potential: ψ

The concept of water potential explains the force that moves liquid water from one location to another. Somewhat surprisingly, it is not the quantity of water present in the soil, but rather the energy required by the plant to extract the water from the soil and then transport this water to the highest leaves that drives the ascent of water in vascular plants.

Work is the energy required to move a mass through a distance, expressed in units of **joules**. The term "work" was likely first used in 1826 by **Gaspard-Gustave Coriolis** (1792–1843; famous for the Coriolis effect) in the context of describing a pump used to raise water in a reservoir (Coriolis, 1829). The work required to move water vertically can be expressed as joules per mass (J kg^{-1}) or per volume of water (J m^{-3}). These units are fundamentally different than those used to specify the quantity of water by mass or volume, since water potential is based on the work (joules) required to displace a unit (a kilogram or cubic meter) of water.

To lift water, suction is required, and the appropriate units for suction are pressure. When using the definition of gauge pressure as the pressure relative to the ambient atmospheric pressure, suction would be represented by negative pressure. Units of pressure are equivalent to units of work per mass or volume of water, as shown below. The unit of work (joules) per mass (kg) is equivalent to units of $m^2\ s^{-2}$:

$$\frac{\mathrm{J}}{\mathrm{kg}} = \frac{mad}{\mathrm{kg}} = \mathrm{kg} \times \frac{\mathrm{m}}{\mathrm{s}^2} \times \mathrm{m} \times \frac{1}{\mathrm{kg}} = \frac{\mathrm{m}^2}{\mathrm{s}^2}$$

since a joule is defined as mass (m) multiplied by acceleration (a) and distance (d). Furthermore, units of work (joules) per volume (m^3) are equivalent to units of pressure (**pascals**):

$$\frac{\mathrm{J}}{\mathrm{m}^3} = \frac{mad}{\mathrm{m}^3} = \mathrm{kg} \times \frac{\mathrm{m}}{\mathrm{s}^2} \times \mathrm{m} \times \frac{1}{\mathrm{m}^3} = \frac{\mathrm{kg}}{\mathrm{s}^2\mathrm{m}}$$

$$P = \frac{F}{A} = \frac{ma}{A} = \mathrm{kg} \times \frac{\mathrm{m}}{\mathrm{s}^2} \times \frac{1}{\mathrm{m}^2} = \frac{\mathrm{kg}}{\mathrm{s}^2\mathrm{m}}$$

Therefore, units of pressure can be used to represent the work performed to move a volume of water through a distance.

To illustrate the concept of work, energy, and water potential, imagine extracting water from a well. An arbitrary point of reference, such as the water table, is helpful to define a coordinate system of positive and negative values. The work (W) required to lift a bucket of water to the surface is the required force (F) that depends on the mass of water (and the bucket and rope; m), the acceleration (a; gravity), and the vertical distance (d). If the bucket filled with water weighs 4 kg and the well is 5 m deep, the work required is:

$$W = Fd = mad = 4\,\mathrm{kg} \times 9.81\frac{\mathrm{m}}{\mathrm{s}^2} \times 5\,\mathrm{m} = 196.2\frac{\mathrm{kg\,m}^2}{\mathrm{s}^2}(\text{or joules}).$$

Once this work is applied to raise the bucket to the surface, the bucket has potential energy relative to the water table that was designated as the reference. By doing work, energy was transferred to the bucket. Even though the bucket has stopped moving at the top of the well, it has potential energy by virtue of its position above the water table reference where the potential energy would be zero. If the rope were to suddenly break, the potential energy would be converted to kinetic energy as the bucket fell.

Knowing the density of water, the volume of water in the bucket can be calculated:

$$\rho = \frac{M}{V}; V = \frac{M}{\rho} = \frac{4 \text{ kg}}{1{,}000 \text{ kg m}^{-3}} = 0.004 \text{ m}^3 \left(4 \text{ litres}\right)$$

and the work required to lift the bucket expressed on a water volume basis is:

$$\frac{196.2 \text{ J}}{0.004 \text{ m}^3} = 49{,}050 \frac{\text{J}}{\text{m}^3} \left(\text{or pascals}\right).$$

As shown, J m^{-3} are equivalent to pascals, so the work required to raise the bucket is equal to roughly 49 kPa. If the bucket and rope were replaced by a pump, a suction of 49 kPa less than the ambient atmospheric pressure (negative gauge pressure) is required. To combine the concepts of work, potential energy, and suction, the term **water potential** (ψ) is used, defined as the potential energy of water per volume (J m^{-3} or pascals) relative to pure water at some reference location. Water always moves from a location of high to low water potential. If ψ is positive, the water is under pressure and will move to areas of lower pressure. For example, a spring or artesian well will expel water. If ψ is negative as is almost always the case in the xylem tissue, the water is under tension and will again move to areas of lower pressure. If $\psi = -10$ kPa at location A and $\psi = -20$ kPa at location B, the water will move from location A to B.

11.6.1 Solute or Osmotic Water Potential: ψ_s

The **solute** or **osmotic water potential** is defined as the water potential that develops because of dissolving solutes in water. Pure water at standard atmospheric pressure has a solute water potential of zero, but when a solute is added, the solute water potential decreases, meaning that the newly created solution has a larger energy deficit than the pure water. This happens because as the solute concentration increases, it becomes much harder for the water within that solution to move out of solution. When the solute molecules bind to water molecules, energy is consumed in the hydrogen bonds that form. This energy is no longer available to perform work, so the potential energy decreases as the solute concentration increases. Another way to think of this is that the negative water potential represents an energy deficit that needs to be fulfilled to bring the water back to zero water potential.

Often a barrier such as a **semipermeable membrane** that allows some molecules to pass through (such as water) but not others (such as the solute) is required to allow a solute concentration difference to develop across the membrane. Therefore, if the cell can develop a solute concentration difference across the semipermeable membrane, water can flow in or out of the cell. This potential for water to perform work due to changes in solute concentration is referred to as the **solute water potential** ψ_S, sometimes referred to as the **osmotic potential.** When a solute is added to pure water, the decrease in water potential relative to pure water permits the movement of water molecules but not the solute across the semipermeable membrane (Figure 11.5).

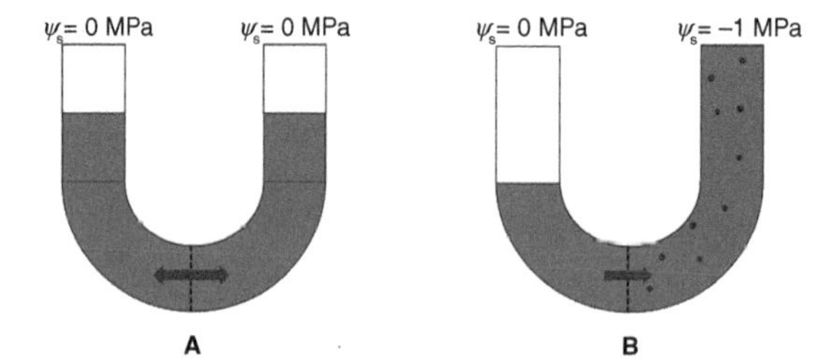

Figure 11.5 (A) With pure water (blue) on each side of the semipermeable membrane (dashed line), the water levels are equal. (B) When a solute that cannot pass through the semipermeable membrane (red spheres) is added to the right-hand side, the solute water potential ψ_s decreases, and water moves (red arrow) from the pure water (left side) to the solute (right side).

The solute water potential is directly proportional to the solute concentration, and can be calculated using Eq. (11.7):

$$\psi_s = -CiRT \tag{11.7}$$

where C is the solute molarity concentration (mol), i is the solute's ionization constant (dimensionless), R is the ideal gas constant ($8.314\,\mathrm{J\,K^{-1}\,mol^{-1}} = \mathrm{m^3\,Pa\,K^{-1}\,mol^{-1}} = \mathrm{kg\,m^2\,s^{-2}K^{-1}\,mol^{-1}}$), and T is temperature in kelvin (°C + 273.15).

The ionization constant is also known as the **van't Hoff factor** and is equal to the ratio of the actual concentration of the dissolved solute in water to the concentration calculated from the solute's mass. If the solute dissociates into cations and anions (an electrolyte) such as salt (NaCl, sodium chloride, i = 2), then the ionization constant is greater than 1. If the solute associates and does not form ions in solution and instead forms other molecules, the ionization constant is less than 1. Lastly, if the solute neither dissociates nor associates in solution, the ionization constant equals 1, as in the case of glucose and sucrose (i = 1 for both).

For pure water, C is zero, so from Eq. (11.7) ψ_s would also be zero. For a 1 mole per liter (1,000 mol m^{-3}) solution of sucrose at 25 °C and standard atmospheric pressure, ψ_s would decrease to roughly −2.5 MPa:

$$\psi_s = -CiRT = -1{,}000\frac{\mathrm{mol}}{\mathrm{m^3}} \times 1 \times 8.314\frac{\mathrm{J}}{\mathrm{K\,mol}} \times 298.15\mathrm{K} = -2{,}479{,}000\frac{\mathrm{J}}{\mathrm{m^3}}(\text{pascals}).$$

If this sucrose solution were placed adjacent to pure water separated by a semipermeable membrane, water would flow from the pure water towards the sucrose solution (Figure 11.5). Through the adjustment of the solute concentration inside compared with outside cells, plants can move water into or out of cells over small horizontal distances. This **osmotic adjustment** can be used to allow water from the soil to enter root cells, to regulate the stomatal cell aperture by controlling the water potential of the adjacent guard cells, and to maintain plant shape and structure for growth. Osmotic adjustment is observed when plants are exposed to either saline or drought conditions (Boyer et al., 2008). In saline conditions, salt entering the roots from the soil lowers the water potential, so the plant must also lower its water potential to maintain water flow into, not out of, the roots (hence the use of the term "adjustment"). Fluid extracted from plant cell tissue shows that the potassium (K^+), sodium (Na^+), and chloride (Cl^-) ions appear to play a major role in osmotic adjustment (Jones and Gorham, 1983; Cusuman, 2001; Boyer et al., 2008). For example, barley and wheat showed an increase in leaf Na^+ and Cl^- as the salinity of the solution they were growing in increased (Boyer et al., 2008), so the plants were able to maintain leaf turgor by dynamically adjusting their internal solute concentration to maintain a constant difference in the water potential inside compared with outside the plant. In drought conditions, the decrease in the plant's water potential increases its ability to extract water as this becomes more difficult as the soil dries.

11.6.2 Pressure or Turgor Water Potential: ψ_p

Fluid pressure within plant cells is known as **turgor pressure**, and this plays a key role in plant form and function. A plant suffering from a lack of water will visibly wilt because of the decreased turgor pressure. As described above, a plant cell can maintain turgor through osmotic adjustment in response to water stress induced by salinity or dehydration. Since plant cells have rigid cell walls, the flow of water into the cell will increase the cell's turgor pressure.

The **pressure or turgor water potential** ψ_p can be directly measured using several techniques. For a healthy plant, ψ_p will be greater than zero, meaning the water inside the cells is under pressure and exerting a force on the cell walls. Inserting a small probe into the cell would then push water into the probe, and the pressure (with a negative sign) required to push the fluid back into the cell equals the turgor pressure. This **pressure probe technique** was developed and tested on tissue slices of pepper fruits (Hüsken, Steudle and Zimmermann, 1978).

Alternatively, the **pressure chamber** (Figure 11.4) can be used to estimate cellular turgor pressure. As described above, the pressure chamber measures the tension (negative pressure) of the water column in xylem tissue, not the turgor pressure inside plant cells. This **balancing pressure** is recorded as the pressure required for water to neither flow into nor out of the cut end of the twig. If this measurement is repeated, and the volume of water expressed by the xylem measured, a **pressure–volume curve** can be plotted showing the relationship between the balancing pressure (P_b) and the volume of water expressed. When plotted as the inverse of the applied incremental pressure ($1/P_b$) on the *y* axis against the volume of water expressed on the *x* axis (both axes linear), the pressure–volume curve decreases rapidly at first but then quickly becomes linear. This means that each subsequent application of pressure results in a greater volume of extracted water from the plant tissue.

Combining theory with the empirical results produced from these pressure–volume curve relationships, Tyree and Hammel (1972) derived several bulk parameters of plant water relations. For example, when the pressure–volume curve becomes linear, the balancing pressure extrapolated to when the extracted water volume is zero (the *y*-axis intercept) is close to the solute (osmotic) water potential. The *x*-axis intercept, when the extrapolated $1/P_b$ is zero, is a measure of the intercellular (symplast) volume. The average turgor pressure of all the cells placed in the chamber (the bulk, or the cell-averaged turgor pressure; Tyree and Hammel, 1972) can be estimated from the difference between the bulk solute (osmotic) water potential and the balancing pressure for each volume of the water expressed. The use of the pressure chamber to derive these plant water parameters has faced both criticism and support (see discussions in Melcher et al., 1998 and Brooks, 2004).

An alternative to using the practical pressure chamber or the complicated pressure probe technique, the turgor water potential can be estimated if the total water potential at the cell level is known. The total water potential in the cell (ψ_c) is equal to the sum of the solute and turgor water potentials (Eq. 11.8):

$$\psi_c = \psi_s + \psi_p. \tag{11.8}$$

The turgor water potential ψ_p could therefore be estimated from the difference between the total cellular and solute water potentials (Eq. 11.9):

$$\psi_p = \psi_c - \psi_s. \tag{11.9}$$

As described, the pressure chamber can be used to measure the total cellular water potential, taken as the average of all the plant tissue inserted into the chamber. Although this measurement is really an integral of all the water in the plant sample, when done carefully, this measurement has been shown to provide a good measure of the total water potential in leaf cells when compared to other methods for measuring the total cellular water potential (Turner, 1981). With ψ_c measured using a pressure chamber (Figure 11.4) and ψ_s calculated using Eq. (11.7), the turgor pressure can then be calculated as the difference between the two (Eq. 11.9).

At the cellular level, the total water potential is equal to the sum of the solute (osmotic) and pressure (turgor) water potential. When the plant cell is in equilibrium with its surroundings, the total water potential is the same throughout the cell walls, cytoplasm, organelles, and vacuoles (Turner, 1981). At the scale of individual plants, however, there are additional water potential components to consider. The **total water potential** ψ_t at any location in a plant is given by Eq. (11.10):

$$\psi_t = \psi_s + \psi_p + \psi_m + \psi_g + \psi_a \tag{11.10}$$

where ψ_m is the **matric water potential** that develops from the surface tension at the menisci between soil particles (Whalley, Ober and Jenkins, 2013), ψ_g is the **gravitational water potential** that develops owing to the vertical position in a gravitational field (small at only 0.01 MPa m^{-1}; see below), and ψ_a is the **liquid–air water potential** that develops owing to the high surface tension across the meniscus interface between liquid water and air.

In the discussion above, $\psi_c = \psi_s + \psi_p$ was used as the total water potential at the cellular level since within a cell ψ_m is not applicable, ψ_g is negligible over a small vertical distance, and ψ_a is not applicable since there is no air present within the cells.

11.6.3 Matric or Soil Water Potential: ψ_m

The **matric** or **soil water potential** (ψ_m) is a measure of the availability of water within the soil. Water contained in the soil must travel through the complex and often tortuous matric of air and/or water-filled spaces to reach the roots. Water can be tightly bound to soil particles owing to the water molecule's electrical attraction to the soil, especially clay particles that have a negative electrical charge and high **cation exchange capacity** across a large surface area. In

Figure 11.6 Two soil tensiometers used for *in situ* measurements of the matric (soil) water potential in a rubber tree plantation in southern Thailand. Photo credit: P. D. Blanken.

addition, the large surface tension that develops across the liquid–air interface within the soil pore spaces creates forces that must be overcome by the plant to extract water.

The **soil water retention curve** shows the relationship between soil water content and the suction (negative soil water potential) required to extract water from within the soil matric. These curves show that the energy required to extract water increases sharply as the soil water content decreases, and depends on soil texture. The soil matric potential in unsaturated soils is always negative since the water is bound to the soil particles (ψ_m approaches zero as the soil water content approaches saturation). When a soil water retention curve for a particular location is developed by measuring the soil's gravimetric or volumetric water content together with the matric potential measured by a **soil tensiometer** (see below), ψ_m can be estimated providing that the *in situ* soil water content is measured. This approach is valid only for soils at the location where the soil water retention curve was developed, since the curve varies with soil properties, and hysteresis occurs depending on if the soil is wetting or drying. Measurements of the soil's water content and water potential can be made in the laboratory using a device known as a **pressure plate**. A soil sample is placed in the chamber of the pressure plate, air pressure is applied, and the volume of water removed as a function of the applied pressure is recorded to yield the relationship between soil water content and soil water potential.

For the plant to transport water from within the soil to inside the root, the water potential inside the root must be less than the water potential in the soil. The water potential inside the root can be decreased by increasing the root cell's solute concentration (Eq. 11.7). In saline soils, the water potential is also decreased by the increase in the soil's solute concentration, hence the plant must develop an even lower internal water potential to prevent water from leaving the roots. Water leaving the roots and desiccating the plant is not beneficial unless preserving vegetables is the goal (placing items in a strong brine solution will extract water and help preservation). Decreasing the water potential inside the roots by increasing the solute concentration requires energy to move ions across cell membranes, so this is costly for the plant.

The soil's *in situ* matric potential is traditionally measured with a **soil tensiometer** (Figure 11.6). This simple but useful instrument, probably developed during the start of the twentieth century, consists of a porous ceramic cup placed at the end of a slender water-filled tube (Or, 2001; Whalley, Ober and Jenkins, 2013). A vacuum gauge is connected to the sealed tube. As water moves from the tube through the porous cup into the soil, suction (tension) is measured by the vacuum gauge. Water can flow from the soil back through the cup until an equilibrium in water potential is reached and the matric potential is read on the vacuum gauge. Automated versions of the tensiometer are available, where electronic pressure transducers are used to measure the tension with the data recorded by electronic dataloggers. As an alternative to the use of porous cup tensiometers, soil properties that vary with water content, such as electrical resistance, have been used to estimate the matric potential, although this approach is questionable because of the variable relationship between water content and water potential (Campbell, 1988).

In Section 11.6.5 we see that the surface tension across the meniscus at the liquid–air interface is related to relative humidity, and this affects water potential. Therefore, if the relative humidity in the air within the soil pore space can be measured, the soil matric water potential can be calculated (see Eq. 11.24). To measure the relative humidity in the soil pore air space, a **thermocouple psychrometer** is commonly used. A psychrometer works on the principle that the difference between the ambient air temperature (the **dry-bulb temperature**; T_a °C) and the temperature of a thermocouple covered by a wick soaked in water (the **wet-bulb temperature**; T_w °C) can be used to calculate the vapor pressure. The **wet-bulb depression** (the difference between the dry- and wet-bulb temperatures) forms the basis of the vapor pressure calculation (e_a; kPa) based on the **psychrometric (or hydrometric) equation**:

$$e_a = e_s - \gamma P(T_a - T_w) \quad (11.11)$$

where e_s is the saturation vapor pressure (kPa), γ is the **psychrometer constant** ($\gamma = C_p / \lambda \sim 6.6 \times 10^{-4}$ °C^{-1}, the ratio of specific heat capacity C_p to the latent heat of vaporization λ), and P is the ambient air pressure (kPa). A **dew-point hygrometer**, where a thermocouple is cooled to the air's dew-point temperature by passing electrical current through the thermocouple junction, can be used to measure humidity within soils. Measurement of the humidity in the soil air space is difficult because of the high humidity, but possible if done carefully.

11.6.4 Gravitational Water Potential: ψ_g

The taller a column of water, the greater the energy required to raise it, known as the **gravitational water potential** ψ_g. From the hydrostatic equation (Eq. 11.2), the gravitational water potential for a 1-m change in height is:

$$\psi_g = P = \rho g h = 1{,}000\ \text{kg m}^{-3} \times 9.81\ \text{m s}^{-2} \times 1\ \text{m} \approx 10{,}000\ \text{kg m}^{-1}\ \text{s}^{-2}\ (0.01\text{MPa}).$$

Compared with the magnitude of the other water potential components, this value is small and therefore often neglected except for tall trees. For example, a tree 100 m tall would have a gravitational water potential of 1 MPa (in reference to the ground), which is about half the magnitude of the total water potential in a plant cell. ψ_g is always positive when height (h) is designated as positive above the reference plane. Thus ψ_g represents the pressure that the column of water exerts relative to the reference plane h. As ψ_g is working against the ascent of water, the other terms in the total water potential equation must offset the magnitude of ψ_g.

11.6.5 Liquid–Air Water Potential: ψ_a

The **liquid–air water potential** ψ_a develops where liquid water meets air within the leaves. Of critical importance to Cohesion–Tension Theory is the surface tension that develops where the liquid water inside the plant contacts the atmosphere to exchange carbon dioxide and water vapor. Located on leaves (typically on the underside) are small pores called **stomata** (Greek for "mouths"; singular **stoma**) where the exchange of gases with the atmosphere occurs (Figure 11.7).

The water potential in the two **guard cells** that flank the stomatal cavity is a fundamental control on carbon uptake and water vapor release between the plant and the atmosphere, and thus these cells play a major role in the biotic regulation of the Earth's climate system. The controls and importance of stomata in transpiration are covered in Chapter 12; here the relevant aspect related to water potential is that water flow through the entire plant vascular system is strongly controlled by the stomatal aperture.

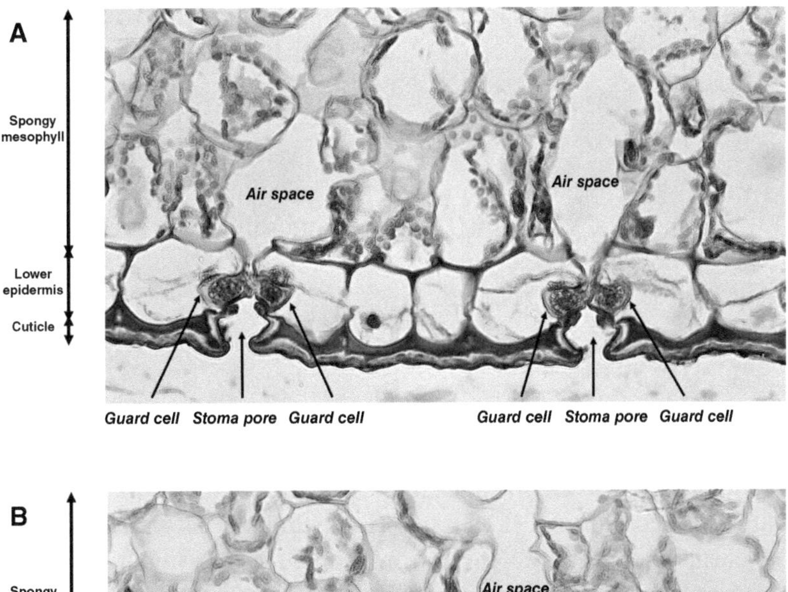

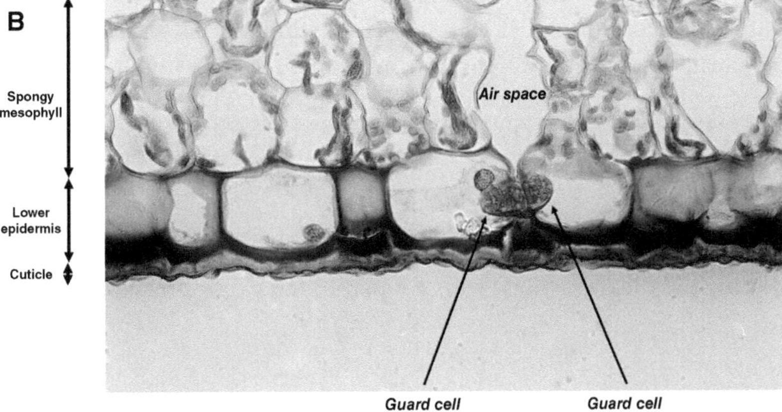

Figure 11.7 Cross-section of open (A) and closed (B) stoma in the epidermis of a succulent xerophyte leaf. Background image source: Berkshire Community College Bioscience Image Library, CC0, via Wikimedia Commons.

Relative humidity (h), the ratio of the actual (ambient) (e_a) to saturation vapor pressure (e_s), $h = e_a/e_s$, is connected to water potential through the **Ideal Gas Law** (Eq. 11.12) and the latent heat of vaporization (λ). An ideal gas is one in which molecules collide but do not interact, so all the internal energy is kinetic energy related to pressure (P; pascals = kg s^{-2} m^{-1}), volume (V; m^3), the number of moles (n; mol), and temperature (T; K). The relationship between these variables is expressed through the Ideal Gas Law and the **Universal Gas Constant** R (8.314 J mol^{-1} K^{-1} = 8.314 kg m^2 s^{-2} mol^{-1} K^{-1}):

$$PV = nRT\,. \tag{11.12}$$

The molar volume, V_m (m^3 mol^{-1}), is defined as the volume occupied by 1 mole of an ideal gas at standard temperature (0 °C; 273.15 K) and pressure (1 atmosphere; 101.3 kPa), and can be calculated by rearranging Eq. (11.12):

$$V_m = \frac{V}{n} = \frac{RT}{P} \tag{11.13}$$

resulting in the calculation that 1 mole of an ideal gas at standard temperature and pressure occupies a volume of 22.4 liters:

$$V_m = \frac{RT}{P} = \frac{8.314\ \text{kg m}^2\text{s}^{-2}\ \text{mol}^{-1}\text{K}^{-1} \times 273.15\,\text{K}}{101{,}300\ \text{kg s}^{-2}\text{m}^{-1}} = 0.0224\frac{\text{m}^3}{\text{mol}}\left(22.4\frac{\text{liters}}{\text{mol}}\right).$$

Assuming water vapor behaves as an ideal gas, substituting the definition of molar volume (Eq. 11.13) into the Ideal Gas Law (Eq. 11.12) and solving for pressure (in units of pascals, kg s^{-2} m^{-1}) gives Eq. (11.14):

$$\begin{aligned} V_m &= V/n;\ V = V_m n \\ P(V_m n) &= nRT \\ P &= \frac{RT}{V_m} \end{aligned}\,. \tag{11.14}$$

The connection between water potential and the latent heat of vaporization can be explained following Campbell and Norman (1998) and Monteith and Unsworth (2013). Imagine the work required to create a volume of water vapor from liquid water without any change in temperature (i.e., all the work is used to increase the volume of the gas). That work is equal to the latent heat of vaporization (λ; J kg^{-1}), and can be expressed as the integral of the change in volume (dV) of vapor pressure (e) from the initial or ambient vapor pressure (e_a) to the saturation vapor pressure (e_s):

$$\lambda = \int_{e_a}^{e_s} e\,dV\,. \tag{11.15}$$

Differentiating the Ideal Gas Law with P equal to the vapor pressure e gives:

$$dV = -\frac{nRT}{e^2}de\,. \tag{11.16}$$

Following conservation of energy, the sum of the work done in the system to change the vapor pressure as the volume increases ($e\,dV$) and the work done to change the volume as the vapor pressure increases ($V\,de$) is zero assuming there is no heat exchange with the surrounding air (an **adiabatic system**):

$$e\,\mathrm{d}V + V\,\mathrm{d}e = 0. \tag{11.17}$$

The differential of the Ideal Gas Law ($\mathrm{d}V$) when multiplied by e gives:

$$e\,\mathrm{d}V = -nRT\frac{1}{e}\mathrm{d}e. \tag{11.18}$$

Linking back to λ, substitution for $e\,\mathrm{d}V$ (Eq. 11.18) in the integral (Eq. 11.15) gives:

$$\lambda = \int_{e_\mathrm{a}}^{e_\mathrm{s}} e\,\mathrm{d}V = -nRT\int_{e_\mathrm{a}}^{e_\mathrm{s}} \frac{1}{e}\mathrm{d}e \tag{11.19}$$

where the values for n, R, and T are considered independent of volume so are moved outside of the integral. Applying the Log Rule for integration to Eq. (11.19) with e expressed as relative humidity gives:

$$\lambda = -nRT\ln\left(\frac{e_\mathrm{a}}{e_\mathrm{s}}\right). \tag{11.20}$$

Note that in this adiabatic system example, conservation of energy requires that $e\,\mathrm{d}V$ and $V\,\mathrm{d}e$ are equal but opposite in sign. Therefore, λ is always positive as it represents the work required to increase the vapor pressure (e) as the parcel expands ($\mathrm{d}V$). Conversely, the water potential across the liquid–air interface (ψ_a) representing the energy required to change state from liquid to gas is an energy deficit, a negative number representing the work required for evaporation to occur. Using the definition of molar volume in the Ideal Gas Law (Eq. 11.14) allows for nRT (kg m^2 s^{-2}) in Eq. (11.20) to be replaced with RT/V_m (kg s^{-2} m^{-1}; pascals) to provide equations to approximate λ in the correct units of J kg^{-1} (equal to m^2 s^{-2}) when divided by the density of liquid water (ρ_w; kg m^{-3}):

$$\lambda = \frac{1}{\rho_\mathrm{w}}\frac{RT}{V_\mathrm{m}}\ln\left(\frac{e_\mathrm{a}}{e_\mathrm{s}}\right) \tag{11.21}$$

and the equivalent liquid–air interface water potential ψ_a in correct units of pressure (kg s^{-2} m^{-1}; pascals):

$$\psi_\mathrm{a} = -\frac{RT}{V_\mathrm{m}}\ln\left(\frac{e_\mathrm{a}}{e_\mathrm{s}}\right). \tag{11.22}$$

Equation (11.22) can be used to calculate the water potential of plant tissue. A thermocouple psychrometer (or any type of hygrometer) can be used to calculate relative humidity ($e_\mathrm{a}/e_\mathrm{s}$) from measurements of the dry- and wet-bulb temperature inside a sealed chamber where the plant tissue sample has had time to equilibrate with the chamber air. The natural logarithm of $e_\mathrm{a}/e_\mathrm{s}$ is negative and approaches zero as $e_\mathrm{a}/e_\mathrm{s}$ approaches 1 (100% relative humidity). Therefore, when the relative humidity is 100% ($e_\mathrm{a}/e_\mathrm{s} = 1$), then $\psi_\mathrm{a} = 0$ Pa since $\ln(1) = 0$, and there is no vapor pressure gradient to drive evaporation. As the air dries and $e_\mathrm{a}/e_\mathrm{s}$ decreases, ψ_a decreases following $\ln(e_\mathrm{a}/e_\mathrm{s})$, and the sharp decrease in ψ_a results in a large decrease in water potential and the final "pull" for water's ascent in vegetation. If the relative humidity is 98% (humid conditions in the plant material), and the air temperature is 20 °C (293.15 K), Eq. (11.22) gives:

$$\psi_\mathrm{a} = -\frac{RT}{V_\mathrm{m}}\ln\left(\frac{e_\mathrm{a}}{e_\mathrm{s}}\right) = \frac{8.314\,\mathrm{kg\,m^2 s^{-2}} \times 293.15\,\mathrm{K}}{1.8 \times 10^{-5}\,\mathrm{m^3\,mol^{-1}}}\ln(0.98) = -2.74\,\mathrm{MPa}.$$

This example shows the important role humidity plays in the water potential, hence water movement and transpiration. The magnitude and controls of transpiration, a major source of water vapor to the atmosphere, are explained in Chapter 12.

Rearrangement of Eqs. (11.21) and (11.22) can yield useful and practical equations. Equation (11.21) can be rearranged to provide an approximation of the vapor pressure, e_a:

$$e_a = e_s \exp\left(\lambda \rho_w \frac{V_m}{RT}\right) \tag{11.23}$$

and Eq. (11.22) can be rearranged to provide an approximation of relative humidity, h:

$$h = \exp\left(\frac{\psi_a V_m}{RT}\right). \tag{11.24}$$

Equation (11.24) is useful to estimate humidity in situations where water potential is easier to measure than humidity. Soil humidity is often much more difficult to measure than the soil water matric potential. If a soil tensiometer (Figure 11.6) is used to measure the soil water matric potential, Eq. (11.24) can be used with ψ_m in the place of ψ_a to calculate the soil humidity. Equation (11.24) also provides an efficient way to calibrate humidity sensors. A solution can be prepared with the solute water potential calculated using Eq. (11.7). The humidity sensor can be placed in a sealed chamber above this solution and calibrated against h calculated using Eq. (11.24), using the ψ_s value in place of ψ_a.

11.7 SUMMARY

The availability of a sufficient, reliable supply of liquid water is critical for terrestrial vegetation's function and form. Photosynthesis, the conversion of solar radiation to chemical energy in the creation of new molecules, requires energy from the Sun and liquid water. Water provides additional energy that is used to capture carbon from the atmosphere, supplies electrons to create new covalent bonds in the formation of carbohydrates, and provides a means for these compounds to be transported through the plant. Water also provides the plant with physical support and the ability to grow tall to gain access to solar radiation. Temperature regulation and freezing tolerance are also provided by internal plant water, made possible by the large specific heat capacity and solubility properties of liquid water.

The unique properties of water coupled with plant physiological adaptations explain the ascent of water in vegetation to heights greater than 10 m. In narrow 20-μm-diameter tubes, the typical diameter of xylem tissue, water can ascend to a height of almost 1.5 m owing to the cohesive and adhesive properties and high surface tension of liquid water alone. Plants have evolved adaptations to xylem tissue such as thick walls, lignin organic polymer cell lining, and tracheid cells, all designed to aid in the ascent of water. The Cohesion–Tension Theory presents a framework for how water ascends to the tops of tall trees. The theory describes how it is not only the cohesion between water molecules within xylem, but also the large surface tension that develops at the liquid–air interface in the stomata, that explains how water can ascend above 10 m. Measurement of xylem water potential made with pressure chambers in tall trees has shown that the required tension sufficient to explain the decrease in water pressure required does exist. Some critics of the Cohesion–Tension Theory, however, point out that xylem water tensions greater than roughly 1 MPa (~10 atmospheres) as indicated by some pressure chamber measurements are not possible since the cohesive force between water molecules in xylem tissue cannot tolerate tensions beyond 1 MPa before cavitation occurs. Theory and laboratory studies, however, show that liquid water at room temperature can withstand cavitation until −25 MPa (−247 atmospheres), so the debate continues.

Water potential indicates the work required or released when water moves from one location to another. The total water potential within cells is equal to the sum of the solute and turgor water potentials. The solute water potential decreases as the solute concentration increases, and it is always negative. Increasing or decreasing the solute concentration inside relative to outside a cell is an effective means for water to move across a semipermeable cell membrane. The turgor water potential is a positive pressure resulting from the flow of water into cells. Turgor pressure gives the plants shape and structure, and wilting is indicative of water stress and the loss of turgor pressure.

At the scale of individual plants, especially tall trees, the total water potential is equal to the sum of the solute, turgor, matric, gravitational, and liquid–air water potential. The matric water potential develops from liquid water being bound to soil particles and within the soil pore space itself. The work required for the plant to lower the water potential in the root cells to a value less than in the soil, so that water can flow into the roots, varies with soil properties such as texture and water content. A soil water retention curve gives the relationship between soil water content and matric potential and is often used to quantify the permanent wilting point. The gravitational water potential refers to the positive water potential that develops between two vertically displaced locations and is relatively small at 0.01 MPa per meter. The liquid–air water potential develops in the stoma where liquid water is first exposed to air, and it is an important component of the Cohesion–Tension Theory. The liquid–air water potential is derived from the connection between potential energy and the latent heat of vaporization and is useful to calculate not only the water potential but also relative humidity.

11.8 QUESTIONS

11.1 Several of the critical roles liquid water serves in terrestrial vegetation's function and form were described. For each of these, describe how one (or more) of the specific unique physical or chemical properties of water makes these critical roles possible.

11.2 In the midlatitudes, the tree line altitude is roughly 3,500 m above sea level where the air pressure is approximately 66 kPa. What is the maximum height of a column of water that atmospheric pressure can support at this altitude? Does your calculation explain the absence of trees at this altitude? Why or why not?

11.3 What are the characteristics of xylem tissue that aid in the ascent of water in tall vegetation?

11.4 The Cohesion–Tension Theory provides a well-known explanation for the ascent of water in tall vegetation. Based on the evidence provided in this chapter and your own research, provide an argument for or against this theory.

11.5 What is the expected relative humidity in the air above a 0.005 mole per liter glucose solution at standard temperature and pressure?

11.6 Explain why a measure of water potential in the soil is a better metric to explain plant–water relations than water content in the soil.

11.7 Of the water potential terms in the total water potential (Eq. 11.10), the liquid–air water potential is often the smallest (i.e., a large but negative number). Explain why.

11.8 Species such as mangroves growing in standing water often have similar very small xylem pressure potentials (i.e., under high tension) compared with species such as mesquite growing in deserts. Explain why.

11.9 How do plants growing in saline soils prevent desiccation? Use an example calculation of the solute water potential for one plant growing in saline soils and one growing in nonsaline soils to support your answer.

11.10 Earth is comprised of both vascular and nonvascular terrestrial plant species and associated ecosystems. As discussed, there are costs associated with the development and energy required to transport water vertically. Is this cost worth it? Discuss in terms of the evolution, adaptation, and resilience of nonvascular compared with vascular species.

REFERENCES

Boyer, J. S., James, R. A., Munns, R., Condon, T. A. G. and Passioura, J. B. (2008) 'Osmotic adjustment leads to anomalously low estimates of relative water content in wheat and barley', *Functional Plant Biology*, 35, pp. 1172–1182. doi: 10.1071/FP08157.

Brooks, J. L. (2004) 'The Cohesion–Tension Theory', *New Phytologist*, 163, pp. 451–452. doi: 10.1111/j.1469-8137.2004.01160.x.

Campbell, G. S. (1988) 'Soil water potential measurement: An overview', *Irrigation Science*, 9, pp. 265–273. doi: 10.1007/BF00296702.

Campbell, G. S. and Norman, J. M. (1998) *An Introduction to Environmental Biophysics*. Second ed. Springer.

Caupin, F. and Herbert, E. (2006) 'Cavitation in water: A review', *Comptes Rendus Physique*, 7, pp. 1000–1017. doi: 10.1016/j.crhy.2006.10.015.

Coriolis, G. (1829) *Du calcul de l'effet des machines ou considérations sur l'emploi des moteurs et sur leur évaluation, pour servir d'introduction à l'étude spéciale des machines*. Paris, Carilian-Goeury, Libraire. Des Corps Royaus des Ponts et Chaussees et des Mines, Quai des Augustins, No. 41.

Cusuman, J. C. (2001) 'Osmoregulation in plants: Implications for agriculture', *American Zoologist*, 41, pp. 758–769. doi: 10.1668/0003-1569(2001)041[0758:oipifa]2.0.co;2.

Davitt, K., Arvengas, A. and Caupin, F. (2010) 'Water at the cavitation limit: Density of the metastable liquid and size of the critical bubble', *EPL* 90. doi: 10.1209/0295-5075/90/16002.

Dixon, H. H. and Joly, J. (1894) 'On the ascent of sap', *Philosophical Transactions of the Royal Society of London B*, 186, pp. 563–576.

Hüsken, D., Steudle, E. and Zimmermann, U. (1978) 'Pressure probe technique for measuring water relations of cells in higher plants', *Plant Physiology*, 61, pp. 158–163. doi: 10.1104/pp.61.2.158.

Jones, R. G. W. and Gorham, J. (1983) 'Osmoregulation', in Lange, O. L. et al. (eds.) *Physiological Plant Ecology III. Encyclopedia of Plant Physiology (New Series)*. 12th ed. Springer, pp. 35–58. doi: 10.1007/978-3-642-68153-0_3.

Lewis, A. M. (1988) 'A test of the air-seeding hypothesis using sphagnum hyalocysts', *Plant Physiology*, 87, pp. 577–582. doi: 10.1104/pp.87.3.577.

Liu, Q., Luo, L. and Zheng, L. (2018) 'Lignins: Biosynthesis and biological functions in plants', *International Journal of Molecular Sciences*, 19. doi: 10.3390/ijms19020335.

Melcher, P. J., Meinzer, F. C., Yount, D. E., Goldstein, G. and Zimmermann, U. (1998) 'Comparative measurements of xylem pressure in transpiring and non-transpiring leaves by means of the pressure chamber and the xylem pressure probe', *Journal of Experimental Botany*, 49, pp. 1757–1760. doi: 10.1093/jxb/49.327.1757.

Monteith, J. L. and Unsworth, M. H. (2013) *Principles of Environmental Physics*. Fourth ed. Academic Press.

Or, D. (2001) 'Who invented the tensiometer?', *Soil Science Society of America Journal*, 65, pp. 1–3. doi: 10.2136/sssaj2001.6511.

Scholander, P. F., Bradstreet, E. D., Hemmingsen, E. A. and Hammel, H. T. (1965) 'Sap pressure in vascular plants', *Science*, 148, pp. 339–346. doi: 10.1126/science.148.3668.339.

Sillett, S. C., Van Pelt, R., Koch, G. W. et al. (2010) 'Increasing wood production through old age in tall trees', *Forest Ecology and Management*, 259, pp. 976–994. doi: 10.1016/j.foreco.2009.12.003.

Simard, M., Pinto, N., Fisher, J. B. and Baccini, A. (2011) 'Mapping forest canopy height globally with spaceborne lidar', *Journal of Geophysical Research: Biogeosciences*, 116, pp. 1–12. doi: 10.1029/2011JG001708.

Sperry, J. S. (2003) 'Evolution of water transport and xylem structure', *International Journal of Plant Sciences*, 164, pp. 115–127.

Tomos, D. (2000) 'The plant cell pressure probe', *Biotechnology Letters*, 22, pp. 437–442. doi: 10.1023/A:1005631921364.

Turner, N. C. (1981) 'Techniques and experimental approaches for the measurement of plant water status', *Plant and Soil*, 58, pp. 339–366. doi: 0.1007/BF02180062.

Tyree, M. T. and Hammel, H. T. (1972) 'The measurement of the turgor pressure and the water relations of plants by the pressure-bomb technique', *Journal of Experimental Botany*, 23, pp. 267–282. doi: 10.1093/jxb/23.1.267.

Wei, C., Steudle, E. and Tyree, M. T. (1999) 'Water ascent in plants: Do ongoing controversies have a sound basis?', *Trends in Plant Science*, 4, pp. 372–375. doi: 10.1016/S1360-1385(99)01466-1.

Whalley, W. R., Ober, E. S. and Jenkins, M. (2013) 'Measurement of the matric potential of soil water in the rhizosphere', *Journal of Experimental Botany*, 64, pp. 3951–3963. doi: 10.1093/jxb/ert044.

Zimmermann, U., Schneider, H., Wegner, L. H. and Haase, A. (2004) 'Water ascent in tall trees: Does evolution of land plants rely on a highly metastable state?', *New Phytologist*, 162, pp. 575–615. doi: 10.1111/j.1469-8137.2004.01083.x.

12 Transpiration

Key Learning Objectives

After reading this chapter you will be able to:

1. Distinguish between transpiration and evaporation, and abiotic and biotic evaporation.
2. Explain how water loss and carbon uptake from microscopic stomatal pores on leaves influence global-scale water and carbon cycles.
3. Describe the techniques used to measure transpiration at the leaf, tree, and canopy scales.
4. Summarize the development and application of modeling techniques used to isolate transpiration from evaporation.
5. Predict how increases in temperature and carbon dioxide concentrations will affect transpiration, plant productivity, and water use efficiency.

12.1 Introduction

As we have seen in Chapter 11, water serves several critical roles in terrestrial vegetation's function, for which water must be stored, replenished, and transported from the roots to the leaves. Water is evaporated within the leaves through the process of transpiration. Knowing the transpiration rate and controls is important not only from a plant physiology perspective, but also from the perspective of the water cycle in terrestrial vegetated ecosystems, including agricultural systems.

The factors controlling the transpiration rate differ from those that control evaporation from other surfaces. Transpiration, or biotic evaporation, is influenced by variables including species, leaf area, and phenology (life cycle timing) and responds to changes in the ambient meteorological conditions (e.g., temperature, solar radiation, wind, humidity) and the availability of water in the soil (e.g., soil moisture and water potential). The location of transpiration water loss and carbon dioxide uptake is in the stomata, microscopic pores located on leaf surfaces. The ability of plants to add water vapor to and remove carbon dioxide from the atmosphere has a significant influence on global water and carbon cycles.

Since evaporation occurs from multiple sources in terrestrial vegetated ecosystems, the transpiration contribution to the total ecosystem evaporation needs to be isolated. Several measurement and modeling approaches have been developed to isolate transpiration from other water vapor sources (abiotic evaporation: soils, streams, rivers, lakes). Transpiration can be measured at the leaf, tree (plant), or stand (forest) scales, and descriptions of these methods are included. The popular eddy covariance method is described, as well as novel approaches used to isolate transpiration using water vapor isotopes.

Theoretical modeling-based approaches to partition transpiration from evaporation originating from the observations of John Dalton and Howard Penman (Chapter 7) for wet, well-watered surfaces are described. John Monteith modified Penman's evaporation equation to include the active role of vegetation in evaporation, resulting in the well-known Penman–Monteith evaporation model. The development, performance, and subsequent modifications of the model are described.

This chapter concludes with a summary of the key variables that control transpiration, with a focus on variables that are changing rapidly, namely temperature and atmospheric carbon dioxide concentration. The atmosphere's demand for water vapor increases with warming, and higher temperatures are often coupled with lower humidity and a decrease in soil moisture, placing additional demands on the transpiration water loss. Since the path of water loss and carbon uptake is through the stomata, how these are coupled through the concept of water use efficiency is discussed. Observations of how water use efficiency has changed with climate change are presented.

12.2 The Process of Transpiration

The terms **evaporation** and **transpiration** both refer to the change of state of water from liquid to gas, so why are different words used to describe the same process? In terrestrial vegetated ecosystems, it is often desirable to know the portion of evaporation that originated from water sources such as lakes or soils relative to the water that has evaporated after moving through vascular plants. The former can be thought of as **abiotic evaporation** and the latter **biotic evaporation**, since the former has no biological control on the evaporation rate whereas the latter does. Since both the source of water for evaporation (e.g., roots have access to deeper water than standing surface water) and the controls of the evaporation rate (e.g., plants can actively regulate water loss) are different for vegetated compared with nonvegetated surfaces, so are the magnitude and timing of the evaporative water loss. Evaporation that occurs from within the live tissue of vascular vegetation is referred to as **transpiration** (T) to distinguish it from **evaporation** (E) that occurs from all other locations including evaporation of intercepted precipitation on (not within) vegetation. The combination of evaporation and transpiration is commonly referred to as **evapotranspiration** (ET), a word some believe should not be used since it neglects the fact that evaporation occurs regardless of location (Miralles et al., 2020). Here, the term transpiration is used to define water that has evaporated from inside the living tissue of vascular plants.

Transpiration occurs within numerous microscopic pores located on the leaf surface called stomata. Most stomata are located on the underside of the leaf and flanked by two **guard cells** that control the aperture through changes in turgor pressure (see Chapter 11 for details on stomatal pores and turgor pressure). **Marcello Malpighi** (1628–1694), who pioneered the use of the microscope, is credited as the first person to observe and describe stomata in 1671, but he did not understand their purpose. **Stephen Hales** (1677–1761) later described the process of transpiration in his book *Vegetable Staticks* (Hales, 1727) where he recognized the importance of the "*…innumerable little pores of the leaves, which are plainly visible with the microscope*" (p. 153). Microscope observations have revealed the characteristics and sheer quantity of stomata. In tomato plants (*Solanum lycopersicum*), using silicon rubber leaf impressions and 1,000× magnification, Fanourakis et al. (2015) measured an average stomatal pore length, depth, and width of approximately 18, 7, and 2 μm, respectively, and an area of 30 μm^2. The **stomata density**, the number of cells per square millimeter of leaf area, varies with species (and therefore location and climate). Tree species such as poplars (*Populus*) have a stomata density of around 150 stomata per mm^2, coniferous pine species (*Pinus*) 430, and deciduous oak species (*Quercus*) 600 (Woodward and Kelly, 1995).

You might think that the water vapor added to the atmosphere from transpiration from such small pores would be negligible. Despite their microscopic size, however, there are many stomata on a leaf, and many leaves on plants, so the number of stomata quickly becomes large at the ecosystem scale. With hundreds of stomata per square millimeter, an oak leaf (for example) with a one-sided surface area of 10 cm^2 would have 600,000 stomata per leaf (600 stomata $mm^{-2} \times 1{,}000\ mm^2$). Estimating that a large oak tree has somewhere between one and two million leaves, there are between 600,000 and 1,200,000 million stomata on one oak tree. When multiplied by the number of trees and other vascular plants on Earth, the number of stomata actively exchanging and regulating water vapor release and CO_2 uptake is astronomical.

The stomata density varies not only with species, but also with climate and location. In particular, the atmospheric CO_2 concentration has been found to influence stomata density. Studies examining changes in stomata density under controlled conditions where the ambient CO_2 concentration was increased, and studies examining macrofossil leaves, have found that stomata density decreases as CO_2 concentrations increase (Woodward, 1987; Woodward and Kelly, 1995; Royer, 2001). Plants can gain the same (or more) CO_2 while losing less water through transpiration, thus maintaining a constant **water use efficiency**, defined as the mass of CO_2 gained relative to the mass of H_2O lost through the stomata. Further evidence of the importance of and connection between stomata and atmospheric CO_2 is apparent in the "sawtooth" pattern of annual changes in atmospheric CO_2 concentration measured at locations such as Mauna Loa, Hawaii. The annual variation in the atmospheric CO_2 concentration is roughly 10 parts per million (ppm) from the minima in October to the maxima in May, illustrating the depletion in the atmospheric CO_2 reservoir from carbon uptake through stomata for photosynthesis.

Several studies have attempted to quantify how much water vapor enters the atmosphere through transpiration. Although transpiration varies with numerous factors including species, location, time, and climate, studies show that in many terrestrial regions most water vapor enters the atmosphere through plants, not open-water sources such as rivers or lakes. As discussed in Chapter 7, an estimated 80% of the 67,900 km^3 of water evaporated from the global land surface is transpired water (Miralles et al., 2011). Estimates of the global terrestrial transpiration relative to the total evaporation range from a conservative 61% (Schlesinger and Jasechko, 2014) to as high as 90% (Jasechko et al., 2013). Some of the variation in these estimates arises from spatial sampling issues and uncertainty with the methods used to partition transpiration from total evaporation, as discussed next. Regardless, transpiration contributes the majority of terrestrial water vapor, and is of critical importance to regional and global water cycling.

12.3 Partitioning Transpiration from Evaporation

As discussed in Chapter 7, natural sources of water for evaporation from terrestrial surfaces include a combination of standing water on or near the surface and water that has been transpired or intercepted by vegetation. There has long been recognition of the need to understand the contribution of evaporation from these sources, and attempts to do so span centuries (Blyth and Harding, 2011), yet the task remains an important challenge to this day as changes in climate, land use, and land cover are affecting the water cycle. Approaches to isolate transpiration from the total ecosystem evaporation can be grouped into measurements made at the leaf, tree, or canopy scale.

12.3.1 Leaf Measurements

Transpiration from individual leaves can be directly measured using an instrument aptly named a **porometer** (Figure 12.1). By placing an area of leaf (a_l; cm^2) in a **cuvette** (enclosed chamber), the transpiration rate (T; $\mu g\ cm^{-2}\ s^{-1}$) can be calculated by measuring the volume flow of dry air (F; $cm^3\ s^{-1}$) required to maintain a constant relative humidity inside the cuvette (LI-COR, 1989). Transpiration is calculated from the dry air flow required to maintain a steady-state condition of constant humidity inside the cuvette, the area of the exposed leaf in the cuvette, and the difference in **water vapor density** (or mole fraction or vapor pressure) between the air inside the cuvette (ρ_c; $\mu g\ cm^{-3}$) and in the dry air stream entering the cuvette (ρ_a; $\mu g\ cm^{-3}$) (Eq. 12.1):

$$T = (\rho_c - \rho_a)\frac{F}{a_l}. \tag{12.1}$$

Since a small area on one side of an individual leaf is measured, using a porometer to measure tall vegetation or ecosystems comprising several species is a challenge. In a forest, transpiration

Figure 12.1 A cuvette placed on Cabernet Sauvignon leaves in an Australian vineyard. By measuring transpiration, the stomatal conductance can be calculated. This instrument can also measure photosynthesis.
Photo credit: CSIRO, CC BY 3.0, https://creativecommons.org/licenses/by/3.0, via Wikimedia Commons.

can vary greatly between species and even within one tree depending on where the leaf is located, since the variables influencing transpiration (see Section 12.5) can vary significantly with height. At a given time, measurements made on leaves at the bottom of the tree will almost certainly be very different than those made at the top, owing to differences in leaf structure, physiology, and ambient conditions such as solar radiation, wind, temperature, and humidity. Canopy access towers are required to reach the different levels in tall forest canopies to properly quantify this variation. As transpiration and its controls vary with species and age, measurements should be made on all the major species and age classes: this quickly becomes a major sampling problem.

Nevertheless, the porometer has been successfully used in many field-based measurement studies. Consistent, multiple measurements under a variety of ambient conditions have been used to develop and calibrate transpiration models. When the **Leaf Area Index** is measured (the area of leaves per area of ground) and the fraction of cover for the dominant species is known, this together with porometer-based field measurements can be used to scale-up and provide an understanding of the ecosystem-scale transpiration magnitude and controls. This approach was used to conclude that transpiration accounted for 80 percent of the total water vapor flux in a **willow–birch shrub forested** Subarctic wetland (Blanken and Rouse, 1994, 1995, 1996).

12.3.2 Tree Measurements

To help overcome the sampling issues surrounding leaf-level measurements, techniques have been developed to provide continuous transpiration measurements at the scale of an individual tree. The measurement principle is based on the application of heat to the xylem tissue and the subsequent measurement of changes in stem temperature from which the transpiration rate can be calculated as the water is transported upwards through that xylem tissue. There are several of these **sap flow measurement** techniques that have been developed and tested based on where stem temperature is measured and how heat is applied (see review by Smith and Allen, 1996). In a database of sap flow measurements made across globally distributed locations and 2,714 vegetation species (the majority were tree species), the **heat (or thermal) dissipation method** was the most common (66%), followed by the **trunk section heat balance** (16%) and the **compensation heat pulse** (8%) methods (Poyatos et al., 2021) (Figure 12.2).

The **heat** or **thermal dissipation method** developed by Granier (1985) measures temperature using two **thermocouple** (or **thermistors**) probes inserted into the stem's xylem tissue, one located roughly 10 cm above the other. In what is known as the **Seebeck effect**, thermocouples produce a small voltage proportional to temperature when the junctions of two dissimilar metals are at different temperatures. Thermistors produce a resistance in a small electric circuit that is proportional to temperature. Located slightly above the lower thermocouple is a heating element. As the sap flows upwards through the xylem tissue, a difference between the sap temperature measured by the lower and the warmer upper thermocouple is created by the warmed sap flowing past it. The temperature difference is dependent on the rate of sap flow, with high sap flow rates corresponding to a decrease in the temperature difference since the heat is dissipated

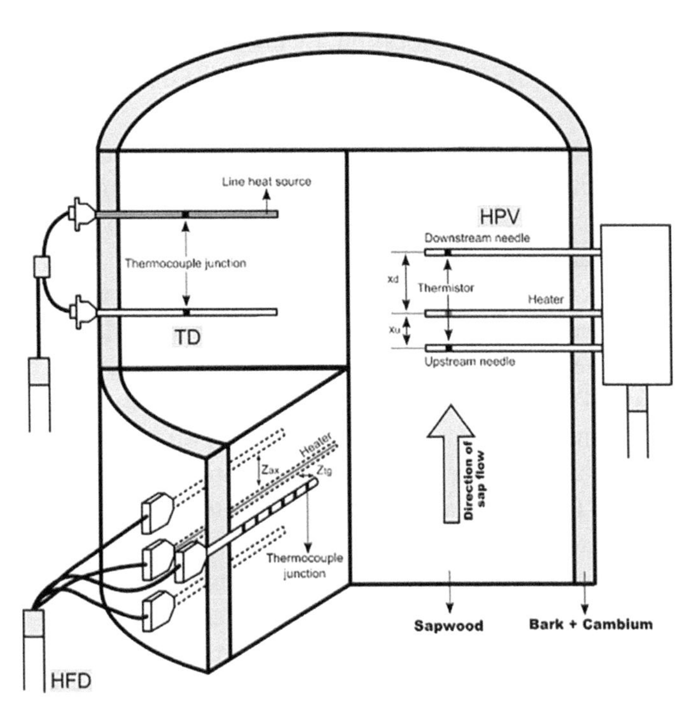

Figure 12.2 Popular methods used to calculate individual-plant (tree) transpiration using the sap flow technique of thermal dissipation (TD), heat field deformation (HFD), and heat pulse velocity (HPV). Distances between the HPV heater and downstream/upstream needle sensors are Xd and Xu, respectively, and distances between the heater and upper and lower thermocouples used in the HFD method are Zax and Ztg, respectively. Reproduced with permission from Steppe et al. (2010).

more rapidly. As with other sap flow measurements, the sap's density and cross-sectional area of sapwood must be known to calculate transpiration. This method is best used in established trees large enough for drilling and inserting probes.

For smaller-diameter trees (as small as 4 mm) and nonwoody herbaceous species where placement of probes in the xylem is not possible, the **trunk section (or trunk sector) heat balance method** can be used (Smith and Allen, 1996). Heat is applied to the circumference of a stem section, and the applied heat is balanced by (equal to) the sum of the vertical and radial heat conducted by the stem and heat uptake by the vertically moving sap. Thermocouples are used to measure radial gradients in temperature away from the heater, and together with the known thermal conductivity of the stem and heater material, this allows the vertical and radial heat flows to be calculated. Sap flow can be calculated as the residual of the heat balance equation.

The **compensation heat pulse method** measures sap flow from the application of a pulse of heat rather than constant heating. Vertically separated thermocouples and a heating element are used, but the temperature is measured at several locations above and below the heater probe. Brief (1–2 second) pulses of heat are applied, and the temperature measurements are used to calculate the velocity of the heat pulse as it is conducted through the sap. Sap velocity is equal to the distance between temperature measurement locations divided by the time required for the heat pulse to reach each location. By integrating the sap's velocity measurements across the cross-sectional sapwood area, the mass flow rate of the sap can be calculated (Smith and Allen, 1996).

Regardless of which sap flow measurement technique is used, the fundamental assumption is that the sap flow rate directly corresponds to the transpiration rate of the entire tree. This assumption is not always true, especially in large trees with a large internal water storage capacity such as those found in tropical forests. As the upper portions of the canopy are exposed to solar radiation at dawn, which triggers the onset of photosynthesis, transpiration begins with water initially supplied from internal storage. Sap flow measured at the base of these trees may therefore be delayed, despite the occurrence of transpiration. This delay can be significant. In a 35-m-tall tropical forest in Panama, the water stored within the trees accounted for roughly 10 percent of the total daily transpiration, and the lag between crown transpiration and basal sap flow varied from 15 minutes after sunrise to 45 minutes during peak transpiration (Meinzer, James and Goldstein, 2004). In a boreal aspen forest in central Canada, Hogg et al. (1997) compared two sap measurement techniques (thermal dissipation and heat pulse) to the eddy covariance method (see below). Owing to water storage, the 20-m-tall thin boreal aspen trees exhibited a 1-hour delay between transpiration measured by the eddy covariance method and the sap flow measurements.

In addition to this potential delay in sap flow and the practical difficulties involved with heating and accurately measuring temperature inside trees, spatial scaling from individual trees to forests is required to understand forest ecosystem response. Typically, fewer than 10 trees at a study site have sap flow measurements, but by careful selection of the trees to be instrumented (e.g., individuals that are representative of the larger forest stand), the errors associated with upscaling can be minimized. Even in a relatively homogeneous forest stand, however, spatial scaling is a concern since the microclimate can vary considerably. In the boreal aspen forest in central Canada, Hogg et al. (1997) found similar diurnal and seasonal trends and relationships between transpiration and the vapor pressure deficit regardless of the method used to measure transpiration. However, there were differences in transpiration rates due to phenology and physiological function variations amongst clonal species.

12.3.3 Forest Measurements: Eddy Covariance

An advantage of a method that samples a broad area of vegetation is that upscaling issues are minimized. Continuous measurements to estimate transpiration across distances spanning hundreds of meters or even kilometers are possible but are not immune to practical limitations and theoretical assumptions. One method that is especially well suited to forests is the **eddy covariance** (**eddy correlation**) method.

The term **eddy** describes the turbulent swirls of various dimensions and characteristics in the atmosphere that can transport energy and mass between the surface and atmosphere. These eddies can vary in diameter from less than millimeters to hundreds of meters. In general, large eddies persist longer but become smaller as they cascade down to smaller sizes as kinetic energy is lost. This understanding is based on early observations of turbulence around cumulus clouds from aircraft flights and ground-based drawings of cloud observations (see page 66 in Richardson, 1922). The term **covariance** is used to note that the method is based on the statistical variance between the measured vertical wind speed and a **scalar quantity** (a quantity that has magnitude but not direction) that is transported by these eddies, such as the mass of water vapor or carbon dioxide.

The eddy covariance method requires accurate high-frequency (10 or 20 Hz) measurements of the **vertical wind speed** (w; m s^{-1}) above the surface of interest. An **anemometer** is an instrument used to measure wind speed, and a traditional **cup anemometer** consists of three or four small hemispherical cups oriented horizontally, connected to a vertical shaft. The rotation rate is proportional to the **horizontal wind speed** (u; m s^{-1}). A **vane anemometer** consists of a propeller rotating vertically when attached to a vane to keep the propeller oriented into the wind. Both the cup and vane anemometers measure horizontal wind speed only; neither the **crosswind speed** component (v; m s^{-1}), perpendicular to the direction of u, nor the vertical wind speed (w; m s^{-1}) required for the eddy covariance method. To measure u, v, and w, cups or vanes can be aligned along all three coordinate planes. Owing to issues including the threshold wind speed required to initiate cup or propeller movement (typically 0.5 to 1.0 m s^{-1}), friction created by bearings, and overestimating wind speed during gusty conditions when momentum keeps the cups or vanes spinning even after the wind has stopped, these mechanical designs lack the fast response time required to resolve the small, quickly moving eddies. Thus, errors in flux measurements are significant (on the order of 25 percent) when using a cup or propeller anemometer unless proper corrections are applied (McBean, 1972).

Motivated by the need to address these limitations, Kaimal and Businger (1963) developed a design based on the transmission of sound waves that could be used for eddy covariance flux measurements, including calculation of the latent heat flux (evaporation). These **sonic anemometers** calculate wind speed by measuring how fast a pulse of **ultrasonic sound waves** travels between pairs of sensors (sonic transducers) placed between 10 and 20 cm apart (Figure 12.3).

Figure 12.3 Sonic anemometer (background instrument with three pairs of opposing transducers) together with closed-path infrared gas analyzer (IRGA) air inlet (air inlet visible in front of the lower arm of the sonic anemometer) and open-path IRGA (foreground) above a subalpine Colorado forest. Photo credit: P. D. Blanken.

The sound wave's transit time (t; s) depends on the sensor separation (d; m), the speed of sound in calm air (C; m s^{-1}), the cosine of the deviation of the path due to horizontal wind (for vertically separated sensors) (α; degrees), and the wind vector component being measured (w for vertically separated sensors) (Kaimal and Businger, 1963) (Eq. 12.2):

$$t = \frac{d}{C\cos(\alpha) + w}. \tag{12.2}$$

For the sound pulse moving downward, w is subtracted, not added, in Eq. (12.2). Based on the difference between the transit times for the sound pulse moving upward (t_u) and downward (t_d), w is calculated as a residual, since this difference depends only on w (primes indicate fluctuations from the means noted by the overbar) (Eq. 12.3):

$$t'_d - t'_u = \frac{2dw'}{\overline{C^2}}. \tag{12.3}$$

Since the speed of sound varies with air temperature, the sum of t_u and t_d can be used to calculate temperature. A pair of sonic transducers oriented vertically is used to measure vertical wind speed (w), or horizontal wind speed (u) when placed in a horizontal orientation. The addition of a second pair of sonic transducers in a horizontal configuration allows for measurement of the crosswind component (v), and when three pairs are used, all three wind components can be measured. These are known as 1D (dimensional), 2D, or 3D depending on whether one, two, or all three dimensions of the wind velocity are measured. In contrast to mechanical anemometers, sonic anemometers can measure wind speed at resolutions of 0.01 m s^{-1} and can sample at 20 Hz or faster, providing the sampling resolution and speed required to resolve fast, small-scale turbulence (eddies) that can transport heat or mass. One significant issue with sonic anemometers is that water on the transducers or precipitation falling in the transducer path often makes the measurements unusable, as the water interferes with the signal between sensor pairs.

In addition to measurements of w (w is a **vector quantity** since it has both magnitude and direction) provided by a sonic anemometer, the scalar quantity of mass (e.g., water vapor or carbon dioxide) or energy (e.g., heat or momentum) the eddy is transporting also needs to be measured at the same location and frequency. As mentioned, air temperature can be calculated from the sonic anemometer's measurements and used together with w to calculate the sensible heat flux (see Eq. 12.6). The transit time for the ultrasonic sound pulses, however, does not affect the measurement of w but does affect air temperature calculations through changes in the ratio of vapor pressure to atmospheric pressure (Kaimal and Businger, 1963). Empirical corrections have been developed to account for water vapor's effect on air temperature calculations, using a fast-response fine-wire thermocouple located in the sonic anemometer's path to directly measure air temperature (e.g., Schotanus, Nieuwstadt and De Bruin, 1983; Burns et al., 2012).

The density of water vapor (ρ_W; kg m^{-3}) and carbon dioxide (ρ_C; kg m^{-3}), required scalar measurements for eddy covariance measurements of the fluxes of water vapor and carbon dioxide respectively, can be measured using an **infrared gas analyzer (IRGA)**. A summary of the eddy covariance method and IRGA operation is provided by Burba (2013). Both H_2O and CO_2 are strong absorbers of **infrared radiation** (both are greenhouse gases) even in small quantities, so measuring the attenuation of infrared radiation in an air sample containing H_2O or CO_2 can be used to measure the gas concentrations. There are two general types of infrared gas analyzers: **open path** and **closed path** (Figure 12.3). In an open-path IRGA, the air is free to pass through the space (typically 15–20 cm) between a broadband infrared radiation source at one end and a detector at the other. Typically, a spinning wheel or "chopper" with filter windows rotates at high speed in front of the infrared radiation source. One of the filters in the chopper allows only wavelengths where CO_2 absorbs strongly, around 4.25 μm, to pass, and one filters to allow wavelengths where H_2O absorbs strongly, around 2.59 μm, to pass. Located between the CO_2 and H_2O filters are reference filters where neither CO_2 nor H_2O is absorbed by the infrared radiation. By comparing the ratio of infrared radiation transmitted between the source and detector as radiation passes through the CO_2 or H_2O filter to the reference filter, we can calculate the densities of CO_2 and H_2O. This open-path design has the advantage of minimal disruption of the air sample and does not require tubing or pumps.

A closed-path IRGA draws an air sample into a small chamber or sample cell where, as with the open-path design, the ratio of the transmission of infrared radiation between the air sample and a reference air sample is used to measure the CO_2 and H_2O densities. The reference cell contains air that has passed through chemicals to remove all the CO_2 and H_2O, and a spinning chopper wheel that allows the infrared radiation to pass through either the sample or reference cell. This closed-path design requires tubing and a pump to draw air into the IRGA, so issues such as the absorption of water vapor inside the tubing, pressure and temperature changes that could promote condensation (heated tubing is often used), and filters to prevent dust and pollen from entering the cells must be considered (Metzger et al., 2016). The closed-path design, however, lends itself to routine automatic calibration, an advantage to ensure quality measurements. A system of solenoid switches connected to air tanks of known concentrations of CO_2 and H_2O can be used to routinely and automatically calibrate the IRGA. Most flux measurement sites use either the open- or the closed-path IRGA with the sensor and/or sample air intake located close to the sonic anemometer (Figure 12.3).

With vertical wind speed measured using a sonic anemometer, and a scalar quantity mole (or mass) fraction in dry air (s; moles or grams of CO_2 or H_2O per mole or grams of dry air) calculated from the CO_2 or H_2O density measured with an IRGA, with both measured simultaneously at the same location, the vertical flux F (kg m^{-2} s^{-1}) can be calculated using the basic equation (Eq. 12.4):

$$F = \overline{\rho_d} \times \overline{w's'} \tag{12.4}$$

where ρ_d is the dry air density (kg m^{-3}), and primes denote instantaneous deviations from the time-averaged mean (typically 10 or 30 minutes). The eddy covariance method is widely used to directly measure evaporation or carbon dioxide exchange from forests. As with any measurement, there are theoretical assumptions that should be recognized to ensure accurate measurements (see Lee, Massman and Law, 2004; Aubinet, Vesala and Papale, 2012).

With an eddy covariance system (a sonic anemometer, a co-located IRGA, and a data acquisition system) placed on a tower above the vegetation (Figure 12.4), measurements of evaporation expressed as latent heat flux (λE; W m^{-2}) can provide a good estimate of forest-scale transpiration (Eq. 12.5):

Figure 12.4 A 26-m-tall tower used to measure fluxes of water and carbon dioxide in an 11.5-m-tall Colorado subalpine forest. Eddy covariance instruments are located near the top, as well as several other meteorological sensors at various heights on the tower. Photo credit: P. D. Blanken.

$$\lambda E = \lambda \times (1 + \mu\sigma) \times \left[\overline{w'\rho_v'} + \left(\overline{\rho_v} / \overline{T} \right) \overline{w'T'} \right] \quad (12.5)$$

where λ is the latent heat of vaporization (J kg^{-1}), μ is the ratio of the molecular mass of dry air (M_a; g mol^{-1}) and water vapor (M_w; g mol^{-1}) ($\mu = M_a/M_w$), σ is the ratio of the mean densities of water vapor (ρ_v; kg m^{-3}) and dry air (ρ_a; kg m^{-3}) ($\sigma = \rho_v/\rho_a$), and T (in kelvin) is the mean absolute temperature at the measurement height. Equation (12.5) contains the fundamental eddy covariance expression for the vertical flux of water vapor $\overline{w'\rho_v'}$ with the water vapor fluctuations expressed as a density. The other terms in Eq. (12.5) correct for the influence of the sensible heat and water vapor on the latent heat flux (Leuning et al., 1982). The **sensible heat flux** (H; W m^{-2}) is contained in the $\overline{w'T'}$ term, and H can be directly calculated using w from the sonic anemometer and T from the sonic air temperature estimate when corrected for humidity effects (Schotanus, Nieuwstadt and De Bruin, 1983), or T can be measured with a fine-wire fast-response thermocouple to calculate the sensible heat flux (Eq. 12.6):

$$H = C_p \times \overline{\rho} \times \overline{w'T'} \quad (12.6)$$

where C_p (J kg^{-1} K^{-1}) and ρ (kg m^{-3}) are the specific heat capacity and mean density of moist air, respectively (Leuning et al., 1982). If the mass rate of evaporation is desired (kg m^{-2} s^{-1}), Eq. (12.5) can be divided by the latent heat of evaporation, and then divided by the density of water if an evaporation rate (e.g., mm per hour or day) is the preferred unit.

The source area, or spatial scale of the eddy covariance flux measurements, is referred to as **fetch** or flux footprint (Chu et al., 2021). This depends on the instruments' height and location, relative wind direction, turbulence characteristics, and the extent of the underlying surface being measured. As a rough rule-of-thumb, the ratio of the instrument height above the surface to the upwind area that is being measured is 1:100 (height above surface: upwind sample distance).

If using the eddy covariance method to measure only transpiration, all other sources of evaporation must be quantified. Most vegetated sites contain some understory vegetation and bare soils, both of which can contribute to the overstory eddy covariance measurements, especially after rainfall. Even if understory vegetation is sparse, the exposed soil (or rock) can contain water that can evaporate. To isolate transpiration occurring from the upper or main canopy, a second eddy covariance system can be placed above the understory, and the difference between the latent heat flux measurements at each level when the canopy is dry should represent the overstory transpiration. The evaporation contribution from bare soil can be measured with **weighing lysimeters**, where changes in mass of soil volume are used to calculate evaporation (Uclés et al., 2013). Since the turbulent flux footprint sampled by the eddy covariance method would be much larger from the above-canopy than the below-canopy measurements, we need an understanding of how these differences in flux footprints could potentially affect the results. In addition, periods during and shortly after precipitation should be avoided to allow time for the intercepted water to evaporate.

Much has been learned about the magnitude and controls of transpiration by using a paired eddy covariance approach to isolate overstory transpiration. Eddy covariance measurements made above and below a **boreal deciduous aspen forest** revealed that, during the summer, roughly 75% of the transpiration originated from the aspen trees, and 25% from the hazel understory (Blanken et al., 1997). During the leafless period, most of the sensible and latent heat production originated from the forest floor, whereas after leaf development, most originated from the aspen overstory (Blanken et al., 2001). Roupsard et al. (2006) reported a similar result with 68% of the transpiration originating from the overstory of a **tropical coconut palm** forest in the South Pacific. Although the magnitudes of the transpiration rates were very different in these two forests (boreal aspen 285 mm yr^{-1}; 0.78 mm day^{-1}; tropical coconut palm 642 mm yr^{-1}; 1.3 to 2.3 mm day^{-1}), the similar overstory contributions to the total forest transpiration implies a **Leaf Area Index** (LAI; m^2 leaf per m^2 ground) control since the values were similar (aspen LAI 2.3; palm LAI 2.95). Indeed, the importance of LAI was demonstrated by Nelson et al. (2020) and Wang, Good and Caylor (2014) who found good relationships between the ratio of transpiration to the total evaporation and LAI across several plant functional types, although additional variables such as the **vapor pressure deficit** (the difference between the saturation and actual vapor pressure) also played an important role.

12.3.4 Forest Measurements: Stable Isotopes

As an alternative to above- and below-canopy eddy covariance measurements, **stable isotopes** of oxygen contained in water vapor can be used to measure transpiration (Wang and Yakir, 2000; Yakir and Sternberg, 2000). This method is based on the understanding that water vapor is depleted in ^{18}O compared with the ^{18}O in the source liquid water. Since ^{18}O is heavier than ^{16}O, ^{16}O will preferentially evaporate over ^{18}O (thus the depletion in the water vapor ^{18}O), and the ratio of $^{18}O/^{16}O$ can be related to evaporation owing to this **isotopic modification** or **fractionation** first quantified by Craig and Gordon (1965). Studies have shown the observed isotopic composition of water found in leaves is usually less than expected from steady-state conditions observed from a freely evaporating water surface, implying that isotopic fractionation does not occur in transpiration.

The isotopic ratio of ^{18}O to ^{16}O in water sampled in soil compared with water sampled inside the xylem and leaves is similar (Wershaw et al., 1970). This means that during the uptake and transport of water by plants, little fractionation occurs, and the composition of water inside the plants reflects the water available to the plant (Flanagan and Ehleringer, 1991). Even with evaporation occurring within the stomata, however, studies have found that when the rates of water uptake and transpiration are equal, the water vapor exiting the leaves has a similar isotopic composition to the water entering the leaves (Flanagan and Ehleringer, 1991; Moreira et al., 1997). This is somewhat contrary to the expectation that since transpiration is evaporation, the lighter ^{16}O isotopes should evaporate preferentially over ^{18}O.

Indeed, fractionation does occur during transpiration. The lighter oxygen isotopes do evaporate more readily than the heavier during transpiration, so leaf water becomes enriched with heavy isotopes (Dongmann and Nürnberg, 1974). As expected, in dry air (low relative humidity), the higher transpiration rate resulted in an enrichment of ^{18}O (depletion of ^{16}O) in leaf water measured in several grass species (Helliker and Ehleringer, 2002). Therefore, the only explanation for why the isotopic compositions of water vapor exiting the leaf and the water entering the leaf are similar is that water in the leaf must be continuously and quickly supplied with water from the soil below that has not undergone any fractionation. This is the steady-state principle that occurs when the rate of transpiration and water uptake are assumed equal, and the isotopic composition of the water vapor from transpired water is similar to the water taken up by the plant's roots.

Water that has evaporated from the surface of water in soils or on top of leaves, however, is much more depleted in the heavier ^{18}O and ^{2}H isotopes. Therefore, the isotopic composition of water vapor from plant transpiration and soil or intercepted water evaporation are distinct and therefore isotopic measurements of water vapor coupled with a mixing model can be used to partition transpiration from other sources of evaporation (Flanagan and Ehleringer, 1991; Moreira et al., 1997; Williams et al., 2004). This approach should be avoided during periods of non-steady-state conditions such as in tall trees in the early morning when transpiration may occur in advance of water intake into the roots (see Section 12.3.2). Wang and Yakir (1995) found that steady-state conditions were rarely attained in a variety of plant species only 2 to 4 months old, so the assumption of steady-state conditions should not be assumed *a priori*.

The stable isotope approach has been used to understand the importance of transpiration in the global water cycle (Yakir and Sternberg, 2000). The method requires precise *in situ* measurements of oxygen isotopes typically using **spectroscopy** (elements and isotopes are identified based on the dispersion of light upon interaction with them). In the Amazon basin, Moreira et al. (1997) found that the ambient water vapor in an **Amazon Rainforest** was mostly, if not completely, generated by transpiration whereas transpiration from a grassy pastureland could not be detected. This is an important finding since this source of water vapor likely plays a large role in the generation of precipitation in the region, thus deforestation will certainly affect the water balance of this major ecological system. In a subalpine forest in the Rocky Mountains of Colorado, Berkelhammer et al. (2016) found strong agreement between isotope and eddy-covariance-based estimates of transpiration, with a daytime growing season average of 61 percent transpiration relative to the total evaporation. In a compilation of several isotope-based partitioning studies, Sutanto et al. (2014) reported that in general, most studies claim that 70 percent or more of the total evaporation flux from vegetated surfaces is from transpiration, and they suggest that these values are biased high because of systematic errors in instrument accuracy, assumptions used in the analysis, and calculation parameters.

12.4 Modeling Approaches

Physically based models used to estimate transpiration are based on simple, yet careful, observations made decades ago. **Howard Penman**'s (1909–1984) research included measurements of evaporation from well-watered surfaces at Rothamsted Research, England. By combining equations based on the amount of energy available for evaporation with the atmosphere's ability to transport water vapor through turbulence, Penman derived a **combination method** or **equation**. Penman showed that with this combination equation the difficult-to-measure surface temperature was not needed, and evaporation from a surface without limits on the availability of water could be calculated based on readily available meteorological data: "*...mean air temperature, mean dewpoint, mean wind speed velocity at the standard height and mean duration of sunshine*" (Penman, 1948).

Penman's equation was developed from observations of evaporation from large cast-iron cylinders (see Figure 7.2). Three cylinders were filled with water, three with bare soil, and three with soil covered by grass (turf). To maintain saturated conditions, the cylinders containing bare soil or grass were connected by a pipe at the base to an adjacent covered cylinder filled with water (referred to as the "minor"). Decreases in water level in the minor and the water-only cylinders indicated daily evaporation rates, and concurrent measurements of meteorological variables were used to examine relationships with the measured evaporation. Penman's work provided a foundation for estimating evaporation, but the equation was developed and tested only for saturated, well-watered surfaces.

To extend Penman's findings to surfaces with a limited water supply, **John Monteith** (1929–2012) incorporated a **bulk surface resistance** term to account for the limitation on evaporation imposed by transpiring vegetation. Penman's equation derived in 1948 with the original units is (Eq. 12.7):

$$\lambda E = \frac{\Delta \times H + \rho c_{\mathrm{p}} \left\{ {(e_{\mathrm{s}} - e_{\mathrm{a}})}/{r_{\mathrm{a}}} \right\}}{\Delta + \gamma} \tag{12.7}$$

where Δ is the rate of change (slope) in the saturation vapor pressure with air temperature (e_s; mm Hg) at a given air temperature (T_a; °F), de_s/dT_a, H is the available energy for evaporation (cal. $cm^{-2}\ s^{-1}$), γ is the constant 0.27, and r_a is "...the time in which 1 cm^3 of air exchanges heat with 1 cm^2 of the surface" (Monteith, 1965), later known as the **aerodynamic resistance**. Note that Penman used the symbol H not for the sensible heat flux, but for the "net radiant energy available at the surface". Based on the energy balance, the energy available at the surface is equal to (Eq. 12.8):

$$R_n - G_0 = \lambda E + H \tag{12.8}$$

where the available energy is the difference between the net radiation (R_n) and ground heat flux at the surface (G_0), and λE and H is the sum of the latent and sensible heat fluxes, respectively (all in units of W m^{-2}). Therefore, "H" as it appears in the early literature has since been replaced with $R_n - G_0$ here as well as in most other sources; care should be taken to avoid confusion.

Monteith's contribution to evaporation studies was to extend the application of Penman's equation (Eq. 12.7) to surfaces not saturated with water. As noted by Monteith (1965), Penman's equation is not valid for surfaces where the vapor pressure is less than the saturation vapor pressure at surface temperature. Using the **Ohm's Law** electrical analogy ($V = IR$; voltage = current × resistance), Monteith modified Penman's equation to account for the resistance imposed on evaporation by the leaf stomata cells (**leaf resistance**) and soil (**soil resistance**) combined in a **crop** or **bulk surface resistance term**. Using evaporation measurements made at Rothamsted, Monteith estimated the magnitude of these resistances and derived Eq. (12.9):

$$\frac{E}{E_0} = \frac{\Delta + \gamma}{\Delta + \gamma\left(1 + {r_s}/{r_a}\right)} \tag{12.9}$$

where E/E_0 is the ratio of the transpiration rate (E) to that from a wet surface (E_0), and r_s is the **surface resistance** of a crop canopy. Monteith (1965) stated:

> When evaporation from the soil surface is negligible r_s is expected to be a plant parameter revealing diurnal and seasonal changes of stomatal resistance. When evaporation from the soil is comparable with transpiration, r_s combines the resistance of crop leaves with capillary resistance to the diffusion of water vapor in the soil pores.

To calculate the transpiration rate (E) from Eq. (12.9) *when the surface is dry*, we multiply both sides by E_0 and substitute for E_0 using Eq. (12.7), yielding Eq. (12.10):

$$\begin{aligned} \frac{E}{E_0} &= \frac{\Delta + \gamma}{\Delta + \gamma\left(1 + {r_s}/{r_a}\right)} \\ E = E_0\left[\frac{\Delta + \gamma}{\Delta + \gamma\left(1 + {r_s}/{r_a}\right)}\right] &= \left[\frac{\Delta H + \rho c_p\left\{{(e_s - e_a)}/{r_a}\right\}}{\Delta + \gamma}\right] \times \left[\frac{\Delta + \gamma}{\Delta + \gamma\left(1 + {r_s}/{r_a}\right)}\right] \\ &= \frac{\Delta H + \rho c_p\left\{{(e_s - e_a)}/{r_a}\right\}}{\Delta + \gamma\left(1 + {r_s}/{r_a}\right)} \end{aligned} \tag{12.10}$$

where in Eq. (12.10) H is not the sensible heat flux but the surface available energy $R_n - G_0$, the net radiation minus the soil heat flux. This notation for H is used here to match the notation used by both Penman and Monteith. Equation (12.10) is the widely known **Penman–Monteith combination equation (model)** since it combines the energy balance with a means to transport the water vapor.

In situations where the surface is not dry, and evaporation from wet soil or leaves is occurring, modifications to the Penman–Monteith equation are required. This is especially likely in situations where bare soil is exposed in sparse canopies such as young crops or in Arctic or alpine regions. To address this issue, Shuttleworth and Wallace (1985) developed a theoretical model framework that explicitly contains a resistance term associated with vegetation and a separate resistance term associated with soil. Shuttleworth and Wallace (1985) express the total evaporation (expressed as latent heat flux, λE) from a surface with vegetation and bare soil using the Penman–Monteith equation (PM) as the sum of evaporation from the canopy (subscript C) and bare soil (subscript S) (Eq. 12.11):

$$\lambda E = C_C \times \mathrm{PM_C} + C_S \times \mathrm{PM_S}. \tag{12.11}$$

In Eq. (12.11) the coefficients C_C and C_S are calculated using an Ohm's Law analogy based on the total resistance to water vapor transfer from the canopy and bare soil, respectively. To use the Shuttleworth–Wallace equation (Eq. 12.11) in a practical sense essentially requires measurement at the canopy and soil of the available energy, meteorological conditions (e.g., air temperature, humidity, wind speed), and water vapor surface and aerodynamic resistances. Measurements or model estimates of these values can be difficult to attain, but this approach has been used in a variety of settings.

The Penman–Monteith (PM) model (Eq. 12.10) has been compared to the Shuttleworth–Wallace (SW) model (Eq. 12.11) in numerous settings. Kato, Kimura and Kamichika (2004) found that the SW model performed better than the PM model in estimates of λE from a sparse sorghum row-crop canopy when compared to measurements of λE made using the **Bowen Ratio Energy Balance** (BREB) method (a trusted method based on profile measurements of air temperature and humidity, and the available energy). The SW model revealed that transpiration showed strong diurnal variation and was as high as 70% of the total evaporation following irrigation. In a sparsely vegetated semi-arid high-elevation rangeland in southern Colorado, Stannard (1993) found the SW model fared better (r^2 = 0.79 for hourly data) than the PM model (r^2 = 0.56 for hourly data) in comparison to eddy covariance measurements of λE. The SW model indicated that 74% of λE was transpiration and 26% soil evaporation.

The SW model has also been tested in boreal environments typified by sparse canopies. In a hummock–hollow Subarctic wetland tundra in the Hudson Bay Lowland of Canada, Wessel and Rouse (1994) found that when compared with λE measured using the BREB method, a PM model weighted by the surface area of the evaporation sources (hummock, hollow, or open water; Chapter 10) outperformed the SW model. In a boreal coniferous forest in central Sweden, Iritz et al. (1999) modified the SW model to include the evaporation of canopy-intercepted water and the soil water regulation on the canopy resistance. The daily total SW estimated evaporation compared well to eddy covariance λE measurements (r^2 = 0.65 and 0.71 in two separate years). Measurements of transpiration using the sap flow technique over the course of a summer were 243 mm, compared with the SW-based estimate of 212 mm. Soil water evaporation measured using chambers was 56 mm compared with an SW estimate of 36 mm, and the measured intercepted evaporation (by collecting the throughfall) was 74 mm compared with 72 mm. Over the summer period, the total evaporation was measured at 322 mm compared with the SW estimate of 320 mm.

12.5 Transpiration Controls

The dilemma faced by vascular plants is the need to sequester carbon dioxide for photosynthesis yet simultaneously face the unavoidable loss of water through transpiration by exposing liquid water to the atmosphere. This dilemma is quantified through **water use efficiency** (WUE), the ratio of the mass of carbon gained to the mass of water lost (Hatfield and Dold, 2019). An efficient plant would gain the most carbon while losing the least amount of water, and the plant's control of the water use efficiency is through active control of **stomatal conductance** (1/resistance), a measure indicative of the stomata pores' aperture (Chapter 11). By opening the stomata during periods when the transpired water loss is low, CO_2 can enter the plant at an optimum time of minimal water loss. For example, in the early morning period when the **photosynthetic active radiation** (400–700 nm) is adequate for photosynthesis (~25 percent of full sunlight), the air is often cool with a high relative humidity and low wind speed; an ideal time to open the stomata to gain CO_2 with minimal water loss. During midday periods when the air is warm, dry, and often windy, transpiration water loss is large, so stomata often close to reduce water stress. Examples of this dynamic diurnal stomatal behavior have been observed in several Subarctic wetland species (Blanken and Rouse, 1996). Even wetland species growing in standing water reached their maximum stomatal conductance with minimal transpiration shortly after sunrise. The stomatal conductance and transpiration both decreased midday during warm and dry conditions. Leaf water potential measurements made with a pressure chamber (Chapter 11) showed that the stomatal closure was triggered by high transpiration rates.

There are different but complementary variables that control transpiration that couple meteorological and soil conditions to plant physiological controls. Stomata opening is triggered by the need for sufficient light for photosynthesis, and the concentration gradient of CO_2 in the atmosphere is greater than that in the leaf so CO_2 diffuses into the leaf (this is seldom an issue). Stomata closure is triggered by excessive water loss, resulting in the water tension in the xylem becoming so large that cavitation xylem embolism is possible (Chapter 11). At a given stoma aperture, this water loss is influenced by the vapor pressure gradient between the leaf and the atmosphere, leaf and air temperatures, and wind speed. Each of these primary controls is described below.

12.5.1 Temperature

Assuming there are no limits on the supply of water to the plant, temperature has a significant effect on the transpiration rate since the saturation vapor pressure varies strongly with temperature through the **Clausius–Clapeyron equation** (Chapter 7). At the leaf level, the transpiration rate (E_l; mol m^{-2} s^{-1}) varies with the vapor pressure gradient between the leaf surface (e_s; kPa) and the atmosphere (e_a; kPa) divided by the ambient barometric pressure P (kPa), but also the plant-regulated stomatal conductance (g_s; mol m^{-2} s^{-1}) (Eq. 12.12):

$$E_l = g_s \times \frac{(e_s - e_a)}{P}. \tag{12.12}$$

If Eq. (12.12) is multiplied by the molecular weight of water (M_w = 18 g mol^{-1}), the units of E_l become g m^{-2} s^{-1}. Since e_s increases exponentially with temperature, any increase in leaf temperature will greatly increase the saturation vapor pressure at the leaf surface since the air inside the stomatal pore is assumed to be saturated (100 percent relative humidity). Even if the vapor pressure in the air outside the leaf (e_a) increases as transpiration continues, the nonlinear increase in e_s relative to e_a means that the leaf–air vapor gradient ($e_s - e_a$) increases as leaf temperature increases, acting to further increase transpiration. As daytime wind and turbulence in the atmosphere generally remove and replace the humidified air with dry air, the leaf–air vapor gradient is maintained or increased allowing transpiration to continue. If, however, the transpiration rate becomes too large relative to the rate of water intake through the roots, approaching the point when the cohesion between liquid water molecules in the xylem could fail, forming an air pocket, the stomata can reduce their aperture (decrease g_s) to decrease transpiration and reduce the xylem water tension. Such behavior was observed by Blanken and Rouse (1996) in several Subarctic willow and birch species.

12.5.2 Soil Moisture

Regarding the supply of liquid water to the roots to meet the atmospheric demand for water vapor, the **soil water retention curve** (Chapter 11) informs how the volume of water stored in soils relates to the energy required by the plant to extract this water. Soil properties such as texture and porosity, rooting depth and extent, and plant species and age all affect the availability of water to the plant. Whether the soil is wetting or drying also affects the soil's matric water potential through a hysteresis effect in the soil water retention curve. Of primary importance is the amount and timing of precipitation to supply water to the soil. Too much water is detrimental to the plant as it limits the availability of oxygen for root respiration. Too little water is detrimental to the plant since, as water intake decreases, stomatal closure or even leaf loss is required to reduce the risk of cavitation. In either case, photosynthesis (and therefore plant growth) is limited.

12.5.3 Carbon Dioxide Concentration

The increasing atmospheric CO_2 concentration affects stomatal conductance, leaf area, and therefore transpiration and water use efficiency on both short and long timescales. The **Free-Air CO_2 Enrichment (FACE) project** was designed to measure the impacts of increased ambient CO_2 concentrations on vegetation growth in outdoor settings. Large vertical-standing pipes emitted CO_2 at roughly 200 ppm above the ambient concentrations, and effects on the vegetation were measured. A summary of nearly 250 observations spanning 14 sites and 5 continents showed that the majority of crop species using the C3 photosynthetic pathway (C3 is the most common photosynthetic pathway in 85% of terrestrial vegetation, including all trees, and is known as the **Calvin cycle**) showed an 18% increase in yield with the higher CO_2 concentration under nonstress conditions, with a similar response during water-stressed conditions (Ainsworth and Long, 2021). This **CO_2 fertilization effect** was also observed at forested FACE sites. The net primary productivity increased by 25% averaged across boreal, temperate, and tropical forests (Hickler et al., 2008). Whether increased CO_2 concentrations increase vegetation productivity or not depends on if there is an adequate water supply.

These increases in productivity in C3 species under higher CO_2 concentrations are due to an increase in photosynthesis concurrent with a decrease in stomatal conductance without any short-term decrease in stomatal density (Ainsworth and Rogers, 2007) and therefore a decrease in transpiration (Manderscheid et al., 2016). However, the effects of the increase in leaf temperature with the decreased transpiration and the coupled vapor pressure gradient effects on stomatal conductance and photosynthesis are unclear from these FACE studies. Several studies conducted under natural conditions concluded that despite increasing CO_2 concentrations, increased water deficits have reduced, not increased, vegetation growth. Using tree-core data from 122 Douglas fir trees at least 125 years old growing across the western United States, Restaino, Peterson and Littell (2016) found strong relationships between the **vapor pressure deficit** ($D = e_s - e_a$) and growth throughout the Douglas fir forests of the western United States. Low growth was associated with a high D, and photosynthesis was inhibited owing to stomatal closure during these dry atmospheric conditions. Tree seedlings also suffer greater transpiration rates and mortality from increased D (Will et al., 2013). Ten seedling species common along the grassland/forest ecotone in the central United States were grown in chambers under non-drought and drought conditions. Seedlings in the drought conditions experience a 40% increase in D compared with the non-drought conditions. Seedlings grown under the higher D conditions had a 14% lower stomatal conductance, a 29% increase in transpiration rate, a 12% decrease in net photosynthesis, and increased mortality compared with those grown in the non-drought conditions (Will et al., 2013).

Outdoors, the many variables that influence transpiration and photosynthesis rates can change rapidly with changes in one variable affecting another. Therefore, many studies isolate one variable at a time to determine its influence, but consequently, the synergetic response of the plant in a natural setting to all variables is not accounted for. On the one hand, plant productivity should increase with increasing CO_2 concentrations but only if there is an adequate water supply. On the other hand, the drier atmosphere that comes with the CO_2-induced warming should decrease plant productivity. Both groups of studies are linked to changes in stomatal conductance associated with transpiration and the water balance. At a global scale, observations and models show that D has remained fairly constant since 1805, but sharply increased in the late 1990s (Yuan et al., 2019) (Chapter 16). Satellite-based **Normalized Difference Vegetation Index**, a measure of plant productivity, showed a pronounced decrease in 59 percent of the world's vegetated region after 1999. Four different LAI datasets showed a similar decreasing trend since the late 1990s, and two global-scale terrestrial productivity models confirmed that the observed decreased productivity starting in the 1990s was due to increased D which offset the positive CO_2 fertilization effect

(Yuan et al., 2019). The interplay of these related variables on transpiration and vegetation productivity is complex and difficult to unwind.

12.6 SUMMARY

Transpiration is the evaporation of water from within living plant tissue under biological control (biotic evaporation) as opposed to evaporation from other sources (abiotic evaporation). Transpiration occurs within microscopic pores or stomata on the leaves, with the number of stomata per leaf varying with species, location, weather, and atmospheric CO_2 concentration. Despite their small size, stomata play a large role in the global water and carbon cycle. The majority of water evaporated from the global land surface is through transpiration, and the seasonal variation in CO_2 concentration in the Northern Hemisphere illustrates the importance of gas exchange through the small but numerous stomatal pores.

Partitioning evaporation into its biotic and abiotic water sources remains a measurement challenge. Transpiration can be measured on individual leaves and then scaled to the plant and canopy with additional measurements of species composition and Leaf Area Index. Transpiration from an individual plant or tree can be determined by sap flow methods based on the application of heat to the stem and use of stem temperature measurements to measure how the sap transfers heat. Stand-level micrometeorological methods such as the eddy covariance method can be used to measure large-scale water and carbon dioxide fluxes based on the covariance between high-frequency atmospheric measurements of the vertical wind speed and scalars such as water vapor or carbon dioxide densities. Such measurements are possible with fast-response sonic anemometers and open or closed-path infrared gas analyzers. These canopy-scale measurements sample the entire surface beneath; hence all sources of water vapor are included, and the partitioning or isolation of the stand's total evaporation into the transpiration component is required. Measurements from eddy covariance instruments placed above and below a forest canopy, or measurements of stable oxygen isotopes, have been used to isolate the transpiration contribution to the total evaporative flux.

Theory supported by observations has been developed and used to separate evaporation from its various sources. Combination methods that combine the energy available for evaporation and the transfer of water vapor through the atmosphere were developed and tested by Penman (1948). Since Penman's equation was developed for surfaces that have no restrictions on water supply, Monteith (1965) modified Penman's equation to develop the Penman–Monteith equation that includes a surface resistance term to account for restrictions on the surface water supply from both vegetation and soil. Subsequent modifications to the Penman–Monteith equation have been developed and tested on sparsely vegetated surfaces allowing for estimates of soil evaporation and transpiration.

The wide variety of transpiration measurements using independent techniques, instruments, and theory, collectively show that transpiration is a major component of the terrestrial water balance, contributing between 60 percent and 80 percent of the total terrestrial water vapor flux. Two of the primary controls on the transpiration rate, and hence the terrestrial (and global) water balance, that are increasing with contemporary climate change, are temperature and the atmospheric carbon dioxide concentration. Increasing temperatures raise the atmospheric demand for water vapor by increasing the vapor pressure gradient between leaf surface and atmosphere. Many studies have found that in response, stomata close to conserve water that is being depleted in drying soils. Since carbon dioxide enters the plant through the same stomata through which water vapor exits the plant, any increase in plant productivity for common C3 species due to higher carbon dioxide concentrations will likely be negated by water-stress-induced stomatal closure.

12.7 QUESTIONS

12.1 What is the difference between abiotic and biotic evaporation? Why is it important to distinguish between the two? List the primary controls for each.

12.2 In the background of the increasing long-term atmospheric CO_2 concentration measured at Mauna Loa, Hawaii, is an annual variation of roughly 10 parts per million (ppm). What could explain this persistent annual variation?

12.3 A typical forest consists of a combination of tall trees, understory shrubs, and small lakes often connected by rivers or small streams. Which method(s) would you use to best measure transpiration from this ecosystem, and why? What are the limitations of your approach?

12.4 Several researchers have compared concurrent sap flow measurements of transpiration to eddy covariance flux measurements of evaporation in forested ecosystems. Needless to say, the two methods sometimes yield different results. Discuss reasons for the possible differences in these measurements. What could be done to maximize the information so that each method provides complementary knowledge to improve our understanding of water cycling through a forest?

12.5 The eddy covariance method has become the standard approach to measure water vapor and carbon dioxide exchange over large areas. Describe the basic principle behind the method, the instruments required, and any advantages and disadvantages of the method.

12.6 How can stable isotopes of oxygen contained in molecules of water be used to distinguish between transpiration and evaporation?

12.7 What motivated Monteith to change the evaporation equation previously developed by Penman? Summarize the changes made by Monteith to the Penman equation (Eq. 12.7) to result in the Penman–Monteith equation (Eq. 12.10).

12.8 Vegetation often will have the highest stomatal conductance (i.e., open stomata) right after sunrise, then reduce the stomatal conductance (i.e., close stomata) during midday. Why would plants exhibit such behavior?

12.9 A group of plants growing in arid conditions (e.g., orchids, cactus, pineapple) use a crassulacean acid metabolism (CAM) photosynthetic pathway where the stomata are closed during the day and open at night. The production of malic acid during the day is used at night as the energy source for photosynthesis. How is this CAM photosynthetic pathway advantageous to these plants to conserve water?

12.10 In a warmer and drier world with increasing atmospheric CO_2 concentrations, some studies have shown that vegetation net primary productivity increased under controlled conditions. Under natural conditions, some studies have shown the contrary: decreased net primary productivity. How does water play a role and explain these apparently contrasting findings?

REFERENCES

Ainsworth, E. A. and Long, S. P. (2021) '30 years of free-air carbon dioxide enrichment (FACE): What have we learned about future crop productivity and its potential for adaptation?' *Global Change Biology*, 27, pp. 27–49. doi: 10.1111/gcb.15375.

Ainsworth, E. A. and Rogers, A. (2007) 'The response of photosynthesis and stomatal conductance to rising [CO_2]: Mechanisms and environmental interactions', *Plant, Cell and Environment*, 30, pp. 258–270. doi: 10.1111/j.1365-3040.2007.01641.x.

Aubinet, M., Vesala, T. and Papale, D. (2012) *Eddy Covariance: A Practical Guide to Measurement and Data Analysis*. Springer.

Berkelhammer, M., Noone, D. C., Wong, T. E. et al. (2016) 'Convergent approaches to determine an ecosystem's transpiration fraction', *Global Biogeochemical Cycles*, 30. doi: 10.1002/2016GB005392.

Blanken, P. D., Black, T. A., Neumann, H. H. et al. (2001) 'The seasonal water and energy exchange above and within a boreal aspen forest', *Journal of Hydrology*, 245, pp. 118–136. doi: 10.1016/S0022-1694(01)00343-2.

Blanken, P. D., Black, T. A., Yang, P. C. et al. (1997) 'Energy balance and canopy conductance of a boreal aspen forest: Partitioning overstory and understory components', *Journal of Geophysical Research Atmospheres*, 102, pp. 28915–28927. doi: 10.1029/97JD00193

Blanken, P. D. and Rouse, W. R. (1994) 'The role of willow-birch forest in the surface energy balance at Arctic treeline', *Arctic and Alpine Research*, 26, pp. 403–411.

Blanken, P. D. and Rouse, W. R. (1995) 'Modelling evaporation from a high subarctic willow-birch forest', *International Journal of Climatology*, 15. doi: 10.1002/joc.3370150110.

Blanken, P. D. and Rouse, W. R. (1996) 'Evidence of water conservation mechanisms in several subarctic wetland species', *Journal of Applied Ecology*, 33, pp. 842–850.

Blyth, E. and Harding, R. J. (2011) 'Methods to separate observed global evapotranspiration into the interception, transpiration and soil surface evaporation components', *Hydrological Processes*, 25, pp. 4063–4068. doi: 10.1002/hyp.8409.

Burba, G. (2013) *Eddy Covariance Method for Scientific, Industrial, Agriculture and Regulatory Applications*. LI-COR Biosciences.

Burns, S. P., Horst, T. W., Jacobsen, L., Blanken, P. D. and Monson, R. K. (2012) 'Using sonic anemometer temperature to measure sensible heat flux in strong winds', *Atmospheric Measurement Techniques*, 5. doi: 10.5194/amt-5-2095-2012.

Chu, H., Luo, X., Ouyang, Z. et al. (2021) 'Representativeness of eddy-covariance flux footprints for areas surrounding AmeriFlux sites', *Agricultural and Forest Meteorology*, 301–302. doi: 10.1016/j.agrformet.2021.108350.

Craig, H. and Gordon, L. I. (1965) 'Deuterium and oxygen 18 variations in the ocean and the marine atmosphere', in Tongiorgi, E. (ed.) *Proceedings of the Third Spoleto Conference: Stable Isotopes in Oceanographic Studies and Paleotemperatures*. V. Lischi & Figli, pp. 9–130.

Dongmann, G. and Nürnberg, H. W. (1974) 'On the enrichment of $H_2{}^{18}O$ in the leaves of transpiring plants', *Radiation and Environmental Biophysics*, 11, pp. 41–52. doi: doi.org/10.1007/BF01323099.

Fanourakis, D., Giday, H., Milla, R. et al. (2015) 'Pore size regulates operating stomatal conductance, while stomatal densities drive the partitioning of conductance between leaf sides', *Annals of Botany*, 115, pp. 555–565. doi: 10.1093/aob/mcu247.

Flanagan, L. B. and Ehleringer, J. R. (1991) 'Stable isotope composition of stem and leaf water: Applications to the study of plant water use', *Functional Ecology*, 5, p. 270. doi: 10.2307/2389264.

Granier, A. (1985) 'Une nouvelle méthode pour la mesure du flux de sève brute dans le tronc des arbres', *Annales des Sciences Forestieres*, pp. 193–200. doi: 10.1007/BF00117583.

Hales, S. and Royal Society, Great Britain (1727) *Vegetable Staticks, or, An Account of Some Statical Experiments on the Sap in Vegetables*. Printed for W. & J. Innys and T. Woodward.

Hatfield, J. L. and Dold, C. (2019) 'Water-use efficiency: Advances and challenges in a changing climate', *Frontiers in Plant Science*, 10, pp. 1–14. doi: 10.3389/fpls.2019.00103.

Helliker, B. R. and Ehleringer, J. R. (2002) 'Differential O-18 enrichment of leaf cellulose in C-3 versus C-4 grasses', *Functional Plant Biology*, 29, pp. 435–442.

Hickler, T., Smith, B., Prentice, I. C. et al. (2008) 'CO_2 fertilization in temperate FACE experiments not representative of boreal and tropical forests', *Global Change Biology*, 14, pp. 1531–1542. doi: 10.1111/j.1365-2486.2008.01598.x.

Hogg, E. H., Black, T. A., den Hartog, G. et al. (1997) 'A comparison of sap flow and eddy fluxes of water vapor from a boreal deciduous forest', *Journal of Geophysical Research Atmospheres*, 102, pp. 28929–28937. doi: 10.1029/96JD03881.

Iritz, Z., Lindroth, A., Heikinheimo, M., Grelle, A. and Kellner, E. (1999) 'Test of a modified Shuttleworth-Wallace estimate of boreal forest evaporation', *Agricultural and Forest Meteorology*, 98–99, pp. 605–619. doi: 10.1016/S0168-1923(99)00127-6.

Jasechko, S., Sharp, Z. D., Gibson, J. J. et al. (2013) 'Terrestrial water fluxes dominated by transpiration', *Nature*, 496, pp. 347–350. doi: 10.1038/nature11983.

Kaimal, J. C. and Businger, J. A. (1963) 'A continuous wave sonic anemometer-thermometer', *Journal of Applied Meteorology and Climatology*, 2, pp. 156–164.

Kato, T., Kimura, R. and Kamichika, M. (2004) 'Estimation of evapotranspiration, transpiration ratio and water-use efficiency from a sparse canopy using a compartment model', *Agricultural Water Management*, 65, pp. 173–191. doi: 10.1016/j.agwat.2003.10.001.

Lee, X., Massman, W. and Law, B. (eds.) (2004) *Handbook of Micrometeorology*. Kluwer Academic.

Leuning, R., Denmead, O. T., Lang, A. R. G. and Ohtaki, E. (1982) 'Effects of heat and water vapor transport on eddy covariance measurement of CO_2 fluxes', *Boundary-Layer Meteorology*, 23, pp. 209–222. doi: 10.1007/BF00123298.

LI-COR (1989) *LI-1600 Steady State Porometer Instruction Manual*. Revision 6. LI-COR Biosciences.

Manderscheid, R., Erbs, M., Burkart, S. et al. (2016) 'Effects of free-air carbon dioxide enrichment on sap flow and canopy microclimate of maize grown under different water supply', *Journal of Agronomy and Crop Science*, 202, pp. 255–268. doi: 10.1111/jac.12150.

McBean, G. A. (1972) 'Instrument requirements for eddy correlation measurements', *Journal of Applied Meteorology and Climatology*, 11, pp. 1078–1084.

Meinzer, F. C., James, S. A. and Goldstein, G. (2004) 'Dynamics of transpiration, sap flow and use of stored water in tropical forest canopy trees', *Tree Physiology*, 24, pp. 901–909. doi: 10.1093/treephys/24.8.901.

Metzger, S., Burba, G., Burns, S. P. et al. (2016) 'Optimization of an enclosed gas analyzer sampling system for measuring eddy covariance fluxes of H_2O and CO_2', *Atmospheric Measurement Techniques*, 9. doi: 10.5194/amt-9-1341-2016.

Miralles, D. G., Brutsaert, W., Dolman, A. J. and Gash, J. H. (2020) 'On the use of the term "Evapotranspiration"', *Water Resources Research*, 56. doi: 10.1029/2020WR028055.

Miralles, D. G., De Jeu, R. A. M., Gash, J. H., Holmes, T. R. H. and Dolman, A. J. (2011) 'Magnitude and variability of land evaporation and its components at the global scale', *Hydrology and Earth System Sciences*, 15, pp. 967–981. doi: 10.5194/hess-15-967-2011.

Monteith, J. L. (1965) 'Evaporation and environment,' *Symposia of the Society for Experimental Biology*, 19, pp. 205–234.

Moreira, M. Z., Sternberg, L., Martinelli, L. et al. (1997) 'Contribution of transpiration to forest ambient vapour based on isotopic measurements', *Global Change Biology*, 3, pp. 439–450. doi: 10.1046/j.1365-2486.1997.00082.x.

Nelson, J. A., Pérez-Priego, O., Zhou, S. et al. (2020) 'Ecosystem transpiration and evaporation: Insights from three water flux partitioning methods across FLUXNET sites', *Global Change Biology*. doi: 10.1111/gcb.15314.

Penman, H. L. (1948) 'Natural evaporation from open water, bare soil and grass', *Proceedings of the Royal Society of London, Series A*, 193, pp. 120–145.

Poyatos, R., Granda, V., Flo, V. et al. (2021) 'Global transpiration data from sap flow measurements: the SAPFLUXNET database', *Earth System Science Data*, 13, pp. 2607–2649.

Restaino, C. M., Peterson, D. L. and Littell, J. (2016) 'Increased water deficit decreases Douglas fir growth throughout western US forests', *Proceedings of the National Academy of Sciences USA*, 113, pp. 9557–9562. doi: 10.1073/pnas.1602384113.

Richardson, L. F. (1922) *Weather Prediction by Numerical Process*. Cambridge University Press.

Roupsard, O., Bonnefond, J. M., Irvine, M. et al. (2006) 'Partitioning energy and evapo-transpiration above and below a tropical palm canopy', *Agricultural and Forest Meteorology*, 139, pp. 252–268. doi: 10.1016/j.agrformet.2006.07.006.

Royer, D. L. (2001) 'Stomatal density and stomatal index as indicators of paleoatmospheric CO_2 concentration', *Review of Palaeobotany and Palynology*, 114, pp. 1–28. doi: 10.1016/S0034-6667(00)00074-9.

Schlesinger, W. H. and Jasechko, S. (2014) 'Transpiration in the global water cycle', *Agricultural and Forest Meteorology*, 189–190, pp. 115–117. doi: 10.1016/j.agrformet.2014.01.011.

Schotanus, P., Nieuwstadt, F. T. M. and De Bruin, H. A. R. (1983) 'Temperature measurement with a sonic anemometer and its application to heat and moisture fluxes', *Boundary-Layer Meteorology*, 26, pp. 81–93. doi: 10.1007/BF00164332.

Shuttleworth, W. J. and Wallace, J. S. (1985) 'Evaporation from sparse crops – an energy combination theory', *The Quarterly Journal of the Royal Meteorological Society*, 111, pp. 839–855. doi: 10.1002/qj.49711146510.

Smith, D. M. and Allen, S. J. (1996) 'Measurement of sap flow in plant stems', *Journal of Experimental Botany*, 47, pp. 1833–1844. doi: 10.1093/jxb/47.12.1833.

Stannard, D. I. (1993) 'Comparison of Penman-Monteith, Shuttleworth-Wallace, and modified Priestley-Taylor evapotranspiration models for wildland vegetation in semiarid rangeland', *Water Resources Research*, 29, pp. 1379–1392.

Steppe, K., De Pauw, D. J. W., Doody, T. M. and Teskey, R. O. (2010) 'A comparison of sap flux density using thermal dissipation, heat pulse velocity and heat field deformation methods', *Agricultural and Forest Meteorology*, 150, pp. 1046–1056. doi: 10.1016/j.agrformet.2010.04.004.

Sutanto, S. J., van den Hurk, B., Dirmeyer, P. A. et al. (2014) 'HESS Opinions "A perspective on isotope versus non-isotope approaches to determine the contribution of transpiration to total evaporation"', *Hydrology and Earth System Sciences*, 18, pp. 2815–2827. doi: 10.5194/hess-18-2815-2014.

Uclés, O., Villagarcía, L., Cantón, Y. and Domingo, F. (2013) 'Microlysimeter station for long term non-rainfall water input and evaporation studies', *Agricultural and Forest Meteorology*, 182–183, pp. 13–20. doi: 10.1016/j.agrformet.2013.07.017.

Wang, L., Good, S. P. and Caylor, K. K. (2014) 'Global synthesis of vegetation control on evapotranspiration partitioning', *Geophysical Research Letter*, 41, pp. 6753–6757. doi: 10.1002/2014GL061439.

Wang, X.-F. and Yakir, D. (1995) 'Temporal and spatial variations in the oxygen-18 content of leaf water in different plant species', *Plant, Cell & Environment*, 18, pp. 1377–1385. doi: 10.1111/j.1365-3040.1995.tb00198.x.

Wang, X. F. and Yakir, D. (2000) 'Using stable isotopes of water in evapotranspiration studies', *Hydrological Processes*, 14, pp. 1407–1421. doi: 10.1002/1099-1085(20000615)14:8<1407::AID-HYP992>3.0.CO;2-K.

Wershaw, R., Friedman, I., Heller, S. J. and Frank, P. A. (1970) 'Hydrogen isotopic fractionation of water passing through trees', in Hobson, G. D. and Speers, G. C. (eds.) *Advances in Organic Geochemistry: Proceedings of the Third International Congress*. Pergamon, pp. 55–67.

Wessel, D. A. and Rouse, W. R. (1994) 'Modelling evaporation from wetland tundra', *Boundary-Layer Meteorology*, 68, pp. 109–130. doi: 10.1007/BF00712666.

Will, R. E., Wilson, S. M., Zou, C. B. and Hennessey, T. C. (2013) 'Increased vapor pressure deficit due to higher temperature leads to greater transpiration and faster mortality during drought for tree seedlings common to the forest-grassland ecotone', *New Phytologist*, 200, pp. 366–374. doi: 10.1111/nph.12321.

Williams, D. G., Cable, W., Hultine, K. et al. (2004) 'Evapotranspiration components determined by stable isotope, sap flow and eddy covariance techniques', *Agricultural and Forest Meteorology*, 125, pp. 241–258. doi: 10.1016/j.agrformet.2004.04.008.

Woodward, F. I. (1987) 'Stomatal numbers are sensitive to increases in CO_2 concentration from the pre-industrial levels', *Nature*, 327, pp. 617–618.

Woodward, F. I. and Kelly, C. K. (1995) 'The influence of CO_2 concentration on stomatal density', *New Phytologist*, 131, pp. 311–327. doi: 10.1111/j.1469-8137.1995.tb03067.x.

Yakir, D. and Sternberg, L. D. S. L. (2000) 'The use of stable isotopes to study ecosystem gas exchange', *Oecologia*, 123, pp. 297–311. doi: 10.1007/s004420051016.

Yuan, W., Zheng, Y., Piao, S. et al. (2019) 'Increased atmospheric vapor pressure deficit reduces global vegetation growth', *Science Advances*, 5, pp. 1–13. doi: 10.1126/sciadv.aax1396.

13 Water and Thermoregulation: Ectotherms

Key Learning Objectives

After reading this chapter you will be able to:

1. Explain how cutaneous resistance to evaporative water loss influences the range of habitats occupied by ectotherms such as frogs, lizards, and insects.
2. Calculate the evaporative water loss from an ectotherm and use this calculation to estimate survival times.
3. Describe how aquatic and terrestrial insects use water and control the rate of water exchange across their bodies to permit habitation across the wettest and driest regions of Earth.
4. Provide examples of behavioral adaptations that insects, including flying insects, have developed to optimize thermoregulation.

13.1 Introduction

Living organisms, like all objects, exchange energy with their environment. These forms of energy include radiation, convection, conduction, and the energy associated with changing states of water. Unlike terrestrial plants, fauna can relocate to seek food or shelter, so movement plays an active role in adjusting energy inputs relative to outputs. The difference between the sum of energy inputs received by the body and the sum of energy outputs lost (plus any release of stored energy) ultimately determines whether the body warms (inputs > outputs), cools (outputs > inputs), or maintains a constant temperature (inputs = outputs). This chapter describes how water is used by ectotherms to help regulate body temperature: endotherms are discussed in Chapter 14.

The animal kingdom can be broadly divided into those animals that maintain a constant body temperature (mammals and birds) and those whose body temperature varies with the ambient environment (arthropoda, fish, amphibia, and reptiles). In both groups, water plays a vital role in regulating the organism's body temperature. This is partly due to the large specific heat capacity of water (to buffer against abrupt changes in the ambient temperature) but most importantly due to the ability of water to cool as a consequence of evaporation. The body's temperature is determined by its energy balance, which in turn is tightly coupled to the body's water balance.

Thermal regulation is a large and complicated subject comprising abiotic controls involving water's thermal properties (Chapter 7), biotic controls on water movement within the body's cells (Chapter 11), and behavioral responses and adaptations (Chapters 13 and 14). This chapter uses examples of amphibian, reptile, and insect species, creatures whose body temperature conforms to the ambient temperature, to illustrate how water is used to control body temperature. Chapter 14 describes how a mammal, *Homo sapiens*, uses water in a different set of processes to maintain a constant body temperature regardless of ambient temperature. In both cases, it will become apparent how the availability of and access to water determines not only the distribution and seasonal range of the organism, but also its behavior and reproduction as water is used to control body temperature and therefore life.

This chapter begins by explaining the concepts that differentiate temperature conformers (**ectotherms**) from temperature regulators (**endotherms**), and the role water plays in this thermoregulation. It describes the group that represents most species on Earth, the ectotherms. The ubiquitous yet threatened amphibian frogs are used to illustrate how water regulates body temperature and function, including arboreal (tree-dwelling) species, in moist environments. At the other extreme of the aridity scale for ectotherms are lizards, capable of inhabiting the driest regions on Earth. The important role of skin in regulating evaporative water loss, and the ways that lizard species have adapted to different habitats, are described. Finally, we examine how insects, which account for roughly 80 percent of known animal species, have adapted to occupy nearly every habitat on Earth from aquatic to extreme desert.

13.2 The Concept of Thermoregulation

Most mammals are **endotherms**, **temperature regulators** that maintain a balance between energy inputs and outputs, and therefore a constant body temperature. In contrast, **ectotherms** (Greek *ecto* for outside, and *therm* for heat), **temperature conformers** (e.g., insects, reptiles, fish), rely on their environment as a source of energy for activity and function (Figure 13.1). As a result, the body temperature of ectotherms varies almost directly with the ambient temperature, whereas that of endotherms does not (Figure 13.2). As opposed to generating heat through metabolism or cooling through sweating as endotherms do, ectotherms rely on **behavioral thermoregulation** (Rozen-Rechels et al., 2019). Both endotherms and ectotherms rely on water to help regulate body temperature.

The term **hydroregulation** has been used to describe the effects of water on thermoregulation, and **thermo-hydroregulation** when both the energy and water balances are integrally combined to regulate body temperature (Rozen-Rechels et al., 2019). This thermo-hydroregulation framework is well suited for ectotherms since body temperature is often dictated by location (specifically the availability of solar radiation), behavior (times of seasonal and daily activity; Stevenson, 1985), and the species' ability to resist water loss based on water availability (Rozen-Rechels et al., 2019). Warm, sunny locations often lack precipitation and accessible fresh water, and have a large vapor pressure deficit (i.e., low humidity), all of which stresses an organism's water balance despite its need to live in such environments to receive adequate thermal energy. For aquatic ectotherms (e.g., amphibians), water is usually continuously available,

Figure 13.1 Ectotherms (left: *Chalcosoma chiron*, 10 cm long) are generally small, with a large surface area to volume ratio, and control energy inputs through moving location and posture. Endotherms (right: standard poodles) are generally large with a small surface area to volume ratio and rely upon metabolism to maintain a constant body temperature. Photo credit: P. D. Blanken.

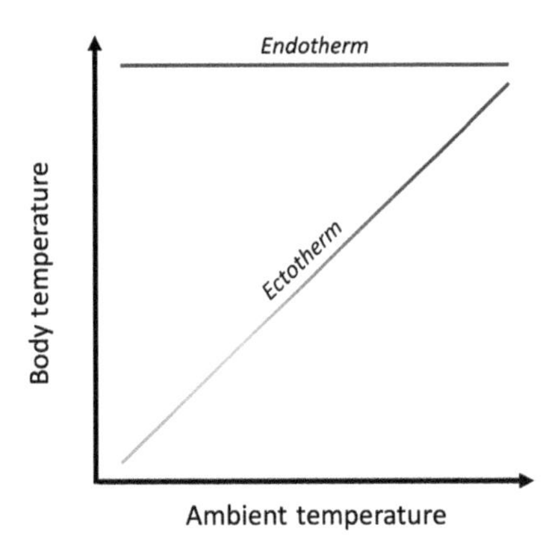

Figure 13.2 General relationship between body (line colors blue: cold; red: warm) and ambient temperature. Endotherms maintain a constant body temperature despite variations in ambient temperature. An ectotherm's body temperature varies directly with ambient temperature.

but those living in hot, dry, desert environments have developed unique strategies to balance water supplies with water demands.

Temperature is determined by the balance, or difference (change in energy storage, ΔE; W m^{-2}), between energy received (E_{in}; W m^{-2}) and energy lost (E_{out}; W m^{-2}) (Eq. 13.1):

$$E_{in} - E_{out} = \Delta E \,. \tag{13.1}$$

Any difference between energy inputs and outputs results in the warming or cooling (dT/dt; °C or K per second) of the body (object) through a layer (z; m) depending on the object's density (ρ; kg m^{-3}) and specific heat capacity (C_p; J kg^{-1} $°C^{-1}$ or K^{-1}) (Eq. 13.2):

Figure 13.3 Two content marine iguanas (*Amblyrhynchus cristatus*) basking in the sun on a warm, dark lava rock surface in the Galapagos Islands, Ecuador. Salt filtered out of their blood and then extruded by the specialized exocrine glands is visible as the white crust on the top of the head and neck. Photo credit: P. D. Blanken.

$$\begin{aligned} \Delta E &= \rho C_\mathrm{p} \frac{\mathrm{d}T}{\mathrm{d}t} z \\ \frac{\mathrm{d}T}{\mathrm{d}t} &= \frac{\Delta E}{\rho C_\mathrm{p} z} \end{aligned} \quad (13.2)$$

If $E_\mathrm{in} = E_\mathrm{out}$, ΔE is zero, meaning that energy is balanced and the body's temperature is constant. If the ectotherm seeks warmth, resting in a sunny location (**basking**) and maximizing the body's surface area contact with a dark, warm surface (through conduction) serves to increase E_in above E_out and therefore increase body temperature (ΔE positive) (Figure 13.3).

Equation (13.1) can be expanded to show the sources of energy an organism typically receives as inputs and outputs (Eq. 13.3):

$$\left(S\downarrow + L\downarrow + M\right) - \left(S\uparrow + L\uparrow + \lambda E + H + G\right) = \rho C_\mathrm{p} \frac{\mathrm{d}T}{\mathrm{d}t} z \quad (13.3)$$

where S is the shortwave radiation incident upon ($S\downarrow$) or reflected ($S\uparrow$) from the body, L is the longwave radiation incident upon ($L\downarrow$) or emitted from the body ($L\uparrow$), M is the metabolic heat production, λE is the latent heat flux (i.e., evaporative water losses), H is the sensible heat flux (i.e., convection), and G is the ground (surface) heat flux (i.e., conduction), with all terms in W m^{-2}. In general, H is negligible for all ectotherms since the body and ambient air temperatures are nearly equal (i.e., there is no temperature gradient), and M is small and poorly regulated. Heat conduction (G) often serves as an energy input term for ectotherms when the surface temperature is greater than the body temperature (Figure 13.3). The primary sources of energy to increase an ectotherm's body temperature are solar radiation and conductive heat inputs, which they gain by changing location or position.

Some of the terms in Eq. (13.3) can be energy inputs or outputs, or negligible, depending on the organism's behavior. The nonradiative energy terms λE, H, and G can be either energy inputs or outputs depending on the vapor or temperature gradients, which depend on the organism's location and activity. For example, the dark-skinned marine iguanas basking in the sun in Figure 13.3 are receiving large energy inputs from the sun and through conduction of heat from body contact with each other and with the dark volcanic rock, so the energy balance in this situation is:

$$\left(S\downarrow + L\downarrow + G\right) - \left(S\uparrow + L\uparrow + \lambda E\right) = \rho C_\mathrm{p}\frac{\mathrm{d}T}{\mathrm{d}t}z\ . \tag{13.4}$$

If the marine iguana dived into the ocean, the shortwave and longwave radiation inputs and outputs would be small and roughly balance, the heat conduction term would become negative as heat was conducted away from the warm body into the cooler water, and evaporation would be negligible (no skin evaporation) so the energy balance in this situation is:

$$M - G = \rho C_\mathrm{p}\frac{\mathrm{d}T}{\mathrm{d}t}z\ . \tag{13.5}$$

Equation (13.5) shows that the swimming iguana's body will cool unless metabolic heat production can offset the body's conductive heat loss to the water.

The state of water influences the magnitude of each of the energy balance terms shown in Eq. (13.3) to various extents. The atmosphere's water vapor concentration and cloud cover influence the amount of incident shortwave ($S\downarrow$) and longwave radiation ($L\downarrow$). The organism's internal water content influences its metabolism and ability to cool through evaporation. Skin temperature is affected by the rate of evaporative water loss on the skin, and therefore water affects the organism's emission of longwave radiation and sensible heat. The quantity of water in the material in direct contact with the organism affects the material's thermal conductivity (Chapter 3), thus directly affecting the heat conduction between the body and ground. Lastly, how fast the organism warms or cools ($\mathrm{d}T/\mathrm{d}t$) depends on the organism's density and specific heat capacity, both of which change with its tissue water content.

Evaporation is often a large term in the energy balance of an organism and is always an energy output term since evaporation requires energy (λ, the latent heat of vaporization); thus, temperature decreases at the location of evaporation on the body. Whether from a lake, soil, stoma (transpiration), or an organism's skin, the difference in water vapor concentration (vapor pressure or density) between the location of evaporation and the air above is one of two crucial variables governing the evaporation rate. The other crucial variable is the **surface resistance** to water vapor transfer, or its reciprocal, a measure of the conductance (sometimes expressed as a transfer coefficient). The surface resistance term often comprises several variable resistances acting in series, involving the surface resistance to water vapor transfer, and another related to wind speed (a **boundary-layer resistance**). This framework, where the flux is equal to a gradient divided by a resistance (or multiplied by a conductance) is captured by the **Ohm's Law** electrical circuit analogy. The dilemma faced by ectotherms is the need to maintain an adequate body temperature with the available energy received in the ambient environment (e.g., solar radiation), given their limited ability to regulate body temperature through metabolism, while concurrently minimizing water loss in environments where the evaporative demand is often high. Compounding this is the fact that most ectotherms are small, so their surface area to volume ratio is large, making them susceptible to desiccation. The development of a large skin resistance to water vapor transfer is the primary means to minimize water loss in dry environments, so the properties of the skin related to water conservation are key.

13.3 Amphibian Species: Frogs

Amphibians are vertebrate ectotherms that live within freshwater aquatic ecosystems. Most amphibians begin life as aquatic larvae with gills and later emerge from the water in adult form with gills transformed into lungs. Amphibians such as **frogs** (order **Anura**) transfer water, oxygen, carbon dioxide, and ions (e.g., sodium, potassium) across their skin (Vanburen, Norman and Fröbisch, 2019), so it is important that the skin remains moist to facilitate this exchange (Figure 13.4). Frogs can breathe through their lungs and across the lining of their mouth, but their skin provides the largest and most effective respiratory surface, especially when submerged.

The abundance of frog species and their ecological significance mean they are used as an example of a "wet-skinned" ectotherm that is highly dependent on water. The modern **Amphibia** class consists of *Anura* (frogs), *Caudata* (salamanders), and *Gymnopiona* (caecilians). Of the roughly 8,600 known amphibian species, nearly 90 percent are frogs, used with other amphibians as sentinel species indicators of aquatic ecosystem health and environmental change. A global-scale synthesis study based on 936 amphibian populations showed that although there is considerable spatial and temporal variability, amphibian populations declined over the years 1950–1997 (Houlahan et al., 2000), with pathogen outbreaks and habitat loss often the cause (Daszak, Cunningham and Hyatt, 2003).

The importance of water and a moist skin for the health of frogs has been known for decades (Mellanby, 1941). Seventy frogs of the species *Rana temporaria* L. ("grass" or "brown frog"), a

Figure 13.4 Yellow-banded dart frogs (*Dendrobates leucomelas*) native to the tropical rainforests of northeastern South America. Their skin excretes toxins and requires moist, humid conditions to minimize evaporative water loss. Photo credit: P. D. Blanken.

common nonarboreal terrestrial frog species found in Great Britain, Europe, and northwestern Asia, were used. Evaporation from the frogs' skin resulted in a decrease in their core temperature by 3 °C over an hour in still air and increased to 5 °C over 5 minutes when exposed to a slight 1.9 m s^{-1} breeze. The evaporation rate, measured by the change in body weight, was 0.3 grams per hour in still air increasing to 3.2 grams per hour when exposed to the breeze. The maximum tolerable evaporative weight loss was roughly 25% of the frog's weight, similar to the 30% value reported in recent studies (e.g., Tracy, Christian and Tracy, 2010).

These studies showed that unlike vascular plants with stomata or humans with sweat glands, frogs and many amphibians cannot dynamically regulate water loss in response to changing ambient conditions. Within a short amount of time, loss of water due to evaporation from skin results in dangerous conditions exacerbated by slight increases in wind speed. Field studies have shown that prolonged periods of drought pose a significant threat to terrestrial frog populations, and that short-term drought relief is insufficient to allow populations to recover (Evans et al., 2020). As maximum temperatures increase and rainfall decreases, the decreased access to moisture, and increased risk of desiccation, is closely associated with frog population declines (Evans et al., 2020).

The need for water begins at the first stage of a frog's life. In the embryonic stage, most frog species require access to liquid water both to deposit eggs and for larvae to develop. Other frog species (e.g., **tree frogs** or **arboreal frogs**) lay their eggs on the underside of leaves located above water. Larvae develop when they fall into the water when hatched. This life strategy, common in many tree frog species, is referred to as **amphibian embryos** since part of their life cycle is completed in and part out of water. In all cases, the dependence of frogs on water availability makes them highly susceptible to environmental change. Climate-induced decreases in wetland water depths have been shown to increase UV-B radiation exposure, resulting in increased embryo vulnerability to pathogen infection (Kiesecker, Blaustein and Belden, 2001). Decreased water levels or the complete elimination of wetlands resulting in habitat loss and fragmentation severely threaten amphibian biodiversity (Sodhi et al., 2008) and lead to overall population decline (Cushman, 2006). As discussed in Chapter 5, wetlands host more than 40% of the world's species, provide traditional medicines for 80% of the world's population (Mitra, Wassmann and Vlek, 2003), and play a large role in the water and biogeochemical cycles. Yet they face constant and continued degradation and loss, with 50% of the world's wetlands having been lost since 1900, and possibly 87% since 1700 (Davidson, 2014).

Proper embryonic development in frogs, especially tree frogs, requires narrow tolerable limits in water temperate, humidity, and dissolved gases such as oxygen. The endangered terrestrial-breeding frog species *Geocrinia alba* and *Geocrinia vitellina*, for example, require stable and cool habitats and cannot tolerate temperatures greater than roughly 30 °C (Hoffmann, Cavanough and Mitchell, 2021). The hydration threshold for these species, defined as the water potential of a substance above which an individual gains water and below which it loses water, is −50 kPa, the lowest recorded for any amphibian species, indicating extreme sensitivity to losing water to dry surfaces (Hoffmann, Cavanough and Mitchell, 2021). Acute oxygen stress can stimulate egg hatching (amphibious species that lay terrestrial eggs hatch aquatic larvae when the eggs are flooded), and chronic oxygen stress can accelerate hatching in several amphibious species, including the red-eyed tree frog *Agalychnis callidryas* (Warkentin, 2002).

The appropriate timing of hatching is important for survival; decreases in dissolved oxygen concentration due to water mixing or eutrophication, or fluctuations in the wetland water level at critical hatching times, can therefore affect population vitality.

Arboreal frogs with amphibious embryos spend a critical portion of the embryonic life exposed to the atmosphere. The timing of hatching and survival of embryos requires high humidity. Premature hatching and reduced hatchling size of the arboreal gliding tree frog *Agalychnis spurrelli* occurred when the relative humidity was decreased from 99% to 96%, with a 98% mortality rate when raised with a 92% relative humidity (González, Warkentin and Güell, 2021). These large impacts on hatchling size and mortality with small decreases in an already high relative humidity show how predicted changes in rainfall variability and increased drought in the tropics where these arboreal frog species reside will likely play a major role in species survival.

Small organisms such as frogs, with a large surface area to volume ratio, are susceptible to heat and water loss. Arboreal frog species tend to be much smaller than nonarboreal species, in part since smaller size and mass makes it easier to move amongst branches and leaves. Arboreal frog length ranges from an average of 7.7 mm for *Paedophryne amanuensis*, discovered in 2009 in Papua New Guinea as the world's smallest known vertebrate (Rittmeyer et al., 2012), to 14 cm for the Australian white-lipped tree frog *Nyctimystes infrafrenatus*, so the surface area to volume ratio for arboreal species is high. Nonarboreal species are generally larger, up to 32 cm and 3.3 kg for the world's largest frog, the endangered *Conraua goliath*, found in central Africa (Nguete Nguiffo, Mpoame and Wondji, 2019). With a smaller surface area to volume ratio and a higher rate of evaporation from the skin than arboreal species, these nonarboreal species find thermoregulation difficult, requiring habitat selection and the use of behavioral posturing to reduce skin water loss (Tracy, Christian and Tracy, 2010). To survive drier conditions in the terrestrial environment, the smaller body size and increased skin resistance to water loss play an important role in avoiding desiccation and regulating gas exchange (Tracy, Christian and Tracy, 2010).

Arboreal frog species are indeed better adapted to regulating evaporative water loss from the skin and are therefore better thermoregulators. By exposing 7 arboreal and 11 nonarboreal frog species to different wind speeds in a wind tunnel, Wygoda (1984) found much lower rates of water loss from the arboreal species. Based on changes in weight, the area-specific cutaneous evaporative water loss from arboreal species was 47 percent lower than for nonarboreal species. Arboreal species had a lower rate of water loss regardless of their weight. The frogs' boundary-layer resistance did not differ between the two or vary with weight, but the skin's resistance to water vapor transfer was 41 times higher in the arboreal species. Wygoda (1984) also found that arboreal-species body and skin temperature averaged 4.4 °C higher in comparison to the nonarboreal species.

A relationship between the rate of water loss through the frog's skin and body temperature and size has been found in many studies, and mathematical models based on physiology agree. Using a biophysical model, Tracy, Christian and Tracy (2010) found both the frog's skin resistance to water vapor transfer and body size had a significant effect on body temperature. Small frogs with a weight less than 10 g could elevate their body temperature while basking even with a small skin resistance, whereas large frogs required a large skin resistance to avoid desiccation. Frogs with a high skin resistance greater than 25 s cm^{-1} were not able to regulate body temperature, suggesting that the primary role of such a large skin resistance is to conserve water rather than regulate body temperature. Overall, frog species with a combination of a skin resistance and body size that promotes slow desiccation rates tend to be the arboreal species (Tracy, Christian and Tracy, 2010). Wet-skinned frog species (the nonarboreal species) had a much smaller skin resistance to water loss, hence were less able to regulate body temperature than the arboreal species that spend much more time away from water.

Overall, studies of the evaporative water loss from several frog species (and most ectotherms, as discussed below) conclude that the skin's resistance is a key variable to regulate cutaneous water loss. Using an **Ohm's Law** analogy, the **cutaneous evaporative water loss** (E_{skin}; kg m^{-2} s^{-1}) can be estimated from the difference in vapor pressure or density between the skin (ρ_{skin}; kg m^{-3}) and the air (ρ_{air}; kg m^{-3}) divided by the total resistance (the sum of the boundary and cutaneous resistances) to vapor loss from the skin (R_{skin}; s m^{-1}) (Eq. 13.6):

$$E_{skin} = \frac{\rho_{skin} - \rho_{air}}{R_{skin}}. \tag{13.6}$$

In practice, R_{skin} is calculated from Eq. (13.6) when the evaporative water loss has been measured based on weight loss and the vapor density difference is measured (e.g., Young et al., 2005).

To calculate the rate of water loss through the skin using Eq. (13.6) requires the calculation of the difference in vapor density between the skin and the atmosphere. The saturation vapor pressures (e_s) can be calculated by using the Teton's solution to the Clausius–Clapeyron temperature–pressure relationship at the appropriate temperature (T) measured in °C (Eq. 13.7):

$$e_s = 0.61121 \exp\left(\frac{17.502T}{240.97 + T}\right). \tag{13.7}$$

If the frog's skin temperature is 20 °C (measured using an infrared thermometer) and the air temperature is 25 °C, the vapor pressures at the frog's skin ($e_{a(skin)}$; kPa) and in the air ($e_{s(air)}$; kPa) are 2.34 kPa and 3.17 kPa, respectively:

$$e_{s(skin)} = 0.61121 \exp\left(\frac{17.502 \times 20}{240.97 + 20}\right) = 2.34 \text{ kPa}$$

$$e_{s(air)} = 0.61121 \exp\left(\frac{17.502 \times 25}{240.97 + 25}\right) = 3.17 \text{ kPa}.$$

These saturation vapor pressures can be converted to units of vapor density (ρ; kg m^{-3}) using the **Ideal Gas Law** with pressure (P) converted from kPa to Pa (1 Pa = 1 J m^{-3}), temperature (T) converted from °C to kelvin (by adding 273.15), and using the gas constant (R) for water vapor (R_v = 461 J K^{-1} kg^{-1}) (Eq. 13.8):

$$\rho = \frac{P}{RT}. \tag{13.8}$$

Using Eq. (13.8) to convert saturation vapor pressure to saturation vapor density at the frog's skin and in the air (respectively) gives:

$$\rho_{s(\text{skin})} = \frac{2{,}340 \text{ J m}^{-3}}{\left(461 \text{ J K}^{-1}\text{ kg}^{-1}\right) \times (293.15 \text{K})} = 0.0173 \text{ kg m}^{-3}$$

$$\rho_{s(\text{air})} = \frac{3{,}170 \text{ J m}^{-3}}{\left(461 \text{J K}^{-1}\text{kg}^{-1}\right) \times (298.15 \text{K})} = 0.0231 \text{ kg m}^{-3}.$$

These saturation vapor densities are then multiplied by the relative humidity at the frog's wet skin (assumed to be 100%) and air (measured at 50%, for example) to provide the actual vapor densities at the skin and in the air:

$$\rho_{\text{skin}} = \text{RH}_{\text{skin}} \times \rho_{s(\text{skin})} = 1.00 \times 0.0173 \text{ kg m}^{-3} = 0.0173 \text{ kg m}^{-3}$$

$$\rho_{\text{air}} = \text{RH}_{\text{air}} \times \rho_{s(\text{air})} = 0.50 \times 0.0231 \text{ kg m}^{-3} = 0.0115 \text{ kg m}^{-3}.$$

Once the vapor densities are known, if the evaporative water loss is determined by measuring the frog's weight loss over time, the skin resistance can be calculated as the residual of Eq. (13.6). Alternatively, the evaporative water loss can be calculated from Eq. (13.6) if a skin resistance value is found from the literature. Based on evaporative weight loss, Young et al. (2005) calculated that the cutaneous skin resistance for the aquatic frog species *Litoria dahlii* was 2.3 s cm^{-1}. Using this resistance value for R_{skin} and the difference in the vapor density (converted from kg m^{-3} to g cm^{-3}) between the frog's skin and the air (calculated above) gives an evaporative water loss per area rate of 2.52×10^{-6} g cm^{-2} s^{-1}, equivalent to 9.07 mg cm^{-2} h^{-1}, comparable to the values measured by Young et al. (2005):

$$E_{\text{skin}} = \frac{\rho_{\text{skin}} - \rho_{\text{air}}}{R_{\text{skin}}} = \frac{1.73 \times 10^{-5} \text{ g cm}^{-3} - 1.15 \times 10^{-5} \text{ g cm}^{-3}}{2.3 \text{ s cm}^{-1}} = 2.52 \times 10^{-6} \text{ g cm}^{-2}\text{s}^{-1}.$$

The skin's surface area is needed to calculate the frog's rate of evaporative water loss. Wygoda (1984) developed an allometric equation to estimate the frog's surface area (cm^2) from mass (g) (Eq. 13.9):

$$A_{\text{frog}} = 9.9 M_{\text{frog}}^{0.56}. \tag{13.9}$$

Using Eq. (13.9) with $M_{\text{frog}} = 25$ g for *Litoria dahlii* gives $A_{\text{frog}} = 60$ cm^2. Wygoda (1984) multiplied A_{frog} by 2/3 to account for only the surface area exposed to the atmosphere, thus reducing A_{frog} to 40 cm^2. This surface area multiplied by E_{skin} gives a rate of water loss of 362.9 milligrams per hour:

$$E_{\text{skin}} = 40 \text{ cm}^2 \times 9.07 \frac{\text{mg}}{\text{cm}^2\text{h}} = 362.8 \frac{\text{mg}}{\text{h}}.$$

The time for the frog's initial 25 g body mass to reach the critical 70% from desiccation water loss (Tracy, Christian and Tracy, 2010) is an estimated 20.7 hours:

$$\frac{\text{Mass} - 0.7 \times \text{mass}}{E_{\text{skin}}} = \frac{25.0 \text{g} - 17.5 \text{g}}{0.363 \text{g h}^{-1}} = 20.7 \text{h}.$$

Based on the conditions used in this example, the frog's critical mass is reached in just under 21 hours, showing how vital the internal water content is. This Ohm's Law framework can be used to estimate the evaporative water loss from frogs or any species if some reasonable estimate of the resistance term is available.

13.4 Desert Species: Lizards

Numerous lizard species inhabit a range of environments and are especially well adapted to arid locations. Currently over 6,000 species of lizards, ranging in size from a few centimeters to 3 m in length, are found in warmer regions of all continents except Antarctica below elevations of roughly 5,000 m. Most inhabit terrestrial regions, but some have adapted to marine environments. For example, the large **marine iguana** (*Amblyrhynchus cristatus*), found only on the Galapagos Islands, eats underwater algae and excretes the ingested salt from specialized exocrine glands in the nostril (Figure 13.3). Most terrestrial lizards are predator carnivores consuming insects, and most species are completely covered in scales, a key adaptation for minimizing water loss in dry environments.

Whereas frogs have a low desiccation tolerance and lack the ability to regulate water loss other than through behavior (e.g., movement to moist habitats), the lizards' skin expands their range to areas far from water. Made from keratin (a protective protein also found in human hair, skin, and nails), these scales prevent oxygen absorption through the skin, so lungs with a muscular diaphragm are required for breathing. The group of *suamate* (scaled) lizards are an example of a large group of species that inhabit a wide range of environments in comparison to the wet-skinned amphibians. There is a thermal advantage also in that the lowering of the skin temperature that occurs with evaporation does not occur in the thick-skinned lizards. Therefore, high body temperatures can be maintained.

Since the **vapor pressure** (or **density**) **deficit** will be large in hot, dry environments, a large skin resistance to water vapor transfer should be required to offset the greater evaporative water demand. This hypothesis has been tested by several studies. Mautz (1982) measured both the respiratory (through breathing) and cutaneous (through the skin) evaporative water losses in dry air with increasing air temperatures from five lizard species in the *Xantusiidae* family spanning a range of mesic and arid environments. The respiratory water loss was roughly a third of the total water loss and did not show consistent patterns with temperature among species. The cutaneous water loss, however, increased substantially for all species with increasing temperature except for desert-dwelling species. The lizard species living in humid caves located in tropical forests of southern Mexico had the highest cutaneous water loss. Since these species were not adapted to arid conditions, their cutaneous water loss increased the most as temperature increased. The differences in water loss through the skin of these species were directly attributed to differences in skin resistance to the water flux created by epidermal lipids, an evolutionary adaptation to arid environments (Mautz, 1982).

Given the range in lizard habitats from moist to dry, an understanding of the tolerance within a single species to changes in aridity and temperature can help us understand their capability to adapt to various changing conditions. The British Virgin Islands are inhabited by the species *Anolis cristatellus*, a relatively small (length from snout to cloacal vent 7.5 cm) arboreal species recognized by the small crest or fin on their back. *Anolis cristatellus* can be found across

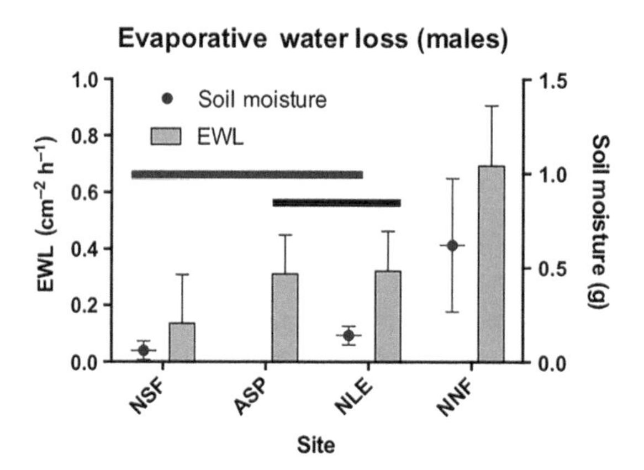

Figure 13.5 Evaporative water loss (EWL) from populations of the small male lizard *Podarcis erhardii* living across a range of moist habitats on and around Naxos, Greece, as shown by changes in soil moisture. See Belasen et al. (2017) for details, including the site descriptions. Reproduced with permission from Belasen et al. (2017).

a wide range of habitats from subtropical moist forests to dry ridges with sparse vegetation. Dmi'el, Perry and Lazell (1997) took advantage of this wide range in habitat to examine how evaporative water loss varied among eight insular *A. cristatellus* populations. Air temperature and relative humidity were measured at the locations inhabited by each population, and a qualitative aridity index was assigned based on the vegetation type and cover. Statistically significant correlations were found between aridity and evaporative water loss (based on lizard weight changes) and the sum of the skin and boundary layer resistances to water vapor loss. Cutaneous water loss was on average 75 percent of the total water loss. As the habitat aridity increased, the cutaneous water loss rate increased from 1.5 to 10.3 mg H_2O per gram body mass per hour. Those populations living in arid habitats had a much higher surface resistance to water loss than those living in humid habitats, showing the capacity within one species to adapt to the ambient conditions. As the habitat aridity increased, the resistance to water loss also increased from 28.5 to 199 s cm^{-1}. As reported in the Mautz (1982) study, Dmi'el, Perry and Lazell (1997) believe that this evolutionary adaptation within one species is an example of phenotypic plasticity.

The finding that an increased skin resistance to water loss in arid environments can evolve within one species of lizards was confirmed by Belasen et al. (2017). The species *Podarcis erhardii* (snout–vent length 7.5 cm) inhabits a broad range of habitats on the south Balkans and central Aegean Islands. Populations in four study sites located on Naxos (Greece) were chosen to represent a wide range in microclimates that varied with elevation and aspect. Evaporative water loss was measured by weight loss, and moisture availability by soil water content measured gravimetrically. As found in previous studies, habitat aridity influenced water loss; as soil moisture increased, so did the evaporative water loss. The *P. erhardii* populations living in dry habitats were better adapted to arid conditions and were better able to limit water loss than the moist-habitat populations (Figure 13.5). The authors stated that the response of *P. erhardii* to changes in temperature across habitats was rigid compared with their response to changes in availability of water.

It is reasonable to assume that the results reported by Mautz (1982), Dmi'el, Perry and Lazell (1997), and Belasen et al. (2017) for three different lizard species in three different locations could be expected for other lizard species living across habitat gradients. *Psammodromus algirus* is widely distributed across a range of habitats from humid to arid in the Mediterranean region of the Iberian Peninsula, southwestern France, and northwestern Africa. Given such a wide distribution, Sannolo et al. (2020) expected that the evaporative water loss from *P. algirus* would vary along an elevational gradient as air temperature decreased and relative humidity increased with elevation. Populations living in the lowlands acclimated to the warm and dry conditions should have lower rates of water loss than those living in the colder, humid higher elevations. The evolutionary development of high skin resistance to water loss in the lowland populations due to constant exposure to warm and dry conditions would be the explanation for the difference in water loss rates within one species of lizards.

Psammodromus algirus showed unexpected responses to air temperature and humidity. Across a 40-km-long elevation transect spanning low, mid, and high elevations (mean elevations 650 m, 850 m, and 1,250 m above sea level), Sannolo et al. (2020) measured the lizard's body temperature and evaporative water loss. Surprisingly, body temperature did not show any significant variation with elevation, but evaporative water loss did. The lack of a significant difference in air temperature over the small 600-m elevation range may explain why the lizards' body temperature did not vary with location. The ambient relative humidity, however, did show a significant increase with elevation. Contrary to expectations, the water loss rates decreased significantly with elevation, with lizards from low-elevation populations losing more water than those from high-elevation populations. Apparently the low-elevation populations had not developed sufficient skin resistance to water vapor transfer to offset the larger vapor pressure gradients experienced at lower elevations. Sannolo et al. (2020) suggested high-elevation populations were more resistant to water loss owing to exposure to more extreme and unpredictable meteorological conditions than low-elevation populations.

13.5 Insects

Arthropoda is the largest animal phylum with over 1.3 million species, accounting for roughly 80 percent of all known animal species. **Insects** (Insecta) are the largest class of Arthropoda and differ from the other arthropods by their segmented exoskeletal body consisting of head, abdomen, and thorax (usually with three pairs of legs and one or two pairs of wings). With over one million described species, insects account for over 80 percent of all arthropods (Zhang, 2013). It is very likely that there are more undiscovered than discovered species of insects. Most insects that have been named and classified are from temperate zones, not the tropics. In the tropics, animal and bird species are much easier to enumerate than in temperate zones since they are generally larger and two to three times more plentiful (May, 1986). If we assume this known-to-unknown ratio of the abundance of animals and birds in temperate zones might apply to the tropics, the total number of insect species would be between two and three million. In combination with other global species extrapolation methods based on assessing the proportion of new species in samples, discovery curves (number of species plotted against time), or changes in body size over time, the potential number of insect species increases to 5.5 million (Stork, 2018).

With a large surface area relative to volume making thermoregulation difficult, one would expect insects to inhabit warm and moist locations, as many indeed do. Aquatic insects require liquid water for part or all of their life cycle, and thus have an obvious requirement for access to liquid water. Most aquatic insects begin life entirely within water (e.g., during the larval stage) before emerging to spend their mature lives as terrestrial adults. Terrestrial insects spend their entire lives on land, but still require water to live, and have evolved strategies to survive even in extremely arid regions. All insects require access to water, and their sheer number, distribution, and length of time inhabiting Earth are testimony to their survival success.

13.5.1 Aquatic Insects

There are approximately 76,000 known aquatic insect species living part or all of their life in water, predominantly in freshwater habitats. There is no distinct taxonomical unit within the *Insecta* class for aquatic insects since, unlike many organisms, aquatic insects likely evolved from terrestrial insects (Derka, Zamora-Muñoz and de Figueroa, 2019). The first evidence of insects in the fossil record dates back to over 400 million years ago, whereas the first fossil evidence of insects that were clearly living exclusively in aquatic habitats dates back to 320 million years ago (Lancaster and Downes, 2013). Unless new fossil findings prove otherwise, there was likely some advantage for terrestrial insects to transition to aquatic environments, as an aquatic origin is unlikely. In addition to the fossil record, there are strong physiological arguments that the respiratory tracheal system more than likely first developed in terrestrial insects before adapting to the aquatic environment (Pritchard et al., 1993). Given the small body size and high surface area to volume ratio for insects, perhaps arid terrestrial conditions, and the inability to conserve water, played a role in this evolutionary trial of going back into the water instead of emerging from it.

Aquatic insects serve as the foundation of the aquatic food chain and are therefore robust indicators of ecological health and diversity. With habitat limitations imposed by dissolved oxygen, water temperature, water chemistry, water flow rates, and turbulence (Hershey and Lamberti, 2001), the diversity and health of aquatic insects at a given location are indicative of water quality and watershed health, an important component of the **River Continuum Concept** (Chapter 5). When submerged, the solute concentration of the ambient water plays a crucial role in maintaining a suitable internal water balance for the aquatic insect. As described in Chapter 11, changes in water potential develop as the solute concentration changes. Water will move into or out of an aquatic insect's cells depending on the gradient in the solute water potential that develops across cell membranes. Using the solute water potential equation (Eq. 11.7), pure water has a solute water potential of zero joules per kg, per m^3, or per mol. If an insect larva or adult aquatic insect were placed in pure water, water would flow into the organism since the solutes within their cells decrease the water potential. In reality, the solute water potential in freshwater aquatic insect habitats is never zero since the water contains dissolved mineral and organic compounds. Living in fresh water as most aquatic insects do, however, can create the situation of gaining too much water and diluting vital body compounds (electrolytes). Each day, the osmotic inflow of water through the body of aquatic insects can amount to 2–10 percent of their total body mass (Noble-Nesbiti, 1990). In contrast, living in brackish or saline water may result in the loss of too much internal water. A higher solute concentration in the ambient water creates a lower solute (osmotic) water potential outside compared with inside the insect and therefore a loss of internal water.

There are two mechanisms available to aquatic insects to control the flow of water into or out of their bodies. One is osmosis (**osmoregulation**) and the other is the skin (**cuticle**) resistance to the flux of water. Both are well represented through **Ohm's Law**; osmoregulation adjusts the gradient in the water potential, analogous to the voltage; the cuticular resistance imposes a limitation to water transfer for a given water potential gradient, analogous to the resistance; and the flux of water across the body is analogous to the current. If the choice is to regulate the water flux through osmosis, such as for **mosquito larvae**, the price to pay for this control is energy. Mosquito larvae have been the subject of several studies to examine the osmoregulation mechanism since they have an expansive habitat range, including several species that are found in both fresh and saline water, and mosquitos are a deadly disease vector. Larvae in general have a poorly developed and highly permeable skin favoring osmoregulation over cuticle resistance to control water flow.

Larvae of the mosquito *Aedes aegypti* (the **yellow fever** mosquito capable of transmitting several viruses and diseases) have been found outdoors in water with salinities up to 16 g of salt per kilogram of water (parts per thousand) (de Brito Arduino et al., 2015). In laboratory settings, *A. aegypti* larvae survived in water containing between 0% and 30% seawater (10.5 parts per thousand salinity, and up to 14 parts per thousand) (Edwards, 1982; de Brito Arduino et al., 2015). For comparison, the average seawater salinity is 35 parts per thousand, so larvae surviving in water with a lower salinity could be found in the brackish waters of estuaries, tidal pools, or small pools of seawater diluted by rainwater.

As with plant cells, mosquito insect larvae can regulate the loss of body water to saline water by increasing the solute concentration inside the body to decrease the water potential gradient. Adjusting the internal solute concentration requires energy to produce ions, so the development of larvae exposed to either very diluted solutions or very saline water can be affected. Edwards (1982) found the oxygen consumption for *A. aegypti* larvae did not vary with changes in salinity, suggesting that the energy cost for osmoregulation was small. De Brito Arduino et al. (2015), however, found that the time for development from larval hatching to adult emergence for *A. aegypti* significantly increased, and adult body size significantly decreased, as salinity increased. Therefore, even if the energy requirement for osmoregulation is small, the development cost for life in the larval stage in waters that are either too fresh or too saline can be large.

The second mechanism to regulate water intake and loss, the development of a large skin (cuticle) resistance, is used by some aquatic insects in the adult life stage. An impermeable cuticle can be especially effective in restricting water inflow in freshwater environments (Noble-Nesbiti, 1990), but also restricts the insect's ability to obtain the required dissolved oxygen. Measurements of the water loss in dry air for 10 species of aquatic insects showed that cuticles for these aquatic insects were more permeable than those for terrestrial species (Holdgate, 1956). Beament (1961) suggested that a thin, monolayer of organized lipid, wax, or grease served to provide the semipermeable cuticle of the adult aquatic insects. Perhaps such a porous cuticle is an adaptation to minimize water exchange but allow for sufficient gas exchange.

13.5.2 Terrestrial Insects

Although common in warm, moist tropical environments, terrestrial insects have adapted to arid regions. Similar to the lizard studies, insect water relations are often based on observations of rates of water loss from populations living along elevational or latitudinal gradients spanning a range in temperature and moisture conditions. As shown in the frog's water loss example, the cuticular resistance to water loss is often calculated by measuring the decrease in mass upon desiccation when the skin-to-air vapor density gradient is measured (Eq. 13.6). The respiratory water loss is often measured as a function of the metabolic rate.

The range in temperature and moisture found in regions that terrestrial insects inhabit is wide, including the exceptionally hot, arid climates of the world's deserts where several species are found. Desert insects are well adapted to tolerate high surface temperatures. In the African Sahara and Namib Deserts, with summertime air temperatures above 40 °C and an annual mean precipitation of less than 100 mm, insects such as ants, locusts, and beetles are well known for their ability to tolerate dry desert conditions. The **Sahara Desert ant** (*Cataglyphis bicolor*) has been observed to forage on surfaces exceeding 70 °C. Under controlled laboratory observations, the desert ant species *Ocymyrmex barbiger* foraged on surfaces with a temperature as high as 51.5 °C (Marsh, 1985). The **desert locust** *Schistocerca gregaria*, responsible for most of the word's crop damage, tolerates a surface temperature of 55 °C before moving off the ground (Maeno et al., 2016). The ground-dwelling desert beetle species can tolerate surface temperatures approaching 50 °C (Edney, 1971). One **desert beetle**, *Onymacris plana*, is the fastest pedestrian insect known, running at an average speed of 90 cm s^{-1} (close to the typical human walking speed) across sand dunes to seek shade provided by the occasional plant (Cloudsley-Thompson, 2001). Each of these desert-dwelling insect species has demonstrated the ability to cope with extreme thermal environments.

Owing to the lack of water vapor, arid environments commonly also experience low nighttime temperatures in addition to the searing daytime heat. When the body temperature of an insect, adult or in the immature larval stage, decreases below the freezing point, the danger of internal ice crystal formation becomes a life-threatening concern, yet insects have adapted to this threat also. Freezing resistance and body temperatures well below 0 °C have been measured for several desert-dwelling insects. For example, freezing points of −7.4 °C, −10.4 °C, and −11.8 °C have been documented for the desert beetles *Adesmia antiqua*, *Trachyderma hispida*, and *Eleodes* species, respectively (Cloudsley-Thompson, 2001). On Marion Island, located in the Subantarctic Indian Ocean, the larva of the flightless moth *Pringleophaga marioni*, showed 100 percent survival of freezing at a body temperature of −6.5 °C (Klok and Chown, 1997).

13.6 Insect Adaptations to Conserve Water

The fact that insects represent by far the majority of the world's species and are found in nearly all terrestrial habits speaks to their ability to conserve water through several physiological and behavioral adaptations. Those insects living in cooler, moister environments passively lose most water through their skin, with smaller amounts lost through respiration (Cloudsley-Thompson, 2001). Some have reported, however, that insects living in damp environments can control the rate of cuticular water loss by adjusting their internal water potential (Galbreath, 1975). Insects living in arid regions (e.g., desert beetles) have a large surface resistance to water vapor transfer provided by a water-impermeable cuticle or a water-resistant superficial epicuticular lipid layer (Noble-Nesbiti, 1990). Some desert insects develop seasonal surface wax deposits, wax blooms, that vary in color to reflect solar radiation and effectively limit water loss (Cloudsley-Thompson, 2001). Evidence of the dominance of respiratory over cuticular water loss in desert-dwelling insects comes from studies finding a strong linear relation between metabolic rate and rate of water loss (Zachariassen, 1996). By reducing their metabolic rate, desert insects can also reduce their water loss. Other water-conserving strategies observed in desert insects include the ability to reabsorb water from their urine (Zachariassen, 1996).

Some insects have developed unique behaviors to collect water. Fog can be common in cool, coastal desert regions, and Hamilton and Seely (1976) observed the Namib Desert beetle *Onymacris unguicularis* employing a unique fog-basking behavior to collect water. These beetles are normally found buried in sand dunes at night, yet when fog is present, they have been observed to emerge from the sand, climb to the crest of sand dunes, and extend their abdomen upwards into the wind, allowing fog to collect on their body and trickle down into their mouth. The water captured by such **fog basking** averaged 12% (with a range from 0% to 34%) of the weight of an individual beetle prior to basking. Several desert beetle species have been observed to collect water that condensed (or collected) on vegetation (Cloudsley-Thompson, 2001).

One small desert beetle species, the **tenebrionid beetles** of the Namib Desert, also used a unique behavior to collect water from fog. Three species of tenebrionid beetles were observed building small trenches on the dunes, perpendicular to the prevailing direction of moisture-laden air flow, that collected water in sufficient quantities to increase the water content by 13.9 percent for one population during one fog event (Seely and Hamilton, 1976).

13.7 Flying Insects: Honeybees

Honeybees (genus *Apis*), through pollination, play a fundamental role in plant ecology and agriculture across the globe where flowering plants are found. Likely native to Europe, the species ***Apis mellifera*** was introduced to the Americas by European explorers and colonists. As with many insect species, bees have developed the means to inhabit a wide range of temperature and moisture environments. Coupled with their high activity levels, bees have several unique adaptations to regulate temperature which are closely coupled to the use of water.

Temperature dictates several important aspects of a bee's life. Foraging and flight take place within an air temperature range from 10 to 40 °C with the highest foraging activity occurring around 20 °C (Abou-Shaara et al., 2017). Since the collection of nectar is restricted beyond this air temperature range, when it is too cold the consumption of stored honey is used to generate body heat through metabolism. Worker bees cluster around the queen bee at the center of the colony, shivering, to generate heat to raise the temperature to between 32 and 36 °C (Jones et al., 2004). Those on the colder periphery will rotate into the center.

When the colony temperature is too high, the fanning of wings is used to help lower the temperature through evaporative cooling. In the North American Sonoran Desert, bees returning to the hive with regurgitated water appeared when the air temperature reached 20 °C, and 40 percent of the returning bees returned with water when the air temperature reached 40 °C (Cooper, Schaffer and Buchmann, 1985). Throughout the year, water is a key requirement for thermoregulation, for honey production (dilution), and to maintain adequate humidity levels for larval development (Visscher, Crailsheim and Sherman, 1996).

The nectar foraging behavior of honeybees is well known, with the famous dance performed to inform the other worker bees of the direction and distance of the nectar source. Perhaps what is less known is that at times of thermal stress, worker bees can specialize exclusively in water, not nectar or pollen, collection. Upon returning to the hive, water is regurgitated, and the bee may perform the same dance to inform other bees of the source of water (Visscher, Crailsheim and Sherman, 1996). In one study, a source of water was gradually moved further from the hive. Under high ambient temperatures (40 °C) and with no other source of water available, bees were able to forage at least 2 km for water. Since these bees were not collecting nectar, the distance they could travel to seek water was limited as they had limited sugar reserves for their return flight (Visscher, Crailsheim and Sherman, 1996).

13.8 SUMMARY

For all terrestrial animals the regulation of body temperature requires the ability to regulate the body's water content, and thus the energy and water balances are closely coupled through the process of thermo-hydroregulation. Species can be divided into two groups: those species that maintain a constant body temperature, the endotherms, and those whose body temperature varies with the ambient temperature, the ectotherms. In both, water is required to protect the body against temperature extremes through evaporative cooling on the skin in addition to providing the basic functions required for life. This chapter focused on thermo-hydroregulation in ectotherms using examples of amphibian species (frogs), desert species (lizards), and aquatic (mosquitos), and terrestrial (beetles) and flying (bees) insects.

Amphibian frogs are fully aquatic in their larval stage and thus are completely dependent on water availability. In adult form, frogs require a wet skin surface for efficient respiration, and when coupled with their large surface area to body volume ratio, they are prone to evaporative water loss and desiccation. Loss of wetland habitats and declines in water quality have had a significant detrimental effect on frog populations worldwide. The primary means for frogs to lessen the loss of water through their skin is through the development of an increased skin resistance to water loss. Several studies have shown that tree-dwelling arboreal frog species have a much larger skin resistance and lower rates of water loss than aquatic frog species, and differences in skin resistance are linked with the frogs' ability to regulate body temperature. Using an Ohm's Law analogy with the concept of skin resistance, evaporative water loss can be estimated.

Lizard species are well known for their ability to inhabit a wide range of environments, especially arid locations. As with frogs, skin resistance to water loss plays an important role in the lizards' ability to conserve water in arid environments. Studies have shown that lizards adapted to arid environments generally have a higher skin resistance and a smaller evaporative water loss than those adapted to cooler, moist environments. Interestingly, field studies have found variations in the evaporative water loss attributable to changes in the skin resistance within the same species of lizards.

Aquatic insect abundance and diversity provide a robust indicator of water quality and overall ecosystem health. Submerged in water, changes in the solute concentration within the body relative to outside the body serve as an important means to regulate water flow into or out of it. Many insects such as mosquitos are fully aquatic in their larval stage, so changes in ambient water solute concentration (e.g., the water's salinity) require changes in the larvae's internal solute concentration. This requires energy and therefore affects the growth and development of the larvae. In adult form, a large cuticular resistance to water loss provided by a semipermeable lipid or wax coating can allow for gas exchange while minimizing water loss.

Terrestrial insects such as ants, locusts, and beetles are found in some of the hottest and driest places on Earth. With heat comes a lack of water, and we saw how these species have developed not only a physiological means to regulate water loss, but also behaviors to minimize water loss, such as fog basking, trench building, or the collection of condensed water on vegetation. Lastly, we examined the importance and use of water by one of the most important insects for agriculture, the honeybee, as well as limitations on their range and foraging behavior due to the availability of water, and how water is used to help regulate the hive temperature.

13.9 QUESTIONS

13.1 Compare and contrast the primary methods used for thermoregulation in ectotherms and endotherms. Are there any advantages for being a temperature conformer over a temperature regulator in terms of habitat range and survivability? What role does water play in the thermoregulation of ectotherms? How do the quantity and state of water influence the terms in the energy balance equation for an ectotherm (see Eq. 13.3)?

13.2 What are the sources (or paths) of water loss from ectotherms? Describe any physiological or behavior controls that ectotherms (e.g., frogs, lizards, or insects) have available to regulate water loss.

13.3 Endotherms such as humans can effectively regulate body temperature through sweating, so why are ectotherms that are not capable of sweating able to inhabit arid regions and tolerate much hotter and drier conditions than humans can?

13.4 Several frog species have evolved to living their adult life in trees (arboreal frog species). For better mobility in the canopy, these arboreal frog species are much smaller than nonarboreal species and are exposed to much drier conditions. How has their smaller size made arboreal frog species more susceptible to desiccation, and how have these species adapted to the drier conditions in the forest canopy?

13.5 The northern dwarf tree frog (sedge frog) *Litoria bicolor* is a small frog native to northern Australia. As expected for an arboreal species, it has a high skin resistance to water vapor transfer to avoid desiccation given its small size and the dry conditions experienced in the forest canopy. Calculate the evaporative water loss from this tree frog in mg per cm^2 per hour with a skin resistance of 65 s cm^{-1}, a skin temperature of 25 °C, and relative humidity at the skin surface of 100%. The air temperature and relative humidity are 30 °C and 25%, respectively. Under these conditions, how many hours would it take for this 0.75-g frog to reach the critical 70% of its mass?

13.6 Imagine that the same dwarf northern tree from Question 13.5 experiences drying of the skin, resulting in a skin temperature increase to 30 °C and decrease in relative humidity at the skin surface to 90%. The air temperature and relative humidity remain at 30 °C and 25%, respectively. Under these different skin conditions, how long would it take this 0.75-g frog to reach the critical 70% of its mass?

13.7 Lizards inhabit a wide range of habitats, and several studies have shown that lizards' evaporative water loss generally decreases as the aridity increases, owing to an increase in the skin's resistance to water loss. Estimate the evaporative water loss (g cm^{-2} s^{-1}) of a lizard by dividing the difference in vapor density (g cm^{-3}) between the skin and atmosphere by a skin resistance of 100 cm s^{-1}, an air and skin temperature of 30 °C, and relative humidity of 20% (air) and 50% (skin). Under drier conditions, the air temperature and lizard skin temperature increase to 40 °C and the atmospheric relative humidity decreases to 10% (the relative humidity at the skin remains at 50%). What value of skin resistance (s cm^{-1}) is required to maintain the same evaporative water loss?

13.8 Are aquatic insects required to regulate the flow of water into or out of their bodies? Why or why not, and under what conditions? If they do have this requirement, describe two general mechanisms available to aquatic insects in either juvenile or adult form to regulate water flow.

13.9 Some insects have been observed displaying unique behaviors in response to extremes in temperatures or arid conditions. Summarize two or three of these behaviors presented here, and search the scientific literature for two or three additional examples not described here and summarize those behaviors also. Do you feel that such behaviors are performed cognitively, or are they simply passive responses?

13.10 Several recent studies have shown that the quantity of flying insects has significantly decreased (e.g., biomass decrease by more than 75 percent over 27 years in Germany; Hallmann et al., 2017). How could changes in the availability of surface water play a role in explaining this troublesome decline? What are the ecological impacts of such decline?

REFERENCES

Abou-Shaara, H. F., Owayss, A. A., Ibrahim, Y. Y. and Basuny, N. K. (2017) 'A review of impacts of temperature and relative humidity on various activities of honey bees', *Insectes Sociaux*, 64, pp. 455–463. doi: 10.1007/s00040-017-0573-8.

Beament, J. W. L. (1961) 'The waterproofing mechanism of Arthropods: II. The permeability of the cuticle of some aquatic insects', *Journal of Experimental Biology*, 38, pp. 277–290. doi: 10.1242/jeb.38.2.277.

Belasen, A., Brock, K., Li, B. et al. (2017) 'Fine with heat, problems with water: microclimate alters water loss in a thermally adapted insular lizard', *Oikos*, 126, pp. 447–457. doi: 10.1111/oik.03712.

de Brito Arduino, M., Mucci, L. F., Nunes Serpa, L. L. and de Moura Rodrigues, M. (2015) 'Effect of salinity on the behavior of *Aedes aegypti* populations from the coast and plateau of southeastern Brazil', *Journal of Vector Borne Diseases*, 52, pp. 79–87.

Cloudsley-Thompson, J. (2001) 'Thermal and water relations of desert beetles', *Naturwissenschaften*, 88, pp. 447–460. doi: 10.1007/s001140100256.

Cooper, P. D., Schaffer, W. M. and Buchmann, S. L. (1985) 'Temperature regulation of honey bees (*Apis mellifera*) foraging in the Sonoran Desert', *Journal of Experimental Biology*, 114, pp. 1–15. doi: 10.1242/jeb.114.1.1.

Cushman, S. A. (2006) 'Effects of habitat loss and fragmentation on amphibians: A review and prospectus', *Biological Conservation*, 128, pp. 231–240. doi: 10.1016/j.biocon.2005.09.031.

Daszak, P., Cunningham, A. A. and Hyatt, A. D. (2003) 'Infectious disease and amphibian population declines', *Diversity and Distributions*, 9, pp. 141–150. doi: 10.1046/j.1472-4642.2003.00016.x.

Davidson, N. C. (2014) 'How much wetland has the world lost? Long-term and recent trends in global wetland area', *Marine and Freshwater Research*, 65, pp. 934–941. doi: 10.1071/MF14173.

Derka, T., Zamora-Muñoz, C. and de Figueroa, J. M. T. (2019) 'Aquatic insects', in Rull, V. et al. (eds.) *Biodiversity of Pantepui: The Pristine 'Lost World' of the Neotropical Guiana Highlands*. Academic, pp. 167–192. doi: 10.1016/B978-0-12-815591-2.00008-2.

Dmi'el, R., Perry, G. and Lazell, J. (1997) 'Evaporative water loss in nine insular populations of the lizard *Anolis cristatellus* group in the British Virgin Islands', *Biotropica*, 29, pp. 111–116. doi: 10.1111/j.1744-7429.1997.tb00012.x.

Edney, E. B. (1971) 'The body temperature of tenebrionid beetles in the Namib Desert of southern Africa', *Journal of Experimental Biology*, 55, pp. 253–272. doi: 10.1242/jeb.55.1.253.

Edwards, H. A. (1982) '*Aedes aegypti*: Energetics of osmoregulation', *Journal of Experimental Biology*, 101, pp. 135–141. doi: 10.1242/jeb.101.1.135.

Evans, M. J., Scheele, B. C., Westgate, M. J. et al. (2020) 'Beyond the pond: Terrestrial habitat use by frogs in a changing climate', *Biological Conservation*, 249, p. 108712. doi: 10.1016/j.biocon.2020.108712.

Galbreath, R. A. (1975) 'Water balance across the cuticle of a soil insect', *Journal of Experimental Biology*, 62: 115–120. doi: 10.1242/jeb.62.1.115.

González, K., Warkentin, K. M. and Güell, B. A. (2021) 'Dehydration-induced mortality and premature hatching in gliding treefrogs with even small reductions in humidity', *Ichthyology and Herpetology*, 109, pp. 21–30. doi: 10.1643/h2020085.

Hallmann, C. A., Sorg, M., Jongejans, E. et al. (2017) 'More than 75 percent decline over 27 years in total flying insect biomass in protected areas', *PLoS ONE*, 12, e0185809. doi: 10.1371/journal.pone.0185809.

Hamilton, W. J. and Seely, M. K. (1976) 'Fog basking by the Namib Desert beetle, *Onymacris unguicularis*', *Nature*, 262, pp. 284–285.

Hershey, A. E. and Lamberti, G. A. (2001) 'Aquatic insect ecology', in Thorp, J. H. and Covich, A. P. (eds.) *Ecology and Classification of North American Freshwater Invertebrates*, 2nd ed. Academic, pp. 733–775. doi: 10.2307/1310031.

Hoffmann, E. P., Cavanough, K. L. and Mitchell, N. J. (2021) 'Low desiccation and thermal tolerance constrains a terrestrial amphibian to a rare and disappearing microclimate niche', *Conservation Physiology*, 9, pp. 1–15. doi: 10.1093/conphys/coab027.

Holdgate, M. W. (1956) 'Transpiration through the cuticles of some aquatic insects', *Journal of Experimental Biology*, 33, pp. 107–118. doi: 10.1242/jeb.33.1.107.

Houlahan, J. E., Findlay, C. S., Schmidt, B. R., Meyer, A. H. and Kuzmin, S. L. (2000) 'Quantitative evidence for global amphibian population declines', *Nature*, 404, pp. 752–755. doi: 10.1038/35008052.

Jones, J. C., Myerscough, M. R., Graham, S. and Oldroyd, B. P. (2004) 'Honey bee nest thermoregulation: Diversity promotes stability', *Science*, 305, pp. 402–404. doi: 10.1126/science.1096340.

Kiesecker, J. M., Blaustein, A. R. and Belden, L. K. (2001) 'Complex causes of amphibian population declines', *Nature*, 410, pp. 681–684. doi: 10.1038/35070552.

Klok, C. J. and Chown, S. L. (1997) 'Critical thermal limits, temperature tolerance and water balance of a sub-antarctic caterpillar, *Pringleophaga marioni* (Lepidoptera: Tineidae)', *Journal of Insect Physiology*, 43, pp. 685–694. doi: 10.1016/S0022-1910(97)00001-2.

Lancaster, J. and Downes, B. J. (2013) 'Evolution, biogeography, and aquatic insect distributions', in *Aquatic Entomology*. Oxford University Press. doi: 10.1093/acprof:oso/9780199573219.001.0001.

Maeno, K. O., Ould Ely, S., Nakamura, S. et al. (2016) 'Daily microhabitat shifting of solitarious-phase desert locust adults: implications for meaningful population monitoring', *SpringerPlus*, 5, pp. 1–10. doi: 10.1186/s40064-016-1741-4.

Marsh, A. C. (1985) 'Thermal responses and temperature tolerance in a diurnal desert ant, *Ocymyrmex barbiger*', *Physiological Zoology*, 58, pp. 629–636. doi: 10.1086/physzool.58.6.30156067.

Mautz, W. J. (1982) 'Correlation of both respiratory and cutaneous water losses of lizards with habitat aridity', *Journal of Comparative Physiology B*, 149, pp. 25–30. doi: 10.1007/BF00735711.

May, R. M. (1986) 'How many species are there?', *Nature*, 324, pp. 514–515. doi: 10.1111/j.1523-1739.1991.tb00145.x.

Mellanby, K. (1941) 'The body temperature of the frog', *Journal of Experimental Biology*, 18, pp. 55–61. doi: 10.1242/jeb.28.3.271.

Mitra, S., Wassmann, R. and Vlek, P. L. G. (2003) *Global Inventory of Wetlands and Their Role in the Carbon Cycle*. ZEF Discussion Paper on Development Policy, Bonn, no. 64, p. 57.

Nguete Nguiffo, D., Mpoame, M. and Wondji, C. S. (2019) 'Genetic diversity and population structure of goliath frogs (*Conraua goliath*) from Cameroon', *Mitochondrial DNA Part A: DNA Mapping, Sequencing, and Analysis*, 30, pp. 657–663. doi: 10.1080/24701394.2019.1615060.

Noble-Nesbiti, J. (1990) 'Insects and their water requirements', *Interdisciplinary Science Reviews*, 15, pp. 264–282.

Pritchard, G., McKee, M. H., Pike, E. M., Scrimgeour, G. J. and Zloty, J. (1993) 'Did the first insects live in water or in air?', *Biological Journal of the Linnean Society*, 49, pp. 31–44.

Rittmeyer, E. N., Allison, A., Gründler, M. C., Thompson, D. K. and Austin, C. C. (2012) 'Ecological guild evolution and the discovery of the world's smallest vertebrate', *PLoS ONE*, 7, pp. 1–11. doi: 10.1371/journal.pone.0029797.

Rozen-Rechels, D., Dupoué, A., Lourdais, O. et al. (2019) 'When water interacts with temperature: Ecological and evolutionary implications of thermo-hydroregulation in terrestrial ectotherms', *Ecology and Evolution*, 9, pp. 10029–10043. doi: 10.1002/ece3.5440.

Sannolo, M., Civantos, E., Martin, J. and Carretero, M. A. (2020) 'Variation in field body temperature and total evaporative water loss along an environmental gradient in a diurnal ectotherm', *Journal of Zoology*, 310, pp. 221–231. doi: 10.1111/jzo.12744.

Seely, M. K. and Hamilton, W. J. (1976) 'Fog catchment sand trenches constructed by tenebrionid beetles, Lepidochora, from the Namib Desert', *Science*, 193, pp. 484–486. doi: 10.1126/science.193.4252.484.

Sodhi, N. S., Bickford, D., Diesmos, A. C. et al. (2008) 'Measuring the meltdown: Drivers of global amphibian extinction and decline', *PLoS ONE*, 3, pp. 1–8. doi: 10.1371/journal.pone.0001636.

Stevenson, R. D. (1985) 'The relative importance of behavioral and physiological adjustments controlling body temperature in terrestrial ectotherms', *The American Naturalist*, 126, pp. 362–386.

Stork, N. E. (2018) 'How many species of insects and other terrestrial arthropods are there on Earth?', *Annual Review of Entomology*, 63, pp. 31–45. doi: 10.1146/annurev-ento-020117-043348.

Tracy, C. R., Christian, K. A. and Tracy, C. R. (2010) 'Not just small, wet, and cold: Effects of body size and skin resistance on thermoregulation and arboreality of frogs', *Ecology*, 91, pp. 1477–1484. doi: 10.1890/09-0839.1.

Vanburen, C. S., Norman, D. B. and Fröbisch, N. B. (2019) 'Examining the relationship between sexual dimorphism in skin anatomy and body size in the white-lipped treefrog, *Litoria infrafrenata* (Anura: Hylidae)', *Zoological Journal of the Linnean Society*, 186, pp. 491–500. doi: 10.1093/zoolinnean/zly070.

Visscher, P. K., Crailsheim, K. and Sherman, G. (1996) 'How do honey bees (*Apis mellifera*) fuel their water foraging flights?', *Journal of Insect Physiology*, 42, pp. 1089–1094. doi: 10.1016/S0022-1910(96)00058-3.

Warkentin, K. M. (2002) 'Hatching timing, oxygen availability, and external gill regression in the tree frog, *Agalychnis callidryas*', *Physiological and Biochemical Zoology*, 75, pp. 155–164. doi: 10.1086/339214.

Wygoda, M. L. (1984) 'Low cutaneous evaporative water loss in arboreal frogs', *Physiological Zoology*, 57, pp. 329–337. doi: 10.1086/physzool.57.3.30163722.

Young, J. E., Christian, K. A., Donnellan, S., Tracy, C. R. and Parry, D. (2005) 'Comparative analysis of cutaneous evaporative water loss in frogs demonstrates correlation with ecological habits', 78, pp. 847–856.

Zachariassen, K. E. (1996) 'The water conserving physiological compromise of desert insects', *European Journal of Entomology*, 93(3), pp. 359–367.

Zhang, Z.-Q. (2013) 'Phylum Arthropoda', *Zootaxa*, 3703, pp. 17–26.

14 Water and Thermoregulation: Humans

Key Learning Objectives

After reading this chapter you will be able to:

1. Describe the energy balance and how water is used by an endotherm to maintain a constant body temperature.
2. Estimate the resting metabolic rate and explain how this changes with activity level.
3. Define conduction and convection and estimate the magnitude of these dry heat terms with changing ambient conditions.
4. Summarize why sweating, and water intake, are so important for the body to remain cool.
5. Calculate sweat rates for a person, and how these rates change with activity level, temperature, and humidity. Estimate your required water intake based on weight, height, activity level, and ambient conditions.

14.1 Introduction

Water plays a vital role in the ability of endotherms to regulate body temperature, especially in hot and dry conditions. All mammals and birds are endotherms, where an endotherm is broadly defined as having the ability to use metabolism to maintain a body temperature that is substantially different than the temperature of the surroundings. Metabolic heat production is especially important to maintain a constant body temperature in cold environments. Evaporative cooling, largely through sweat production, is especially important to maintain a constant body temperature in hot environments. An adequate supply of water is required to produce sweat, and the rate of cooling achieved by the evaporation of sweat on the skin depends on the complete energy balance of the body and the gradient in humidity between skin and air. This chapter focuses on how humans balance energy inputs with outputs through evaporation on the skin surface to maintain a constant body temperature. These concepts and calculations could be applied to any endotherm.

We begin by examining how the energy balance of an endotherm differs from that of an ectotherm. Beginning with shortwave and longwave radiation, each term in the energy balance equation for a human is described. There is a summary of how the net radiation received by a person can be estimated based on measurements of a person's albedo, surface area, surface temperature, and location, and sample calculations are provided. Next, we look at how to estimate the rate of metabolic heat production when a person is at rest or active and examine results from classic research that are still used today.

Additional exchanges of energy that a person experiences that do not directly involve the gains or losses of water within the body include conduction and convection, both of which will be explored in this chapter.

Wet heat exchange from a body, the rate of evaporation of sweat on the skin, is used as the primary means to remain cool under periods of high energy inputs. Sweating is described last because the other energy balance terms need to be known prior to estimating how much sweating is required to offset all energy gains. After a brief review of how sweat is produced, we describe why sweating is so effective in thermoregulation, and how it is distinct from respiration cooling. We will examine how evaporation rates are calculated under various conditions, as well as how a person's required water intake varies with activity level and relative humidity.

14.2 An Endotherm's Thermoregulation

In sharp contrast to **ectotherms**, whose body temperature varies with the ambient temperature, **endotherms** strive to maintain a constant body temperature despite exposure to a wide range of ambient conditions. Although the same basic concepts and principles for thermoregulation explained in Chapter 13 apply, a fundamental distinction between endotherms and ectotherms is that endotherms (Greek *endo* for inside, and *therm* for heat) can regulate body temperature by increasing their metabolic rate. The energy balance equation for an endotherm is the same as for an ectotherm, but an endotherm is much less reliant on position and posture to regulate body temperature. When an endotherm's body temperature is decreasing, the metabolism is used to generate body heat; when the body's temperature is increasing, evaporation (sweating) is used to cool it.

The body's core temperature is determined by the difference in energy received (E_{in}) and energy lost (E_{out}), equal to the change in internal energy storage (ΔE; J m^{-2} s^{-1}):

$$E_{in} - E_{out} = \Delta E. \tag{14.1}$$

When there is a difference between energy inputs and outputs, the body's rate of warming or cooling ($\mathrm{d}T/\mathrm{d}t$; °C or K per second) can be calculated knowing the body's density (ρ; kg m^{-3}) and specific heat capacity (C_p; J kg^{-1} °C^{-1} or K^{-1}) over the thickness (layer) of warming or cooling (z; m) (Eq. 14.2):

$$\begin{aligned} \Delta E &= \rho C_p \times \frac{\mathrm{d}T}{\mathrm{d}t} \times z \\ \frac{\mathrm{d}T}{\mathrm{d}t} &= \frac{\Delta E}{\rho C_p z}. \end{aligned} \tag{14.2}$$

Ideally, $E_{in} = E_{out}$ and ΔE is zero, meaning that the body is neither losing (cooling) nor gaining (warming) energy, and the body's temperature is constant. Except when inside in a controlled environment, this condition is rare since solar radiation, wind, temperature, and humidity are constantly changing through time or as one moves through the environment. When walking from a shady to a sunny location, the increase in solar radiation received on our skin would result in $E_{in} > E_{out}$ and therefore an increase in ΔE, resulting in warming. Walking back into a cool, shady location would result in $E_{in} < E_{out}$ and therefore a decrease in ΔE, resulting in cooling. In either case, our body responds with unconscious, imperceivable adjustments to maintain a constant body temperature.

To do this, a person must constantly work to balance the energy emitted through internal means and/or regulate the energy received by changing location. Equation (14.1) can be written to include the important energy input and energy output terms (Eq. 14.3):

$$\left(S\downarrow + L\downarrow + M\right) - \left(S\uparrow + L\uparrow + \lambda E + H + G\right) = \rho C_p \times {}^{dT}\!/_{dt} \times z \quad (14.3)$$

where S is the shortwave radiation incident upon ($S\downarrow$) or reflected ($S\uparrow$) from the body, L is the longwave radiation incident upon ($L\downarrow$) or emitted from the body ($L\uparrow$), M is the metabolic heat production, λE is the latent heat flux (i.e., evaporation), H is the sensible heat flux (i.e., convection), and G is the ground (surface) heat flux (i.e., conduction), with all terms in W m^{-2}. In contrast to ectotherms, the metabolic term (for heat production) and latent heat flux (for heat loss) terms in Eq. (14.3) can be used by endotherms to keep body temperature constant across a wide range of ambient temperatures. Both M and especially λE are highly dependent on the organism's access to water, but the balance between all the energy inputs and outputs determines the extent to which both M and λE are utilized to regulate body temperature.

14.3 Estimating the Energy Balance Components

An endotherm has the ability to increase its metabolic rate to increase body temperature when cold, or produce sweat to reduce body temperature when warm. The extent by which M or λE are varied, however, depends on the magnitude of all of the energy balance terms shown in Eq. (14.3). Measuring or estimating the magnitudes of these terms is difficult. A multitude of factors such as age, sex, weight, height, activity level, and health affect these energy balance terms even when the ambient conditions are constant. Ambient conditions can vary greatly, and these conditions, as well as other factors such as clothing and activity level, need to be considered for an accurate estimation of the complete energy balance.

14.3.1 Shortwave and Longwave Radiation

The first two sources of energy are the incident shortwave and longwave radiation. The **net radiation** (R_n; all energy terms in W m^{-2}) provides a measure of the radiative energy available (if positive) or energy deficit (if negative) and is calculated as the sum of the difference between the incident (incoming) shortwave (or solar) radiation ($S\downarrow$) and reflected (outgoing) shortwave radiation ($S\uparrow$), and the difference between the incident (incoming) longwave radiation ($L\downarrow$) and emitted (outgoing) longwave radiation ($L\uparrow$) (Eq. 14.4):

$$R_n = \left(S\downarrow - S\uparrow\right) + \left(L\downarrow - L\uparrow\right). \quad (14.4)$$

Shortwave radiation is composed of two components, the **direct-beam radiation** coming directly from the sun, and the **diffuse-beam radiation** coming from solar radiation that has changed direction and scattered after passing through the atmosphere. The **global shortwave** radiation is the sum of the direct and diffuse components. This diffuse radiation could be thought of as "indirect" radiation since it appears not to originate from a direct point (object). The **albedo** is the ratio of the reflected to incident shortwave radiation, $\alpha = S\uparrow / S\downarrow$, and substitution of albedo into Eq. (14.4) gives (Eq. 14.5):

$$R_{\mathrm{n}} = S\downarrow(1-\alpha) + (L\downarrow - L\uparrow). \tag{14.5}$$

The albedo for an endotherm varies with the type and concentration of **pigments** (**chromophores**) in the epidermis cells and hair or fur. In humans, the pigments **melanin**, **carotene**, and **hemoglobin** determine the amount of shortwave radiation reflected from the skin, and thus the color the human eye perceives. Hair or fur color is determined from the pigment melanin, with greater quantities of melanin resulting in dark hair and lower quantities of melanin resulting in lighter-colored hair. In humans, skin color has been found to be strongly correlated with UV-A shortwave radiation (ultraviolet radiation with wavelengths between 315 and 400 nm), with dark pigmentation increasing in populations living near the Equator and at high elevations (Jablonski, 2004). Melanin provides sunburn protection from excessive skin exposure to ultraviolet radiation, yet high melanin concentrations decrease the body's efficiency of vitamin D_3 synthesis (Jablonski, 2004). The albedo measured at a wavelength of 685 nm (red) for *Homo sapiens* on the upper inner arm was on average 46%, ranging from 19% to 67%, with females having a significantly higher reflectance than males (Jablonski and Chaplin, 2000). Skin albedo measured across the full solar radiation spectrum (roughly 0.1 to 4.0 μm) generally varies from 20% to 45%, and 37% is the recommended value to use for a person with average skin wearing typical clothing under outdoor conditions (Kenny et al., 2008).

The radiation absorbed by a person outside during the daytime can be a large, important energy input term. For surfaces such as a forest or lake, R_{n} is represented as the rate of energy exchange (W or J s^{-1}) passing through an imaginary horizontal plane (m^2). A person's body, however, is not well represented by a horizontal plane. A simple cylindrical shape has been used to better approximate the surface area of a person for radiation exchange, and small (e.g., 11 cm long by 1.3 cm diameter) cylindrical radiation thermometers have been constructed to estimate R_{n} by direct measurements. Taken as a miniature analog of a standing person, R_{n} can be calculated by measuring the temperature inside a cylinder and then solving the energy balance equation using Ohm's Law and heat transfer theory (Brown and Gillespie, 1986). Different paint colors and coatings can be used to simulate various skin and clothing properties.

Outdoors, estimates of the net radiation received by a person (i.e., the total radiation absorbed assuming no transmission of radiation) based on a cylindrical radiation thermometer compare well with other methods. Kenny et al. (2008) calculated R_{n} for a cylindrical body using a cylindrical radiation thermometer placed outside during clear or overcast conditions. They found that estimates of R_{n} compared well (within 3–8 percent error) with measurements using a traditional net radiometer or calculated using theoretical estimates. During clear-sky conditions in early June at a mid-latitude North American location, all three methods showed that R_{n} increased with solar altitude, reaching a maximum value of roughly 600 W m^{-2} by mid-morning, decreased 50 W m^{-2} at solar noon, then increased back to roughly 600 W m^{-2} in the late afternoon before decreasing as the sun set. The decrease in R_{n} around solar noon was due to the decrease in the exposed surface area of the cylinder to direct-beam shortwave radiation (the area exposed, or **view factor**, by a cylinder shape is roughly half the cylinder's surface area). During overcast conditions, this same diurnal pattern was observed although dampened because of the decrease in the direct-beam shortwave radiation. When R_n measured or calculated for the cylinder was converted to more accurately account for the irregularities of the human body and radiation exchanges between body parts that vary when standing or seated, Kenny et al. (2008) found that the total shortwave and longwave radiation inputs were large compared to metabolic heat production. Outdoors under clear-sky conditions, the radiation absorbed by the average person was roughly between 400 and 500 W m^{-2}, four times the metabolic heat input whether sitting, standing, or performing mild to moderate physical activity (Kenny et al., 2008).

In the absence of exposure to a large amount of incident solar radiation, such as when someone moves into a shady location, under overcast conditions, at night, or indoors, the longwave radiation received from the surroundings is often equal to that emitted by the person. The **Stefan–Boltzmann Law** can be used to estimate the longwave radiation emitted by a person ($L\uparrow$; W m^{-2}) (Eq. 14.6):

$$L\uparrow = \varepsilon \sigma T_0^4 \tag{14.6}$$

where ε is the **emissivity** (the ratio of the energy emitted by the object to the energy emitted if the object was a perfect emitter; dimensionless), σ is the **Stefan–Boltzmann constant** 5.67×10^{-8} W m^{-2} K^{-4}, and T_0 is the **skin temperature** (**surface** or **radiant temperature**) (K). Since the human body has a constant skin surface temperature of roughly 35 °C (308.15 K), using Eq. (14.6) a person emits a nearly constant 500 W m^{-2} assuming an emissivity of 0.98 (Eq. 14.7):

$$L\uparrow = 0.98 \times 5.67 \times 10^{-8}\ \text{W m}^{-2}\text{K}^{-4} \times (308.15\ \text{K})^4 = 501.02\ \text{W m}^{-2}. \tag{14.7}$$

The Stefan–Boltzmann Law can be used to calculate the energy emitted by objects surrounding the person, and is especially relevant when indoors and surrounded by warm surfaces. The emissivity and surface temperatures of the objects, and the position of the objects relative to the person, are needed to calculate the relative radiation contributions the person receives. Since most building materials have a high emittance, the emissivity is often assumed close or equal to 1. The mean surface temperature to which a person is exposed can be calculated from the sum of the measured surface temperature of the walls and surfaces and their position relative to the person. For example, in a simplified case assuming all surfaces are at the same surface temperature equal to the room temperature of 20 °C (293.15 K) with an emissivity of 0.98, from the Stefan–Boltzmann Law (Eq. 14.6) the objects in the room are emitting 410.36 W m^{-2} (Eq. 14.8):

$$L\downarrow = 0.98 \times 5.67 \times 10^{-8}\ \text{W m}^{-2}\text{K}^{-4} \times (293.15\ \text{K})^4 = 410.36\ \text{W m}^{-2}. \tag{14.8}$$

A person with a surface temperature of 35 °C would be emitting 501 W m^{-2} (Eq. 14.7) and receiving 410 W m^{-2} from the objects in the room (Eq. 14.8). Therefore, under these conditions, the person experiences a loss of 91 W m^{-2} of longwave radiation ($410 - 501 = -91$ W m^{-2}).

As an alternative to using the Stefan–Boltzmann Law, an Ohm's Law analogy can be used to calculate the **radiative heat loss** (the heat lost by energy transfer in the form of radiation) from a person. As in many applications, instead of dividing the difference of the appropriate scalar (e.g., temperature) by a resistance to calculate the flux, a conductance expressed as a **scalar transfer coefficient** can be used. This approach can be used to calculate the radiative heat loss for a clothed person (American Society of Heating, Refrigerating and Air-Conditioning Engineers Inc., 2017) (Eq. 14.9);

$$R = f_{cl} h_r \left(T_{cl} - T_r \right) \tag{14.9}$$

where R (W m^{-2}) is the radiative heat loss for a person (positive for heat loss; negative for heat gain), f_{cl} is the **clothing area factor** (the ratio of the area of a clothed to nude human body; 1 for nude), h_r (W m^{-2} °C^{-1} or K^{-1}) is the **radiative heat transfer coefficient**, T_{cl} is the clothed-body mean surface temperature (°C or K), and T_r is the mean radiative surface temperature (°C or K) of the environment as perceived by the person's body, as described above.

Use of Eq. (14.9) requires an estimate of an appropriate value for the radiative heat transfer coefficient. Using heated mannequins to represent the size, shape, and volume of a person, De Dear et al. (1997) measured h_r = 4.5 W m^{-2} K^{-1} whether seated or standing, close to the generally accepted value of 4.7 W m^{-2} K^{-1}. Using Eq. (14.9) to estimate the radiative heat loss from a person with f_{cl} = 1.15 (trousers, short-sleeved shirt; American Society of Heating, Refrigerating and Air-Conditioning Engineers Inc., 2017), h_r = 4.7 W m^{-2} K^{-1}, T_{cl} = 35 °C, and T_r = 20 °C gives R = 81.1 W m^{-2} (Eq. 14.10):

$$R = 1.15 \times 4.7 \text{ W m}^{-2}\text{K}^{-1} \times \left(35\ °\text{C} - 20\ °\text{C} \right) = 81.1 \text{W m}^{-2}. \tag{14.10}$$

This value is comparable to the 90.7 W m^{-2} radiative heat loss based on the net longwave radiation received by the person as calculated using the Stefan–Boltzmann Law at the body and room temperatures.

14.3.2 Metabolism

The third term in Eq. (14.3) represents a source of energy for endotherms through the production of heat by metabolism (M). Given the importance of metabolism as a biological process to actively regulate body temperature, and as a key distinguishing feature separating endotherms from ectotherms, research aimed to understand the metabolic process has a long and colorful history.

As described in Chapter 2, Lavoisier's combustion experiments led him to realize that if oxygen played an important role in material combustion, it may also play an important role in respiration (respiration is essentially a form of internal combustion). Lavoisier measured the amount of CO_2 and heat produced when burning carbon in a device he invented, called a **calorimeter**. He then placed a live guinea pig in the calorimeter and measured the amount of CO_2 and heat produced. By comparing the CO_2 generated and heat produced by the carbon and the guinea pig, Lavoisier concluded that respiration is a slow form of combustion requiring O_2 and generating heat. After Lavoisier's research ended abruptly upon his execution, **Carl von Voit** (1831–1908) showed that the O_2 consumed by an organism is a result of **cellular metabolism**, and his student **Max Rubner** (1854–1932) demonstrated that heat is generated as a product of metabolism (Da Poian, El-Bacha and Luz, 2010). Subsequent studies identified the molecule responsible for generation of heat, and the term **respiration** is now used to refer to the production of energy in the form of **adenosine triphosphate** (ATP) and the release of CO_2 in cells through the **oxidation** (combustion) of **glucose**. In addition to the generation of 36 ATP molecules, 6 CO_2 and 6 H_2O molecules are produced by the respiration reaction (Eq. 14.11):

$$C_6H_{12}O_6 + 6O_2 = 6CO_2 + 6H_2O + 36ATP. \tag{14.11}$$

Respiration (Eq. 14.11) is one of several chemical reactions required for life referred to as **metabolic reactions** or simply metabolism. In addition to the conversion of food to energy (respiration), other broad categories of metabolic reactions include growth and reproduction, cellular maintenance and repair, and waste elimination.

Early studies showed that the greater the mass of the organism while at rest, the greater the **basal** or **resting metabolic rate**. These studies often use a method known as **allometry**, where an empirical relationship between body size, shape, or mass (variable x) and physiology, functioning, or behavior (variable y) is derived. Many such relationships are represented by an **allometric equation** often best fitted by a power law (Eq. 14.12):

$$y = kx^a \tag{14.12}$$

where k and a are empirically derived coefficients experimentally determined by the relationship between variables x and y. Kleiber (1932) summarized several studies that measured the basal metabolic rate and mass for mammals ranging in size from a dove to a cow. The relationship between weight (w; kg) and the basal metabolic rate (M_0; watts) was linear when plotted on a log–log axis, resulting in the allometric equation known as **Kleiber's Law** (Eq. 14.13):

$$M_0 \cong w^{3/4} = 3.52\, w^{0.74}. \tag{14.13}$$

Using Kleiber's Law, the basal metabolic rate for a person weighing 80 kg is:

$$M_0 = 3.52 \times (80 \text{ kg})^{0.74} = 90.1 \text{ W}.$$

The greater the animal's mass, the greater the number of cells that require energy, and the greater the volume of liquid water within the organism. Much of this mass is due to the large volume of liquid water, although water content typically decreases with age, as the body's energy requirements decrease as growth slows. In humans, for example, the proportion of body weight made up by fluids is over 70% at birth, decreases to 60% in childhood, and then further decreases to 50% in adulthood (Hooper et al., 2014).

Kleiber (1947) and others (e.g., McMahon, 1973) wondered why the exponent in Eq. (14.13) was close to 3/4, and not 2/3 as expected based on the heat loss calculated from the body's surface area to volume ratio. One explanation is that the circulatory systems found in both mammals and in plant xylem tissue have been shown to scale with weight to the 3/4, not 2/3, power (West, Brown and Enquist, 1997; Smil, 2000). Since most mammals including humans have a consistent liquid water content at a given age (plasma is 93 percent water; Nguyen et al., 2007), Eq. (14.13) could conceivably be written in terms of water content instead of body weight, although the former is much harder to measure than the latter.

Kleiber's allometric relationship between weight and metabolism neglects to explicitly recognize the surface area of the animal in relation to its volume, and expressing the basal metabolic rate as W m^{-2} rather than watts allows scaling by surface area and direct comparison with the other energy balance components in Eq. (14.3). The volume of a person or any object with a complex shape can be determined by measuring the volume of water displaced when the object is submerged (the **water displacement method**). The surface area of a person (S; cm^2) can be estimated from the allometric equation based on body weight (w; kg) and body length (L; cm) known as the **du Bois Formula** (du Bois and du Bois, 1916) (Eq. 14.14):

$$S = 71.84\, w^{0.425} L^{0.725}. \tag{14.14}$$

Although Eq. (14.14) is widely used, it was based on measurements of the surface area of only nine individuals. Nearly 100 years later, a reanalysis of the data used by du Bois and du Bois (1916) when combined with data from several subsequent studies resulted in a slightly modified form of Eq. (14.14) now recommended for use (Shuter and Aslani, 2000) (Eq. 14.15):

$$S = 94.9\, w^{0.441} L^{0.655}. \tag{14.15}$$

Using Eq. (14.15), a person weighing 80 kg and 180 cm tall would have a surface area of nearly 2 m^2:

$$S = 94.9 \times (80\ \text{kg})^{0.441} \times (180\ \text{cm})^{0.655} = 19{,}667\ \text{cm}^2 \left(=1.97\ \text{m}^2\right).$$

Using the estimated body surface area of ~2 m^2, the resting metabolic rate for this 80-kg person is about 45 W m^{-2} (Eq. 14.16):

$$\frac{M_0\ (\text{W})}{S\left(\text{m}^2\right)} = \frac{90\ \text{W}}{2\ \text{m}^2} = 45\ \text{W m}^{-2}. \tag{14.16}$$

Water plays a vital role in metabolism, as demonstrated by the relationship between the resting metabolic rate and mass, with water contributing most of this mass in humans. Water is required in the metabolic process to transport the oxygen required for respiration and to remove CO_2 as the waste product. Water serves as the solvent allowing the body to use blood and the circulatory system to transport dissolved gases, energy (ATP), hormones, and other molecules. Given the critical importance of water in metabolism, including respiration, dehydration and other heat-related injuries have serious health consequences. Definitions of dehydration based on rapid changes in body weight show that a decrease in body weight of only 4 percent within seven days is considered a clear sign of dehydration in humans (Hooper et al., 2014). Small changes in water consumption have been found to have a large impact on metabolism. A 30 percent increase in both men and women's metabolic rate was sustained for more than an hour after drinking only 500 ml of water (Boschmann et al., 2003).

14.3.3 Conduction

Energy output pathways used by an endotherm to lose energy and dissipate heat include the body's reflected shortwave radiation ($S\uparrow$), the emitted longwave radiation ($L\uparrow$), the latent heat flux λE (evaporative water losses), the sensible heat flux H (convection), and the ground (surface) heat flux G (conduction). The radiative terms $S\uparrow$ and $L\uparrow$ were discussed in Section 14.3.1 in the context of the body's albedo (Eq. 14.5) and emitted longwave radiation calculated using the Stefan–Boltzmann Law (Eq. 14.6). The nonradiative energy loss terms due to the transfer of heat through conduction (G), convection (H), and evaporation (λE) are described next. Although a body may gain or lose energy through any of these energy terms, under typical conditions they usually represent a loss of energy from the body to the ambient environment.

Jean-Baptiste Joseph Fourier (1768–1830) was a mathematician known for his studies of the harmonic patterns found in time-series data which became known as Fourier series and the Fast Fourier Transform (FFT). He also formulated the relationship between temperature and the thermal properties of materials now used to calculate thermal **conduction**, referring to the transfer of energy from one object to another when at different temperatures due to direct contact between them. In 1824 and 1827, Fourier published papers describing the energy balance of the Earth. He realized that the Earth's surface temperature was much higher than he calculated based on the Earth–Sun distance and incident solar radiation, laying the foundation for our understanding of the **greenhouse effect** by recognizing the importance of the atmosphere in Earth's energy balance. In addition, Fourier described how the conduction or transfer of energy (G, W m^{-2} or W) between two locations varies with the ability of the object to transfer energy (the **thermal conductivity** k; W m^{-1} K^{-1} or °C^{-1}) and the temperature difference (dT; K or °C) measured at two locations separated by distance dz (m). The relationship between these variables is known as **Fourier's Law** (Eq. 14.17):

$$G\left(\mathrm{W\ m^{-2}}\right) = -k\frac{\mathrm{d}T}{\mathrm{d}z}. \tag{14.17}$$

Equation (14.17) can be expressed in watts when the surface area A (m^2) of contact between the objects is known:

$$G(\mathrm{W}) = -kA\frac{\mathrm{d}T}{\mathrm{d}z}. \tag{14.18}$$

Water has a significant effect on the magnitude of conductive heat loss because of its large thermal conductivity (Table 14.1). The units of thermal conductivity describe the rate of energy transfer (power; joules per second) required to change the temperature by 1 °C (or K) over a 1-m distance of the material with a surface area of 1 m^2. In general, the denser the material, the larger the value of k, such as with metals. Natural materials such as soils are comprised of a mixture of solid minerals and organics, air, and water (sometimes in all three states), so the value of k is the sum of the k values from each constituent weighted by the volume fraction of each, with variations due to changes in the liquid water content of the soil. Frozen water has a high k, which is somewhat unexpected since ice has a lower density than liquid water. The integral lattice connections between water molecules when frozen result in the large thermal conductivity. Water vapor, with the large spacing between molecules, has a small k value comparable to that of air.

The surface area in contact between one object and another directly affects the conductive heat transfer. A person standing has a small surface area in contact with the ground compared with someone sitting or lying. Ectotherms such as lizards often choose a posture with a large area of ground contact to maximize conductive heat transfer into their bodies. The difference in temperature, or the temperature gradient, across the material determines the magnitude of conductive heat transfer. If there is no temperature gradient, there is no conductive heat transfer.

Table 14.1 Thermal conductivities for various materials

Sources: Jacobsen et al. (2003); Engineering ToolBox (2005); Vogel (2005).

Material	Thermal conductivity (k) ($\mathrm{W\ m^{-1}\ K^{-1}}$)
Water	
Solid (0 °C)	2.18
Liquid (25 °C)	0.60
Gas (1 atm; 127 °C)	0.016
Natural materials	
Air (10 °C)	0.026
Dry sand	0.15–0.25
Moist sand	0.25–2.00
Dry soil	0.25
Moist soil	2.0
Clay	1.28
Sandstone	1.60–2.10
Granite	1.70–4.00
Wood (moist)	0.12–0.17
Wood (dry)	0.20
Human skin	0.50
Animal fur	0.024–0.063
Metals (0 °C)	
Aluminum	236
Copper	401
Gold	318
Steel (wrought carbon)	59

Fourier's Law (Eq. 14.17) can be used to illustrate the impact of thermal conductivity on the rate of conductive heat loss or gain (depending on the temperature gradient) from a person. Imagine a person standing in air, and another submerged in water. Both have a skin temperature (T_s) of 35 °C, and the air (T_a) and water temperature (T_w) is 20 °C. From Table 14.1, k_{air} is 0.026 and k_{water} is 0.60 W m^{-1} °C^{-1} (approximately). There is no wind or water current. The person's surface area can be calculated using Eq. (14.15) (1.96 m^2 for an 80-kg, 180-cm-tall person) to calculate the energy exchange in watts.

For the person exposed to air:

$$G(\text{air}) = -k\frac{\mathrm{d}T}{\mathrm{d}z} = -k\frac{T_a - T_s}{\mathrm{d}z} = -0.026\frac{\mathrm{W}}{\mathrm{m\ °C}} \times \frac{20\ °\mathrm{C} - 35\ °\mathrm{C}}{1\mathrm{m}} = 0.39\frac{\mathrm{W}}{\mathrm{m}^2} \quad (14.19)$$

or submerged in water:

$$G(\text{water}) = -k\frac{\mathrm{d}T}{\mathrm{d}z} = -k\frac{T_w - T_s}{\mathrm{d}z} = -0.60\frac{\mathrm{W}}{\mathrm{m\ °C}} \times \frac{20°\mathrm{C} - 35°\mathrm{C}}{1\mathrm{m}} = 9.00\frac{\mathrm{W}}{\mathrm{m}^2}. \quad (14.20)$$

Since we subtracted T_a or T_w from T_s, the positive sign indicates the person is losing heat to the adjacent air or water (a negative value would indicate a heat gain). The loss of heat from the person to the water is 23 times larger than the heat loss to the still air. If the water was cooler, at 5 °C, G would double to 18 W m^{-2}. This example shows that although conductive heat loss is small when compared with the body's absorbed radiation or resting metabolic rate, water has a large impact on the magnitude of the heat loss through the thermal conductivity term.

14.3.4 Convection (Sensible) Heat Flux

The **sensible heat flux** is the transfer of energy from one location to another due to the mass movement of the fluid material (e.g., air). It is "sensible" in that it can be calculated from a measurable difference in temperature between the object and its surroundings. The sensible heat (H; watts) or sensible heat flux (H; W m$^{-2)}$ is often referred to as **convection**, or **convective currents**, that occur when heating a parcel of air results in the vertical displacement of the parcel through the surrounding air. As the parcel warms, it expands and the volume increases. The **Ideal Gas Law** ($PV = nRT$; pressure × volume = the number of moles of gas × the gas constant × temperature) shows how these variables are related. Thus, with warming and expansion, the pressure within the air parcel will be less than in the surrounding air. This is referred to as an **unstable condition**, and the parcel will rise. Conversely, when the air parcel is at a lower temperature than its surroundings, its volume will decrease and its pressure will increase, causing the parcel to descend and return to a location where the pressure inside the parcel relative to the air outside is equal. This is referred to as a **stable condition**. The initial displacement of the air parcel could be due to the air being forced to rise over a barrier such as a mountain range (**mechanical** or **forced convection**), or due to surface warming heating of the parcel (**free convection**). During **neutral stability** conditions, the air parcel remains in position or returns to its initial position when displaced.

Often the parcel of air is treated as an **adiabatic** system, meaning that there is no transfer of heat or mass between it and the surrounding environment. This adiabatic assumption works well in the atmosphere where changes in volume and pressure of an air parcel compare well with changes in temperature calculated using the Ideal Gas Law. In other words, the work required to expand or contract the air parcel comes from within the parcel itself as expressed by changes in internal temperature. When there is external energy exchange, the process is **diabatic**.

The body's rate of convective heat loss (H; W) can be calculated from the temperature difference between the body's surface (T_s; °C or K) and the surrounding air (T_a; °C or K), the body's surface area exposed to heat loss (A; m^2), and a **convective heat transfer coefficient** (k_H; W m^{-2} K^{-1} or °C^{-1}) (Eq. 14.21):

$$H = k_H A(T_s - T_a). \quad (14.21)$$

When the human body's surface temperature exceeds the ambient air temperature, H will be positive, indicating convective heat loss from the body. With high ambient air temperatures that exceed the body's surface temperature, H will be negative, indicating that the body is gaining heat (Chapter 16).

The appropriate k_H for a person (sometimes noted as h_c in the literature) is required to calculate H. Based on measurements from 112 experiments using 16 human-replica heated nude mannequins in still air with a total surface area of 1.471 m^2, De Dear et al. (1997) measured a k_H of 3.4 W m^{-2} K^{-1} (standing) and 3.3 W m^{-2} K^{-1} (seated). As wind speed increases, k_H increases, thus increasing the convective heat loss. Indoors, wind speed is generally negligible, but even so De Dear et al. (1997) derived an empirical relationship between indoor wind speeds U_i (m s^{-1}) between 0.2 m s^{-1} and 0.8 m s^{-1} and k_H (W m^{-2} K^{-1}) for general indoor applications regardless of posture (Eq. 14.22):

$$k_H\left(\text{W m}^{-2}\,\text{K}^{-1}\right) = 10.3U_i^{\,0.6}. \tag{14.22}$$

Since most people wear clothing even when indoors, it makes sense to calculate H when the surface area and temperature are representative of a clothed rather than naked body, and express H in W m^{-2} instead of watts. This requires slight modifications of Eq. (14.21). First, a **correction factor** f_{cl} based on the ratio of the clothed to naked body surface area replaces A in Eq. (14.21). Values for f_{cl} typically range from 1.10 for someone lightly clothed to f_{cl} = 1.46 for someone heavily clothed. Second, the body's skin surface temperature T_s in Eq. (14.21) is replaced by a measure of the temperature at the clothing's surface, T_{cl}. Lastly, the heat transfer coefficient k_H in Eq. (14.21) is referred to as h_c (the heat transfer coefficient) and the symbol C in units of W m^{-2} is often used instead of H for the convective heat loss (American Society of Heating, Refrigerating and Air-Conditioning Engineers Inc., 2017). These modifications result in Eq. (14.23):

$$C = f_{cl}h_c\left(T_{cl} - T_a\right). \tag{14.23}$$

Using typical clothed person values of f_{cl} = 1.15, h_c = 3.4 W m^{-2} K^{-1}, T_{cl} of 35 °C, and T_a of 23 °C to calculate C using Eq. (14.23) gives 46.92 W m^{-2}, or roughly 94 W based on a body surface area of 2 m^2 (Eq. 14.24):

$$C = 1.15 \times 3.4\text{W m}^{-2}\text{K}^{-1} \times \left(35\ ^\circ\text{C} - 23\ ^\circ\text{C}\right) = 46.92\text{W m}^{-2} \times 2\text{m}^2 = 93.84\,\text{W}. \tag{14.24}$$

Water affects convective heat loss by decreasing the skin's surface temperature, and a simple experiment can be used for illustration purposes. Based on measurements of the author's bare arm (f_{cl} = 1) in 24 °C air after being sprayed with water, the skin temperature reduced from 35 °C to 26 °C measured with an infrared thermometer. This resulted in a large decrease in C from 37.4 W m^{-2} when dry to 6.8 W m^{-2} when wet (Eq. 14.25):

$$\begin{aligned} C(\text{dry}) &= 1.0 \times 3.4\ \text{W m}^{-2}\text{K}^{-1} \times \left(35\ ^\circ\text{C} - 24\ ^\circ\text{C}\right) = 37.4\ \text{W m}^{-2} \\ C(\text{wet}) &= 1.0 \times 3.4\ \text{W m}^{-2}\text{K}^{-1} \times \left(26\ ^\circ\text{C} - 24\ ^\circ\text{C}\right) = 6.8\ \text{W m}^{-2}. \end{aligned} \tag{14.25}$$

14.3.5 Evaporative Cooling: Sweating

The energy exchange of the human body is rarely balanced, especially when outdoors. Energy received from solar radiation can overwhelm the body's ability to dissipate energy by conductive and convective heat losses, and evaporation must be used to maintain an energy balance and prevent overheating. To produce **sweat**, which is mostly water with some dissolved salt, and move it from sweat glands beneath the skin to the surface requires energy and water, so the heat lost by evaporation is referred to as the **wet heat exchange**. All of the **dry heat exchanges** (radiation, convection, and conduction) depend on the difference between skin temperature and the surroundings, whereas the evaporative heat flux depends on the difference in vapor pressure (or vapor density) between the skin's surface and the surrounding atmosphere, divided by an appropriate resistance to vapor transfer or multiplied by an appropriate vapor transfer coefficient. Evaporative cooling can occur through the generation of sweat on (or near) the skin's surface (**cutaneous cooling**), and through breathing (**pulmonary cooling**). Since sweating is ineffective in furry creatures, pulmonary cooling (e.g., panting) is much more effective for them. As humans evolved mostly hairless bodies, sweating became an effective cooling mechanism (Everts, 2021).

Sweating is an involuntary complex response to produce liquid water on the skin surface in an attempt to lower the body's temperature. The high latent heat of vaporization, the energy required to evaporate water, means that evaporation is a highly effective cooling mechanism if liquid water is available. Evaporation of water from the skin's surface depends on many of the same factors that affect evaporation from well-watered surfaces (Chapter 7), including the body's availability of water and the vapor pressure in the atmosphere. A key difference, however, between sweating and open-water evaporation from the landscape is that with sweating (as with transpiration) there is a strong and variable biological control.

When a person sweats for thermoregulation is controlled by temperature, location and number of sweat glands, and the water available to the body. When the body's core and skin temperatures exceed a set-point temperature, typically 37.1 °C and 31 °C respectively (Campbell et al., 1994), the hypothalamus triggers a suite of chemical responses resulting in sweat production (Baker, 2019). Most sweat is produced by the **eccrine glands** distributed throughout the body, with the highest density located on the palms and soles (~250–500 glands per cm^2), possibly to supplement friction as a fight-or-flight response. Over the entire human body, there is a total of between 2 and 4 million eccrine sweat glands (Sato et al., 1989) that produce sweat in response to both thermal and emotional stimuli (Baker, 2019).

The evaporation rate required to prevent overheating and to maintain thermal regulation can be calculated following the methods and equations outlined below (summarized by the American Society of Heating, Refrigerating and Air-Conditioning Engineers Inc., 2017). It is important to note that these calculations represent conditions experienced indoors in the context of human comfort. The **maximum evaporation rate** possible, E_{max} (W m^{-2}), is calculated using the same framework used to calculate the convective flux; the appropriate gradient (for evaporation this is the saturation vapor pressure minus the ambient vapor pressure; kPa) multiplied by a **vapor transfer coefficient** (h_e; W m^{-2} kPa^{-1}) with a surface area adjustment for clothing (f_{cl}) (Eq. 14.26):

$$E_{max} = f_{cl} h_e \left(e_s - e_a\right). \tag{14.26}$$

The saturation vapor pressure in Eq. (14.26) (e_s; kPa) should be calculated at the skin's surface temperature, and the ambient vapor pressure (e_a; kPa) can be calculated by multiplying the saturation vapor pressure calculated at the ambient temperature by the ambient relative humidity. The saturation vapor pressure can be calculated at any temperature over liquid water using Eq. (7.3), repeated here for convenience:

$$e_s = 0.61121 \exp\left(\frac{17.502T}{240.97 + T}\right)$$

which calculates e_s in kPa with temperature (T) in °C (Buck, 1981).

The evaporative heat transfer coefficient h_e can be estimated using the **Lewis number** (Le), equal to the heat transfer coefficient divided by the vapor transfer coefficient (Le = h_c/h_e) (Lewis, 1922). Under typical indoor conditions, Le = 0.0606 kPa K^{-1}, so h_e (W m^{-2} kPa^{-1}) can be calculated when h_c is known (Eq. 14.27):

$$h_e = \frac{1}{\text{Le}} \times h_c. \tag{14.27}$$

The clothing area factor f_{cl} used in the calculation of the radiative and convective heat losses can differ for water vapor transfer depending on the clothing material used and can be adjusted accordingly. For example purposes, however, the same value of f_{cl} is used for the radiative, convective, and evaporative fluxes.

The maximum evaporation rate is seldom equal to the **actual evaporation rate** from the skin (E_{sk}; W m^{-2}) since the body is neither able to produce enough sweat to cover the entire skin surface, nor is such a high evaporative rate required to maintain a constant body temperature under normal conditions. The **required evaporation rate,** the rate required to maintain a constant body temperature (E_r; W m^{-2}), can be calculated from the energy balance equation (Eq. 14.3) by determining the energy required to balance energy inputs (M; metabolic heat production) and energy outputs ($R + C$; radiative and convective heat loss), and neglecting G (conduction) since this term is small (<1 W m^{-2}) under typical indoor conditions (Eq. 14.28):

$$E_r = M - (R + C). \tag{14.28}$$

The required evaporation rate is the rate of evaporation needed to prevent overheating, so the body needs to produce enough sweat to meet this evaporative demand. This rate has several variables such as activity level, conditioning, age, sex, genetics, and fitness level, so is difficult to estimate. Therefore, empirical relationships between the total metabolic rate (M) and E_r have been developed that vary with acclimation level, activity level, clothing ensembles, or other variables (Shapiro, Pandolf and Goldman, 1982).

Once sweat appears on the skin's surface, the proportion of the body that must be wetted to regulate body temperature through evaporation, the **required skin wetness fraction** (w_r; dimensionless; 0–1) is equal to the ratio of required evaporation rate E_r (W m^{-2}) to the maximum possible evaporation rate E_{max} (W m^{-2}) (Eq. 14.29):

$$w_r = \frac{E_r}{E_{max}}. \tag{14.29}$$

To account for the small proportion of evaporation that results from the diffusion of water through the skin (6 percent under normal conditions), the total proportion of the body that is wet, the **actual skin wetness fraction** (w_{sk}; dimensionless; 0–1) can be expressed as (Eq. 14.30):

$$w_{sk} = 0.94 w_r + 0.06. \tag{14.30}$$

The **actual rate of evaporation from the skin** E_{sk} (W m^{-2}) can then be calculated by multiplying w_{sk} and E_{max} (Eq. 14.31):

$$E_{sk} = w_{sk} E_{max}. \tag{14.31}$$

Equations (14.26) through (14.31) can then be used to calculate the evaporative heat loss from a person's skin. As mentioned, this calculation requires assumptions and generalizations; however, these

Table 14.2 Parameters, variables, and equations used in an example calculation of the energy required for evaporative sweat loss under indoor conditions for a person at the resting (basal) metabolic rate

The conductive heat loss is assumed to be negligible.

Parameter or variable	Symbol	Value used	Equation used	Value calculated
		Surface area and basal metabolic rate		
Body weight	w	80 kg	–	–
Body length	L	180 cm	–	–
Resting metabolic rate	M_0	–	14.13	90 W
Body (nude) surface area	S	–	14.15	1.97 m^2
Resting metabolic rate	M_0/S	–	14.16	**45.7 W m^{-2}**
		Radiative heat		
Clothing area factor (clothing/naked surface area)	f_{cl}	1.15 (ratio)	–	–
Radiative heat transfer coefficient	h_r	4.7 W m^{-2} per °C or K	–	–
Clothing surface temperature	T_{cl}	30 °C	–	–
Radiative surface temperature	T_r	32 °C	–	–
Radiative heat loss	R	–	14.9	**−10.8 W m^{-2}**
		Convective (sensible) heat		
Clothing area factor (clothing/naked surface area)	f_{cl}	1.15 (ratio)	–	–
Convective heat transfer coefficient	h_c	3.4 W m^{-2} per °C or K	–	–
Clothing surface temperature	T_{cl}	30 °C	–	–
Air temperature	T_a	35 °C	–	–
Convective heat loss	C	–	14.23	**−19.6 W m^{-2}**
		Evaporative heat		
Clothing area factor (clothing/naked surface area)	f_{cl}	1.15 (ratio)	–	–
Vapor transfer coefficient	h_e	–	14.27	56.1 W m^{-2} kPa^{-1}
Saturation vapor pressure at skin temperature × 100% relative humidity	e_s	T = 32 °C RH = 100%	7.3	4.76 kPa
Saturation vapor pressure at air temperature × 50% relative humidity	e_a	T = 35 °C RH = 50%	7.3	2.81 kPa
Maximum evaporation rate	E_{max}	–	14.26	125.8 W m^{-2}
Required evaporation rate	E_r	–	14.28	76.1 W m^{-2}
Required skin wetness fraction	w_r	–	14.29	0.60
Actual skin wetness fraction	w_{sk}	–	14.30	0.62
Actual skin evaporation rate	E_{sk}	–	14.31	**78.0 W m^{-2}**

equations can be used to illustrate the importance of evaporative cooling for a person under the conditions summarized in Table 14.2. Under the conditions in this example, the person is indoors, with the air temperature and surface temperature of the surrounding objects higher than the clothing surface temperature. These conditions result in the person gaining radiative and convection heat (both R and C are negative), so the body must produce sweat to cool through evaporation (positive E_{sk}).

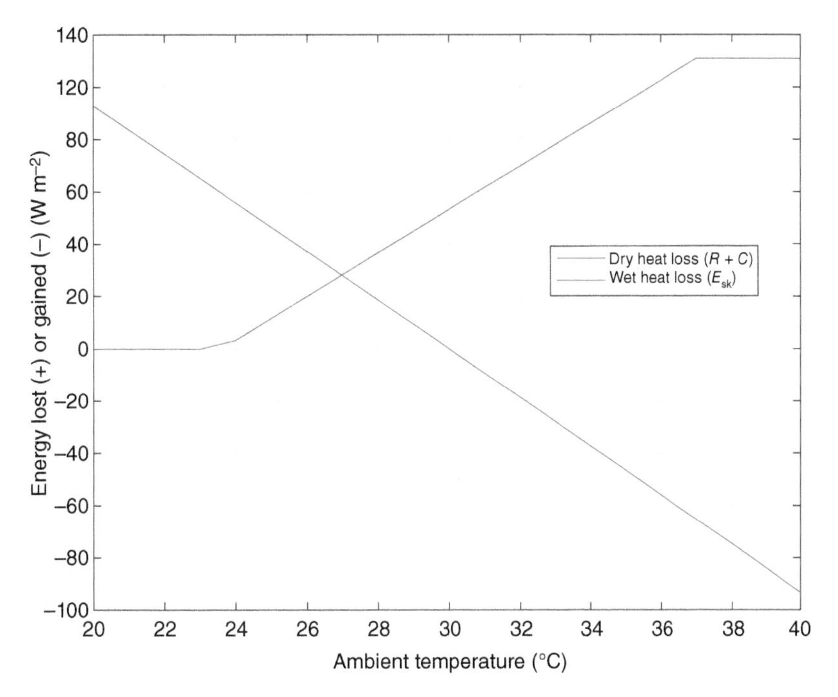

Figure 14.1 Changes in the dry heat (radiative, R, and convective heat, C) and wet heat (evaporative cooling from sweat, E_{sk}) for a subject at rest indoors as the ambient temperature varies, calculated using the equations provided in this chapter. Positive values on the y-axis indicate a loss of energy from the subject, negative values an energy gain.

Table 14.2 shows that a substantial amount of sweat evaporation, 78 W m^{-2}, is required to maintain an energy balance, and thus a stable body temperature. Without sufficient water within the body, an energy balance could not be maintained, and the body temperature would rise quickly to dangerously high temperatures, leading to possible heat stress or heat stroke.

Ambient conditions can change rapidly, often with changes in one variable affecting another (e.g., air temperature and relative humidity). To illustrate how the dry heat terms and the wet heat term vary as conditions change, $R+C$ and E_{sk} are plotted against a range of ambient temperatures (Figure 14.1). The subject's weight and height were 80 kg and 180 cm, respectively; they were indoors, and at rest so the basal metabolic rate was used. Ambient temperature was used as an indicator of both the air and radiative surface temperatures. A 30 °C and 1.15 clothing surface temperature and clothing factor (respectively), and 50 percent relative humidity, were used. Heat loss coefficients for radiation, convection, and evaporation were 4.7 W m^{-2} K^{-1}, 3.4 W m^{-2} K^{-1}, and 56.1 W m^{-2} kPa^{-1}, respectively. For the calculation of the wet heat loss (E_{sk}), a lower limit of 0 W m^{-2} was set since heat gain from condensation on the skin is not likely under the conditions used in the calculation. An upper E_{sk} limit was determined by setting E_{sk} equal to the required evaporation rate (E_r) when the actual skin wetness fraction (w_{sk}) exceeds 1.

Figure 14.1 shows that the dry heat terms ($R + C$) decrease linearly as the ambient temperature increases. For every 1 °C increase in the ambient temperature, the dry heat loss decreases by roughly 9 W m^{-2}, and shifts to a heat gain when the ambient temperature exceeds the body's surface temperature. Evaporative cooling by sweating begins when the ambient temperature exceeds 23 °C, and plateaus at a maximum of roughly 111 W m^{-2} at an ambient temperature of 37 °C. Sweating begins at 23 °C since above this temperature the basal metabolic rate, which does not vary with the ambient temperature, exceeds the dry heat loss which is decreasing as the ambient temperature increases. Between 23 °C and 37 °C, E_{sk} increases at a rate of nearly 11 W m^{-2} for every 1 °C increase in the ambient temperature, 1.2 times the rate of decrease in the dry heat term, showing the importance of evaporative cooling. Heat stress and stroke are likely when the ambient temperature exceeds 37 °C (in this example) since the body is no longer able to produce enough sweat to meet the evaporative demand required to maintain a constant body temperature.

14.4 Requirement for Evaporative Cooling during Activities

The ability to sweat is critical for the body to maintain an energy balance, and the body must have a sufficient internal water supply to meet the evaporative demand. When a person is active, their metabolic heat increases, adding energy that needs to be balanced by energy output (sweating) to prevent overheating. The calculations so far have been based on a person at rest. Let's now look at the sweating and hydration requirements when a person is active.

The additional energy required for various activities is expressed through the metabolic heat generated relative to a typical sedentary person. Referred to as the **Metabolic Equivalent of Task (MET)**, these estimates of the increased metabolic rate can be used to calculate the sweat cooling requirements for the activity being performed. A MET value of 1 defined as equal to 58.1 W m^{-2} (1 kcal kg^{-1} h^{-1} or 50 kcal m^{-2} h^{-1}) is the generally accepted reference value based on the metabolism of a seated, quiet person. A MET value of 2 indicates that the task being performed requires double the metabolic rate compared with a resting, seated person.

Caution should be used when applying published MET values for an individual since their sedentary resting metabolic rate may not be 58.1 W m^{-2}. The MET value of 1 was originally derived from the resting O_2 consumption of 3.5 ml O_2 kg^{-1} min^{-1} for one 70-kg, 40-year-old male subject. Based on a study of 642 women and 127 men ranging in age from 18 to 74 years and weight from 35 to186 kg, Byrne et al. (2005) found that the average resting metabolic rate was 2.6 ml O_2 kg^{-1} min^{-1}, roughly 26 percent less than the standard 1 MET based on 3.5 O_2 kg^{-1} min^{-1}. In addition to individual variations in the resting metabolic rate used to define MET = 1, the task-specific metabolic equivalent will vary based on several factors such as age, health, and task-conditioning (skill). Passmore and Durnin (1955) found a large range in the subject's metabolic rate due to both subtleties in conditions of the activity being performed (e.g., type of surface being walked or run upon) and the subject's age, sex, and weight.

Although we must recognize the limitations of the MET values, these values are still useful for a more appropriate calculation of the energy balance, thus evaporative cooling requirements, for an active person. Various MET values from 1 (resting), 2 (normal walking speed on a level surface), 3 (fast walking or light household chores), 4 (light jogging or jobs with some

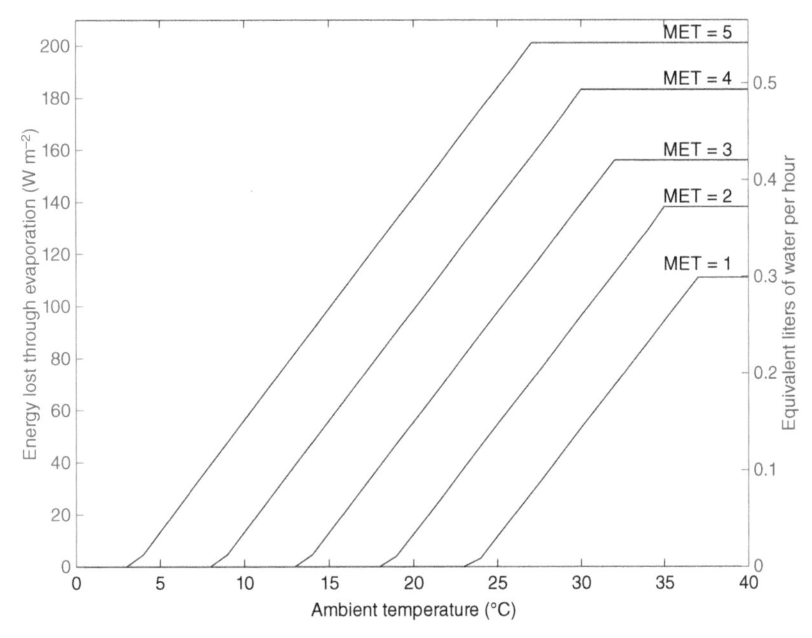

Figure 14.2 Influence of activity (MET) on the evaporative heat loss and the required water intake.

lifting required) through 5 (moderate to vigorous activities such as sports such as tennis or physical labor jobs) were used to calculate the energy lost through sweating required to maintain an energy balance (Figure 14.2). The procedures and conditions used to generate the plots shown in Figure 14.2 were the same as used in Table 14.2, except the resting metabolic rate was multiplied by MET values ranging from 1 (resting) through 5 (vigorous activities).

Figure 14.2 shows that as the MET increases, sweating begins and reaches the maximum rate at a lower ambient temperature. In this example, sweating is not required at a low ambient temperature of 10 °C for MET activities of roughly less than 3, but is required for moderate activity levels (MET > 4). At a higher ambient temperature, say 30 °C, sweating is required even when at rest (MET = 1) and the evaporative cooling requirement increases as the activity level increases. As the ambient temperature increases, the ability for the body to provide enough sweat for the evaporative demand decreases with any further increase in the activity level. For example, at a MET of 2, Figure 14.2 shows that the subject would be able to maintain a constant body temperature by increasing E_{sk} until the ambient temperature reaches 35 °C. At a MET of 3, 32 °C is the maximum tolerable ambient temperature for effective sweating. At a MET of 5, 27 °C is the maximum tolerable ambient temperature for effective sweating. As the metabolic rate increases with increasing activity level, evaporative cooling begins and stops at lower ambient temperatures.

Also included in Figure 14.2 is the equivalent rate of water intake in liters per hour required to replace water lost through sweating. The conversion of the latent heat flux (λE; W m^{-2}) to liters of water per hour requires multiplying λE by the body's surface area (S; m^2), dividing by the latent heat of vaporization (λ: J kg^{-1}), the density of water (ρ_w; kg m^{-3}), conversion from m^3 to liters (1,000 cm^3), and conversion from seconds to hours (Eq. 14.32):

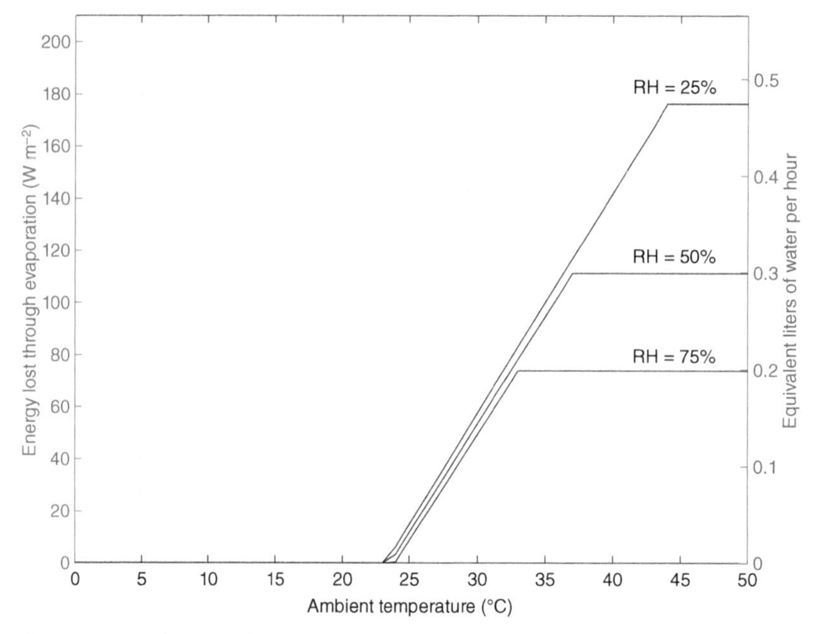

Figure 14.3 Influence of relative humidity on the evaporative heat loss and required water intake for person at rest (MET = 1).

$$E\left(\text{liters per hour}\right) = \lambda E\left(\text{J m}^{-2}\,\text{s}^{-1}\right) \times S\left(\text{m}^2\right) \times \frac{1}{\lambda\left(\text{J kg}^{-1}\right)} \times \frac{1}{\rho_\text{w}\left(\text{kg m}^{-3}\right)} \times \frac{1{,}000\ \text{liter}}{\text{m}^3} \times \frac{3{,}600\ \text{s}}{\text{h}}. \tag{14.32}$$

As the ambient temperature increases, so does the water intake required to balance the water lost to evaporation for all activity levels. Also, as the intensity of the activity increases, so does the water intake required to replenish the water lost. Keep in mind some water loss constantly occurs through non-sweating mechanisms including respiration and the diffusion of water through the skin, so ample hydration is required even when at rest.

The rate of water intake to replenish water lost through sweating depends not only on ambient temperature and activity level, but also on atmospheric humidity (as well as the other variables already mentioned). Changes in humidity directly affect the sweat evaporation rate since the difference in the water vapor density (or vapor pressure) between the skin and the air directly affects the evaporation rate. Since the skin (or clothing) surface temperature is fairly constant in the range of 30 to 35 °C, the saturation vapor pressure also remains fairly constant, whereas the atmospheric vapor pressure (humidity) can change rapidly.

Figure 14.3 shows how changes in relative humidity affect the required evaporative cooling rate and the equivalent water intake for a person at rest (MET = 1) using the same equations and constants shown in Table 14.2. Evaporative cooling through sweating is much less effective as the relative humidity increases. As the relative humidity increases, the decrease in the skin-to-air vapor pressure gradient decreases the evaporation rate, so E_{sk} reaches a lower maximum value at a lower ambient temperature, making one feel "hot and clammy" during hot and humid conditions. The equivalent water intake required to supply the sweat lost also decreases as the relative humidity increases. At a high ambient temperature, the effects of increasing relative humidity become especially apparent. For example, with an ambient temperature of 35 °C and a relative humidity of 25%, Figure 14.3 shows that E_{sk} is roughly 100 W m^{-2}, requiring roughly 0.3 liters of water per hour. Sweating can increase and effectively cool the body until an ambient temperature of 44 °C at a relative humidity of 25%, decreasing to 37 °C at a relative humidity of 50%. At a high 75% relative humidity, E_{sk} reaches its maximum at an ambient temperature of 33 °C, indicating that E_{sk} is not able to effectively cool the body at that combination of temperature and humidity. The impacts of the increasing frequency and magnitude of heatwaves on human health are discussed in Chapter 16.

14.5 SUMMARY

This chapter described how an endotherm maintains a constant body temperature through adjustment of energy inputs relative to energy outputs. Water, in the production and evaporation of sweat on the skin surface, plays a major role in cooling the body during periods of excessive heat or activity. The body receives energy from shortwave radiation, longwave radiation, and metabolic heat production. When outdoors, the quantity of shortwave radiation received can be large during the daytime, even under cloudy conditions, owing to diffuse-beam shortwave radiation. When indoors, longwave radiation inputs dominate over shortwave, with the surface temperature and emissivity of the surrounding walls and objects determining the quantity of longwave radiation the body receives. Radiation losses from the body include a proportion (typically 37 percent) of the incident shortwave radiation reflected, and the emitted longwave radiation which depends on the body's surface temperature and emissivity. Estimates of the net radiation absorbed can be made assuming the body's shape is represented by a cylinder. Estimates are complicated not only by variations in incident radiation, but also the body's surface area and the characteristics of clothing worn.

Metabolic heat production is critical when the body is cold. The identification of metabolism and respiration as a process involving the consumption of oxygen and the production of carbon dioxide and heat by living organisms was proposed by studies performed by Lavoisier, von Voit, and Rubner in the late nineteenth and early twentieth centuries. An allometric relationship between an endotherm's weight and metabolic rate when at rest was developed by Kleiber in the 1930s, as was a relationship between a person's weight, length, and body surface area by du Bois and du Bois in the 1910s. Together, these relationships are still used today to estimate the resting metabolic rate, which can then be used to determine the required rate of evaporative cooling through sweating.

The body loses energy from nonradiative conductive and convective energy exchanges. Energy transfer via conduction depends on the temperature difference between the body and objects the body touches, the surface area of the body in contact with the object, and the thermal conductivity of the object. Generally, the conductive energy term is small since the surface in contact with objects is small (e.g., the soles of the feet when standing). Convective heat exchange depends on the difference in temperature between the body's surface and surrounding air. Since an endotherm strives to maintain a constant body temperature, the rate of convective heat loss varies with the ambient air temperature. The human body generally loses heat through convection when the ambient air temperature is less than the body's surface temperature of between roughly 30 and 35 °C (depending on clothing), and gains heat through convection when the ambient air temperature is greater than the body's surface temperature.

The dominant energy loss term in the energy balance of an endotherm is the evaporative heat loss achieved through the process of sweating, a critical mechanism to maintain a constant body temperature and prevent overheating during periods of excessive energy inputs. Although some cooling is achieved through breathing, humans use cutaneous cooling from sweat produced from sweat glands as the primary means for evaporative cooling. The rate of evaporative cooling required to maintain a constant body temperature depends on all the terms that appear in the energy balance equation, so example calculations of these terms were provided prior to the calculation of the required rate of evaporative cooling. The difference in vapor pressure (or vapor density) between the skin and the air above is the primary determinant of the maximum evaporation rate, and the required evaporation rate depends on the magnitude (and whether a gain or loss of energy from the body) of all the energy balance terms. The rate of cooling achieved by the evaporation of sweat also depends on the characteristics of the individual (weight, height, sex, age, conditioning, etc.), but also whether at rest or active (expressed using the Metabolic Equivalent of Task), and the ambient conditions (e.g., air temperature and humidity).

14.6 QUESTIONS

14.1 How does water influence the ability of an endotherm to maintain a constant body temperature? To answer this question, describe the role of water in each term in the energy balance equation (Eq. 14.3) and how a person outdoors and/or indoors depends on the availability of water to maintain a constant body temperature.

14.2 We have all experienced the sensation of feeling much cooler as we move from a sunny to a shady location. Explain why we experience this sensation if there was no change in air temperature between the sunny and shaded locations.

14.3 Use the Stefan–Boltzmann Law to calculate the longwave radiation gained or lost by a person inside a walk-in freezer where the average surface temperature of the room is −10 °C. The person then relocates to a sauna where the average surface temperature of the room is 38 °C. Calculate the longwave radiation gained or lost by a person in the sauna. Assume the person's surface temperature is 35 °C and the emissivity for the person and the rooms is 0.98. Under what condition(s) would a person not gain or lose any longwave radiation to the environment?

14.4 Using your weight and height, calculate your basal (resting) metabolic rate in units of watts and watts per square meter. To illustrate how mass (weight) of an organism affects the basal metabolic rate (M_0), calculate M_0 for a mouse (w = 25 g) and an elephant (w = 5,250 kg). Divide M_0 by mass to obtain the mass-specific M_0 (W kg^{-1}). Explain the reason(s) why M_0 expressed in watts and M_0 in watts per kilogram are so different for the mouse and elephant.

14.5 Imagine sitting on a bench made of aluminum; then, after feeling cold, you move to an adjacent bench made of wood. If both benches have the same temperature and your body temperature has not changed, explain why you feel colder on the aluminum bench. Calculate your heat loss due to conduction from both benches to support your explanation.

14.6 Some studies report that women are more comfortable at a higher room air temperature (25 °C) than men (22 °C) (Kingma and Lichtenbelt, 2015). Estimate the convective heat loss using Eq. (14.23) (e.g., when indoors) at each of these room temperatures assuming clothing temperatures of 28 °C for each. Does your calculation help provide an explanation for the difference? Which term(s) in the energy balance equation (Eq. 14.3) must differ between females and males to account for the differences in the convective heat loss?

14.7 Calculate your maximum, required, and actual evaporation rates for any MET based on activity level and ambient conditions that interest you. State the conditions and the parameters you used. Use Table 14.2 as a guide.

14.8 Estimate your intake of water required to meet the evaporative sweat demand that you calculated in Question 14.7.

14.9 The rate of heat illness among high school student football athletes in the United States during 2005–2009 was 4.5 per 100,000, 10 times higher than the average rate for eight other sports (Yard et al., 2010). Using Figure 14.2 as a guide, state the maximum ambient temperature that vigorous athletic activities should be restricted to. Alternatively, recreate Figure 14.2 using parameters and conditions applicable to your own situation, keeping in mind that the calculations presented in Table 14.2 are based on indoor conditions.

14.10 In addition to activity level (MET), humidity has a large effect on a person's ability to effectively use sweat to cool (see Figure 14.3). Provide an explanation for why the evaporation of sweat on the skin is so sensitive to changes in humidity. Create the plots shown in Figure 14.3 for a person not at rest (i.e., use an MET > 1) to examine how both increased activity level and humidity affect the body's ability to sweat and prevent heat illness.

REFERENCES

American Society of Heating, Refrigerating and Air-Conditioning Engineers Inc. (2017) 'Thermal Comfort', in *ASHRAE Handbook – Fundamentals 2017: SI Edition*. American Society of Heating, Refrigerating and Air-Conditioning Engineers. Available at: https://books.google.com/books?id=6VhRswEACAAJ.

Baker, L. B. (2019) 'Physiology of sweat gland function: The roles of sweating and sweat composition in human health', *Temperature*, 6, pp. 211–259. doi: 10.1080/23328940.2019.1632145.

du Bois, D. and du Bois, E. F. (1916) 'A formula to estimate the approximate surface area of height and weight be known', *Archives of Internal Medicine*, 17, pp. 863–871.

Boschmann, M., Steiniger, J., Hille, U. et al. (2003) 'Water-induced thermogenesis', *Journal of Clinical Endocrinology and Metabolism*, 88, pp. 6015–6019. doi: 10.1210/jc.2003-30780.

Brown, R. D. and Gillespie, T. J. (1986) 'Estimating outdoor thermal comfort using a cylindrical radiation thermometer and an energy budget model', *International Journal of Biometeorology*, 30, pp. 43–52. doi: 10.1007/BF02192058.

Buck, A. L. (1981) 'New equations for computing vapor pressure and enhancement factor', *Journal of Applied Meteorology and Climatology*, 20, pp. 1527–1532.

Buono, M. J. (1999) 'Sweat ethanol concentrations are highly correlated with co-existing blood values in humans', *Experimental Physiology*, 84, pp. 401–404. doi: 10.1111/j.1469-445X.1999.01798.x.

Byrne, N. M., Hills, A. P., Hunter, G. R., Weinsier, R. L. and Schutz, Y. (2005) 'Metabolic equivalent: One size does not fit all', *Journal of Applied Physiology*, 99, pp. 1112–1119. doi: 10.1152/japplphysiol.00023.2004.

Campbell, A. B., Nair, S., Miles, J. and Webbon, B. (1994) 'Modeling the sweat regulation mechanism', *SAE Technical Papers*, 103, pp. 459–469. doi: 10.4271/941259.

Da Poian, A. T., El-Bacha, T. and Luz, M. R. M. P. (2010) 'Nutrient utilization in humans: Metabolism pathways', *Nature Education*, 3, p. 11.

de Dear, R. J., Arens, E., Hui, Z. and Oguro, M. (1997) 'Convective and radiative heat transfer coefficients for individual human body segments', *International Journal of Biometeorology*, 40, pp. 141–156. doi: 10.1007/s004840050035.

Engineering ToolBox (2005) *Thermal Conductivity of Metals, Metallic Elements and Alloys.* Available at: www.engineeringtoolbox.com/thermal-conductivity-metals-d_858.html (Accessed: September 22, 2021).

Everts, S. (2021) *The Joy of Sweat: The Strange Science of Perspiration*. W.W. Norton & Company.

Hooper, L., Bunn, D., Jimoh, F. O. and Fairweather-Tait, S. (2014) 'Water-loss dehydration and aging', *Mechanisms of Ageing and Development*, 136–137, pp. 50–58. doi: 10.1016/j.mad.2013.11.009.

Jablonski, N. G. (2004) 'The evolution of human skin and skin color', *Annual Review of Anthropology*, 33, pp. 585–623. doi: 10.1146/annurev.anthro.33.070203.143955.

Jablonski, N. G. and Chaplin, G. (2000) 'The evolution of human skin coloration', *Journal of Human Evolution*, 39, pp. 57–106. doi: 10.1006/jhev.2000.0403.

Jacobsen, R. T., Lemmon, E. W., Penoncello, S. G. et al. (2003) 'Thermophysical properties of fluids', in A. Bejan and A. D. Krause (eds.) *Heat Transfer Handbook*. John Wiley and Sons.

Kenny, N. A., Warland, J. S., Brown, R. D. and Gillespie, T. G. (2008) 'Estimating the radiation absorbed by a human', *International Journal of Biometeorology*, 52, pp. 491–503. doi: 10.1007/s00484-008-0145-8.

Kingma, B. and Lichtenbelt, W. V. M. (2015) 'Energy consumption in buildings and female thermal demand', *Nature Climate Change*, 5, pp. 1054–1056. doi: 10.1038/NCLIMATE2741.

Kleiber, M. (1932) 'Body size and metabolism', *Hilgardia*, 6, pp. 315–353.

Kleiber, M. (1947) 'Body size and metabolic rate', *Physiological Reviews*, 27, pp. 511–541.

Lewis, W. K. (1922) 'The evaporation of a liquid into a gas', *Transactions of the American Society of Mechanical Engineers*, 44, 325–340.

McMahon, T. (1973) 'Size and shape in biology', *Science*, 179, pp. 1201–1204.

Nguyen, M. K., Ornekian, V., Butch, A. W. and Kurtz, I. (2007) 'A new method for determining plasma water content: Application in pseudohyponatremia', *American Journal of Physiology – Renal Physiology*, 292, pp. 1652–1656. doi: 10.1152/ajprenal.00493.2006.

Passmore, R. and Durnin, J. V. G. A. (1955) 'Human energy expenditure', *Physiological Reviews*, 35, pp. 801–840. doi: https://doi.org/10.1152/physrev.1955.35.4.801.

Sato, K., Kang, W. H., Saga, K. and Sato, K. T. (1989) 'Biology of sweat glands and their disorders. I. Normal sweat gland function', *Journal of the American Academy of Dermatology*, 20, pp. 537–563. doi: 10.1016/S0190-9622(89)70063-3.

Shapiro, Y., Pandolf, K. B. and Goldman, R. F. (1982) ‘Predicting sweat loss response to exercise, environment and clothing’, *European Journal of Applied Physiology*, 48, pp. 83–96.

Shuter, B. and Aslani, A. (2000) ‘Body surface area: Du Bois and Du Bois revisited’, *European Journal of Applied Physiology*, 82, pp. 250–254. doi: 10.1007/s004210050679.

Smil, V. (2000) ‘Laying down the law’, *Nature*, 403, p. 597.

Sunkpal, M., Roghanchi, P. and Kocsis, K. C. (2018) ‘A method to protect mine workers in hot and humid environments’, *Safety and Health at Work*, 9, pp. 149–158. doi: 10.1016/j.shaw.2017.06.011.

Vogel, S. (2005) ‘Living in a physical world IV. Moving heat around’, *Journal of Biosciences*, 30, pp. 449–460. doi: 10.1007/BF02703717.

West, G. B., Brown, J. H. and Enquist, B. J. (1997) ‘A general model for the origin of allometric scaling laws in biology’, *Science*, 276, pp. 122–126. doi: 10.1126/science.276.5309.122.

Yard, E. E., Gilchrist, J., Haileyesus, T. et al. (2010) ‘Heat illness among high school athletes – United States, 2005–2009’, *Journal of Safety Research*, 41, pp. 471–474. doi: 10.1016/j.jsr.2010.09.001.

15 Water Quality

Key Learning Objectives

After reading this chapter, you will be able to:

1. Summarize the mechanisms behind, and provide examples of, how salt can be detrimental to vegetation growth, productivity, and survival.
2. Describe the connection between water availability, fertilizer use, harmful algal blooms, and eutrophication.
3. Identify the environmental and socioeconomic costs associated with the degradation of water quality due to algal blooms.
4. Explain the scale and scope of microplastics in the environment.
5. State the current and anticipated health consequences for the ingestion of microplastics.

15.1 Introduction

One in four people do not have access to safe drinking water, and consuming unsafe water is responsible for 1.2 million deaths annually (Ritchie and Rose, 2021). Water quality is subjective; a condition or contaminant considered harmful to one organism may not be harmful to another. Threshold concentrations or exposure levels are often arbitrarily determined, and can differ between or within species depending on age, size, health, and other factors. Regardless, since the properties of water permit it to act as the universal solvent, and water is required for plant, ectotherm, and endotherm form, function, and thermoregulation, the state of water and any compound dissolved within it has the potential to alter water quality.

Water quality can be broadly defined as how the characteristics of water make it suitable for a particular use. These characteristics can include any chemical, physical, or biological properties, depending on the water use. Thus, water use defines water quality parameters and limits. Here, **water use** is defined as the use of water by an organism to adequately fulfill its life functions, and **water pollutant** describes any substance or condition that has detrimental effects on an organism, thus lowering water quality and limiting water use. For example, the release of warm water into a river from industry could be considered polluting if it has a negative effect on local fish populations that normally rely on cooler water for foraging.

Water is the universal solvent, and changes in the solute concentration can affect vegetation growth and crop productivity. Since saline water dominates the surface of the Earth and saline soils are found worldwide, especially in arid regions where irrigation is common, high soil water salinity can hinder vegetation growth and productivity, or even kill crops and forests. Irrigation in arid and semi-arid regions can dissolve salts as the water moves down through the soil; then, with high evaporation rates and capillary rise, salts may precipitate out of solution at the soil surface, harming or killing crops. The use of groundwater for irrigation can exacerbate the problem, especially in coastal regions where saline coastal waters may replace groundwater that has been removed for irrigation purposes. Rising sea levels compound the problem.

Water-soluble nutrients applied as chemical fertilizers to agricultural crops and urban landscapes continue to be a major source of water pollution with detrimental effects on both freshwater and marine ecosystems worldwide. The reliance on only a few crops to feed most of the world's population and increased livestock production has resulted in decreased biodiversity and greater susceptibility to disease, all in the context of population increase and climate change. As the application of fertilizers containing nitrogen, phosphorus, and potassium has continued to increase, so have their impacts on water quality.

Excessive growth of algae and bacteria in both freshwater and marine aquatic environments occurs when runoff from agricultural or urban sources contains excessive amounts of dissolved nutrients, especially phosphates. When the algal concentration creates conditions harmful to animals and people, they are termed **harmful algal blooms**, or **HABs**, harmful from the production of toxins with the tell-tale appearance of **green tides** in fresh water, or **red tides** in salt water. Even if the algal blooms are not immediately harmful, water quality decreases as the excessive aquatic growth, or **eutrophication**, results in depletion of dissolved oxygen and hypoxic or anoxic conditions as the plant matter decomposes. Lake Winnipeg is used as an example of a large lake with increased occurrences of eutrophication events. Other examples are used to illustrate the social and economic costs of algal blooms.

Microplastics are quickly becoming a major water quality, ecosystem, and human health concern. In this chapter we will examine the history, uses, and sources of plastics in the environment, and the various mechanisms by which microplastics enter soils. We will then look at how these pollutants affect the properties of soils, especially those that affect water movement through soils and into vegetation, as well as microplastics that have relocated into rivers and streams, and finally into the oceans. Lastly, although the research is in its infancy, we will review the known and probable effects of microplastics ingestion through air and food to human health.

15.2 Vegetation and Salinity

When is water available but not suitable for use by vegetation? As we saw in Chapters 11 and 12, water is required by vegetation for photosynthesis, to transport and remove compounds, for physical support, and for temperature regulation. Soil water may become polluted and unavailable to vegetation when the solute water potential outside the roots is less than the solute water potential inside the roots. Water may then move out of the plant's roots if there is insufficient resistance to this transfer of water across the root–soil interface. The resulting dehydration and desiccation of the plant tissue can lower the xylem water pressure potential below the permanent wilting point (Chapter 11).

Figure 15.1 Surface salt deposits in a low-lying area near Ceylon, southern Saskatchewan, Canada. Photo credit: P. D. Blanken.

Salt is prevalent on Earth's surface primarily in the world's oceans owing to the erosion of minerals containing salts from the continents, made possible by the water cycle (Chapter 5), and the dissolution of salts such as **sodium chloride** in water due to the bipolar nature of the molecular bond between hydrogen and oxygen atoms in the water molecule (Chapter 3). Coupled with these natural properties are the anthropogenic pressures on the landscape and atmosphere that have increased the exposure of vegetation to pollution in the form of increases in soil salinity, especially in arid, semi-arid, and coastal regions. When soils become wet from precipitation or irrigation, minerals and salts can easily dissolve. As the soil water moves to the surface by capillary rise or accumulates in low-lying areas, salt crusts can form on the surface as minerals precipitate out of solution when evaporation rates are high (Figure 15.1). Increases in the withdrawal of groundwater for irrigation that often already contains high concentrations of dissolved salts, coupled with high surface evaporation rates, further exacerbate the salinity increase in surface soils.

As the concentration of dissolved salts increases, the plant's ability to access usable fresh water and nutrients decreases, resulting in decreased growth and crop yield. Relationships between soil water salinity measured by the **electrical conductivity** of soil water extracts (EC_e measured in units of siemens, or decisiemens, per m, dS m^{-1}) and a crop yield index (Y, with 100 the maximum) for specific species have been developed. These indicate that crop yield decreases linearly with salinity after a species-specific salinity threshold (EC_t) is exceeded (Maas and Hoffman, 1977) (Eq. 15.1),

$$Y = 100 - S\left(EC_e - EC_t\right) \tag{15.1}$$

where S is the slope of the relationship between the reduction in crop yield and increase in salinity above the species-specific threshold. Most crops have low salinity thresholds with large decreases in yield resulting from small increases in salinity (Panta et al., 2014). Typical values of S (% per dS m^{-1}) range from 1.0 (carrot) to 6.0 (wheat), and EC_t (dS m^{-1}) from 14.0 (carrot) to 7.1 (wheat) (Corwin, 2019). At present, only 30 crop species provide 90 percent of plant-based human food, and the majority of these vegetable crops are highly salt-sensitive (Zörb, Geilfus and Dietz, 2019).

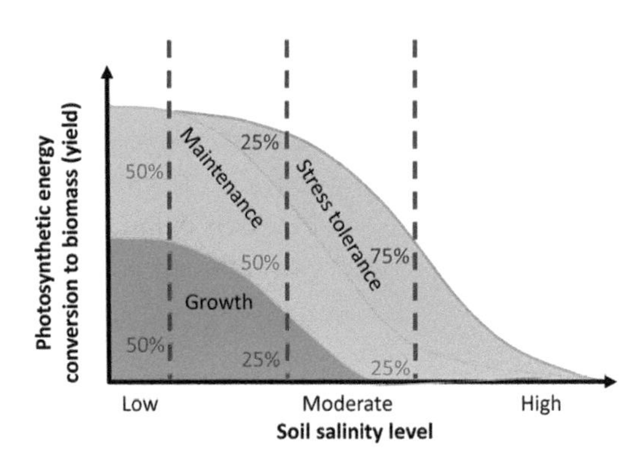

Figure 15.2 Relative allocation of photosynthetic energy resources to growth, maintenance, and stress tolerance as a function of soil salinity. Reproduced with permission from Zörb, Geilfus and Dietz (2019).

As soil salinity increases, the increase in sodium (Na^+) and chloride (Cl^-) ions, in addition to other ions such as magnesium (Mg^{2+}), decreases plant growth and crop yields by decreasing the rate of photosynthesis, closing stomata, and causing cellular damage (Zörb, Geilfus and Dietz, 2019). As soil salinity increases, vegetation shifts resources that were primarily dedicated to growth to the mitigation of stress (Zörb, Geilfus and Dietz, 2019) (Figure 15.2). Any decrease in crop yield will translate directly to decreased profits for farmers and therefore increased costs to consumers. Economic costs for agriculture vary widely, but Munns and Gilliham (2015) provide a useful example of the economic costs of salinity. In large-scale dryland farming, production costs may be \$300 ha^{-1} for a crop, yielding a gross income of \$600 ha^{-1} for a profit of \$300 ha^{-1}. If increased salinity resulting from increased irrigation in dry years decreases crop yield from 3 tonnes ha^{-1} in a normal year to 2 tonnes ha^{-1} in a dry year, gross income decreases to \$400 ha^{-1} for a profit of only \$100 ha^{-1}. Globally, the cost of salt-induced land degradation in 2013 was estimated at \$441 ha^{-1} (US dollars), equal to an annual economic loss of US\$27.1 billion (Shahid, Zaman and Heng, 2018).

Inundation of **saline water** in coastal estuaries and wetlands due to sea level rise or the lowering of the water table from groundwater extraction has direct negative consequences for vegetation growth. **Nuisance flooding**, defined as widespread low levels of flooding between 3 and 10 cm with a negative but nondestructive impact on transportation and public health and safety, can occur because of a combination of increased vulnerability (urbanization), changes in the water balance (precipitation amounts and timing), and rising sea levels (Moftakhari et al., 2018). Along the eastern coast of the United States, six tidal gauges have recorded an average mean sea level rise of 35 ± 15 cm since the 1920s, and are predicted to experience an additional rise of 34 ± 17 cm by 2050, with a corresponding 55 percent predicted increase in the occurrence of nuisance flooding (Moftakhari et al., 2015). Events such as these nuisance floods may not destroy property, but they can harm vegetation because of exposure to saline water.

Exposure to saline water results in decreased plant growth and crop yield, and can also affect established forested ecosystems. In coastal regions of the United States, established pine, maple, and cypress forests are dying and being replaced by saltmarshes. Ury et al. (2021) found that these **ghost forests** containing the remnants of standing dead trees and fallen trunks have increased in area by 11 percent between 1985 and 2019 in coastal eastern North Carolina. Similar ecosystem shifts from coastal forests to saltmarshes are expected in other low-lying coastal regions worldwide. This ecosystem shift from freshwater mature hardwood forests to saltwater marshes will affect water and carbon balances, storm surge vulnerability, nuisance flooding occurrences, and overall terrestrial and aquatic ecology in these regions.

15.3 Agriculture and Urban Fertilizer Runoff

Today's agricultural crop production lacks biodiversity and therefore is dependent on the application of **fertilizers** containing water-soluble nutrients. Fertilizers can degrade water quality with excessive application or application at inappropriate times. Modern agriculture has altered land use and cover, with recent estimates indicating that 12% of the global land surface area is used for crops and 22% for pasture and rangelands. Of the crop land area, 60% is used to grow five crops: wheat (22%), maize (13%), rice (11%), barley (9%), and soybeans (5%) (Leff, Ramankutty and Foley, 2004). Rice feeds roughly half of the world's population, and together with wheat and maize, supplies roughly half of the world's daily calories (International Development Research Centre, 2010).

In many regions, these crops require irrigation and fertilizer application, especially when grown at large scales. Fertilizers containing nitrogen (N), phosphorus (P), and potassium (K) are commonly applied, and in increasing quantities. Plants require **nitrogen** for growth, so the application of N increases crop yield. **Potassium** is required for the plant to metabolize the nitrogen and to control stomata aperture by changing the guard cell turgor pressure through changing the solute water potential, so it too affects growth and crop yield. **Phosphorus** is a major component of adenosine triphosphate (ATP), so affects nearly all aspects of plant growth and vitality. Each of these elements provides plant nutrients that are applied in water-soluble forms (e.g., nitrogen in the form of nitrate NO_3, ammonia NH_3, or ammonium NH_4; potassium in the form of potash; phosphorus in the form of orthophosphoric acid) and readily used by plants, so any excess, unabsorbed fertilizer is likely to enter local streams, rivers, lakes, and ultimately the oceans. The lack of diversity and crop rotation, the presence of large fields, and economic pressure to increase crop yield in the face of a changing climate and increased population result in pressure to apply greater quantities of fertilizers. In 1961, 12.68, 11.14, and 9.37 million tonnes of N, P, and K (respectively) were produced worldwide; by 2019, this had risen to 122.97, 42.86, and 43.64 million tonnes respectively (Ritchie, Rose and Rosado, 2013). These increases do not reflect an increase in arable land area, since fertilizer use per hectare of cropland worldwide has nearly doubled from 70.95 kg per hectare in 1976 to 136.82 kg per hectare in 2018 (Ritchie, Rose and Rosado, 2013).

The demand for irrigation is expected to increase. In the United States in 2007, 7.5% of all croplands and pasturelands were irrigated, with three-quarters of the irrigated agricultural land occurring in the arid and semi-arid western United States (Schaible and Aillery, 2012). Roughly half of the irrigation water originated from surface water sources, and half from regional aquifers (Schaible and Aillery, 2012). Worldwide, 15–25% of arable land is irrigated, two-thirds in

Asia, and this irrigated land produces a disproportionate 30–40% of the world's food calories (Portmann, Siebert and Döll, 2010). The per capita irrigated land area of 0.045 ha per person has remained relatively constant since the 1960s (Howell, 2001), so without accounting for a changing climate, increases in soil salinity, improvements in plant water use efficiency and so on, the land area under irrigation will likely increase as population increases.

The fraction of arable land that is not irrigated depends on water supplied by rain or melting snow. Snowmelt contributes more than 50% of surface water runoff over 26% of the global land area, and 40% of the world's rye and barley production is in snowmelt-dominated basins (Qin et al., 2020). Snowmelt as a source of water is likely to become less reliable with climate change, as the timing and quality of water provided by melting snow is changing, with melting events occurring more often in wintertime. Rain-on-snow events can result in a pulse of water through already saturated soils, mobilizing nitrogen and phosphorus that has been applied as fertilizers, often over many years. Seybold et al. (2022) found that 53% of the area in the contiguous United States experiences rain-on-snow events, with the runoff created by these winter events responsible for placing an estimated 43% of the total nitrogen and phosphorus pools at risk for export to groundwater and surface water.

15.4 Harmful Algal Blooms

Runoff from agricultural and urban regions containing dissolved nitrogen and phosphate eventually enters streams and rivers, which ultimately reach lakes and oceans. When the fertilizer-enriched runoff enters aquatic systems, algae and other aquatic plant species are enriched with these dissolved minerals and respond just as their terrestrial counterparts do: they grow. The estimated annual costs associated with decreased water quality resulting from excessive algal growth in the United States (1987 to 2000) and the European Union (1989 to 1998) were US$82 million (1987 to 2000) and US$813 million, respectively, in cost categories of commercial fishing, public health, recreation/tourism, and monitoring/management (Hoagland and Scatasta, 2006). In the Canadian portion of Lake Erie alone, the projected estimated annual costs of algal blooms over the period 2015 to 2045 are $272 million (2015 Canadian dollars) (Smith et al., 2019). There are two major and related detrimental impacts resulting from the runoff of water abnormally rich in dissolved nutrients into rivers, lakes, and oceans: harmful algal blooms and eutrophication.

Harmful algal blooms, or **HABs**, occur when colonies of algae grow rapidly, creating harmful or toxic conditions that affect the environment and human health. Algal blooms give the surface water a bright green or red appearance. In the shallow, warm waters of the Gulf of Mexico, HABs known as **red tides** are created when the single-celled dinoflagellate algae ***Karenia brevis*** produces a potent class of neurotoxins known as **brevetoxins**. Harmful algal blooms of *Karenia brevis* have resulted in shellfish contamination, massive fish deaths, and deaths in marine mammals. The toxin has been found to bioaccumulate and transfer to upper tropic levels (Steidinger, 2009). Thirty-four endangered Florida manatees (*Trichechus manatus latirostris*) and 107 bottlenose dolphins (*Tursiops truncatus*) died off the coast of Florida at a time when extensive water surveys revealed low concentrations of *Karenia brevis*. High toxin concentrations of brevetoxins in the stomach contents of manatees and dolphins indicated that the source of the toxins was from the consumption of contaminated seagrass for the manatees, or seafood for the dolphins (Flewelling et al., 2005).

Fleming et al. (2011) reviewed the human health impacts resulting from exposure to brevotoxins through food, water, or air. The consumption of shellfish or fish contaminated with *Karenia brevis* was not fatal as it was for marine mammals. Emergency room visits for gastrointestinal illness did increase significantly during active red tide events in Florida, so the acute illness effects are documented but the chronic effects are not yet known. Exposure to aerosolized red tide toxins led to acute and possibly chronic respiratory illness, especially in those with preexisting respiratory illness such as asthma. Swimming in waters containing the toxins has caused eye and skin irritation and increases the chances of inadvertently swallowing contaminated water.

In fresh water where blooms of the photosynthetic **cyanobacteria** occur, the water has a cyan or blue-green appearance, so cyanobacteria are often referred to as **blue-green algae** despite not being algae. Although cyanobacteria once released enough oxygen from photosynthesis to change life on Earth (Chapter 4), they also produce **cyanotoxins** such as **microcystins** that are harmful if ingested. When freshwater HABs occur (**green tides**), the toxins can enter the body through water or contaminated food. When ingested, significant gastrointestinal distress is common, with respiratory distress and/or multiple organ failure and death possible (Pulido, 2016). The World Health Organization defined concentrations of 1 μg of microcystin per liter as the limit for safe life-long drinking water, with 10 and 20 μg/L for recreational water use with mild and moderate probabilities of health effects, respectively (Ho and Michalak, 2015).

Recognizing when HABs are occurring, and the ability to forecast when they are likely to occur, are vital for issuing health warnings and appropriate water resource management. The occurrence of a HAB can be defined based on the abundance (by mass or volume) of the total phytoplankton, cyanobacteria, or individual species of concern (Ho and Michalak, 2015). Water samples collected at locations near municipal water intakes can be monitored for chlorophyll *a* concentration, and satellite-based remote sensing can be used to detect areas of large algal blooms (Figure 15.3). These measures provide an indicator of abundance but do not distinguish toxic from nontoxic species. Further testing, such as DNA analysis, is required to determine the presence of harmful species.

Periods of heavy rain resulting in high surface runoff that flushes nutrients into rivers and lakes, high water temperatures that promote algal and bacterial growth, or calm winds and water currents that do not disperse the blooms, all act to promote HAB formation. Since accurately forecasting HAB events depends on hydrologic and meteorologic factors (and climatologic on longer timescales), lake or ocean internal processes, aquatic biology, land use and land cover, and rates of fertilizer application (therefore economic factors), it is complicated and challenging (Glibert et al., 2010).

Water quality can be affected by excessive plant and algal growth resulting from enrichment by nutrients even if not classified as a HAB. The source of the nutrients is typically runoff from agricultural sources (fertilizer or manure), municipal sources (sewage, fertilizer, or industrial waste), or aquaculture (regular fish feeding). This overproduction of organic material in aquatic systems due to anthropogenic activities on short timescales (years) is referred to as

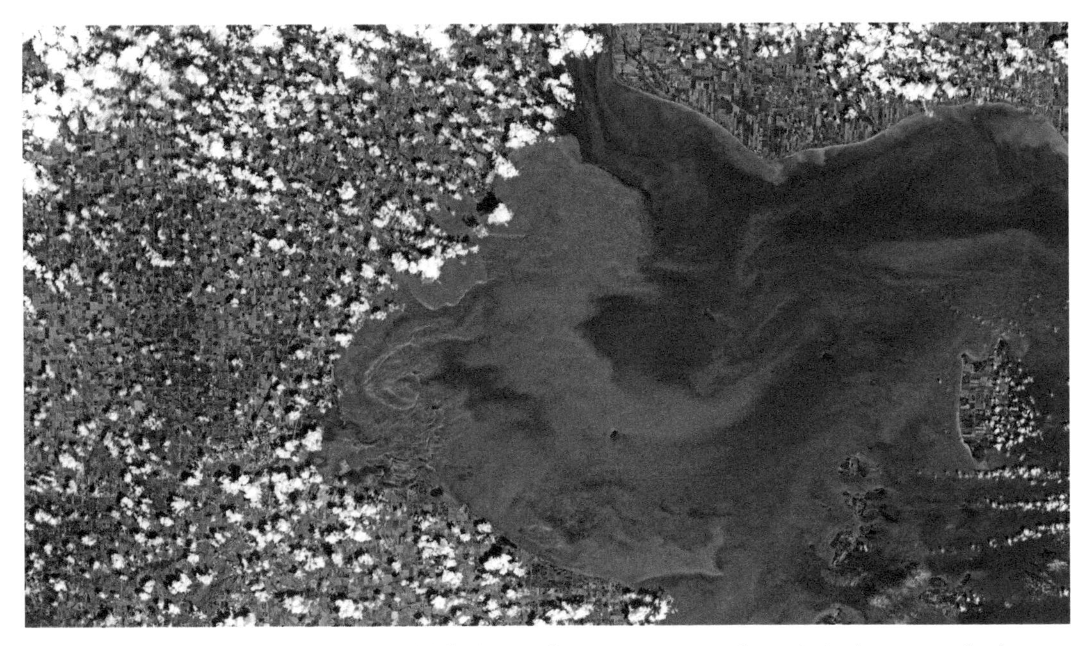

Figure 15.3 A severe toxic algal "green tide" bloom of *Microcystis* cyanobacteria in the western basin of Lake Erie on July 30, 2019, taken by the Operational Land Imager on the Landsat 8 satellite. Image vertical and horizontal distances are approximately 60 and 100 km, respectively. Image credit: NASA's Earth Observatory. https://earthobservatory.nasa.gov/images/145453/eerie-blooms-in-lake-erie.

anthropogenic eutrophication to distinguish it from the natural, long-term (centuries) **eutrophication** process that results in the infilling of small lakes or wetlands (Le Moal et al., 2019).

Lake Winnipeg, in Manitoba, Canada, serves as an example of how nutrient inputs into lakes can change over time. Located in the flat Canadian Prairie region, Lake Winnipeg is a postglacial lake, the third-largest freshwater lake entirely within Canada. The lake has a large surface area of 24,515 km^2, a shallow mean depth of 12 m (maximum 36 m), and a volume of 284 km^3. Land cover around the lake is mainly agricultural, with boreal forests on the northern and eastern shores. Most of the water entering the lake comes from the Saskatchewan, Dauphin, Assiniboine, and Red Rivers located to the west and south. These river basins are all located in agricultural regions with springtime flooding from snowmelt a common occurrence. Although soils in these river basins are phosphorus-rich, twentieth-century increases in the clearing of land for agriculture, increases in the inputs of synthetic fertilizers and manure, and increases in population (Figure 15.4) have largely coincided with a sudden increase in springtime snowmelt-driven discharge and phytoplankton concentration after the mid-1990s (Schindler, Hecky and McCullough, 2012) (Figure 15.5).

Schindler, Hecky and McCullough (2012) summarized how these anthropogenic changes in the terrestrial environment coupled with hydroclimatic changes have affected the frequency of eutrophication and HABs events in Lake Winnipeg. The near-doubling of the size of cyanobacteria blooms (which favor P over N) since the mid-1990s coincides with increased inputs of phosphorus entering the lake from the land. With continued warming

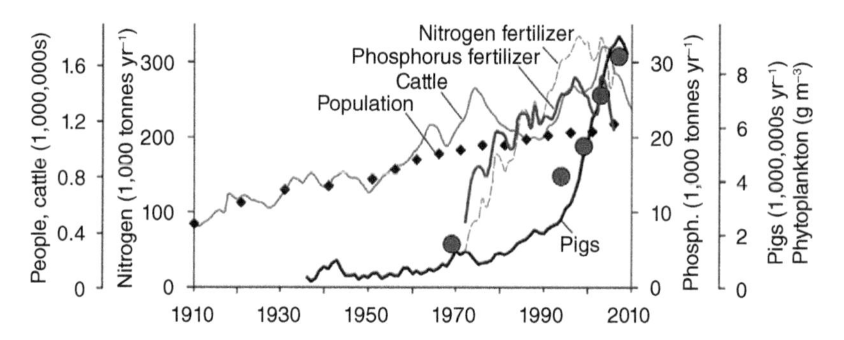

Figure 15.4 Changes in human, cattle and pig populations, and nitrogen and phosphorus fertilizer applications, in the Manitoba portion of the Lake Winnipeg watershed over the years shown on the *x*-axis. Circles denote the mean midsummer phytoplankton biomass. Reproduced with permission from Schindler, Hecky and McCullough (2012).

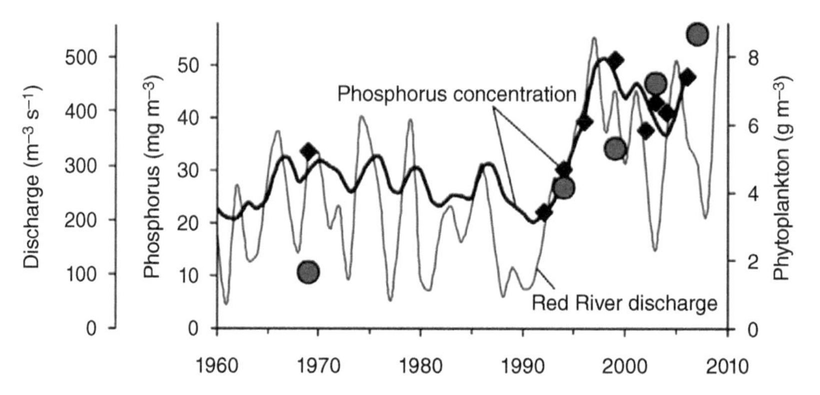

Figure 15.5 Changes in the discharge of the Red River (3-year running mean; thin black line) and the measured (black diamonds) and modeled (thick black line) mean midsummer phosphorus and phytoplankton (circles) in Lake Winnipeg. Reproduced with permission from Schindler, Hecky and McCullough (2012).

and changes in precipitation expected for the next several decades, eutrophication and HAB events are expected to continue and worsen unless phosphorus applications and inputs are reduced.

Water quality degradation from eutrophication such as that occurring in Lake Winnipeg impacts drinking water supplies, commercial and recreational fishing, recreation, and tourism, and even home and property values. In England and Wales, Pretty et al. (2003) estimated that the damages from freshwater eutrophication cost between US$105 and US$160 million per year. The top three costs were for drinking water treatment to remove nitrogen ($28.1 million), drinking water treatment to remove algae ($26.6 million), and reduced values of waterside properties ($13.76 million). The sewage treatment cost to remove phosphates was $70.4 million per year. In the United States, the estimated annual costs associated with losses in recreational water usage, waterfront real estate, spending on the recovery of threatened and endangered species, and drinking water, was US$2.2 billion (Dodds et al., 2009). In the North American Great Lakes region, where decades of population and industrial growth have resulted in decreased water quality, increasing eutrophication and HABs events have been a consequence and remediation costs are high. Garcia-Hernandez, Brouwer and Pinto (2022) estimated that the annual cost required to reduce the total phosphorus emissions in and around the Great Lakes by a government-set target level of 40 percent relative to the 2008 levels entering shallow and warm Lake Erie is $3 billion (Canadian dollars).

15.5 Microplastics

Beginning in the 1950s, the increased production of **plastics** and the continued production of single-use plastic materials, together with the long residence time of plastics in the environment, has led to the point where plastic material is found in almost every location on Earth. Of the currently 320 million tonnes of plastic produced annually, over 40% is single-use, and 74–94% is released into the environment (Wright and Kelly, 2017; Alimi, Hernandez and Tufenkji, 2018).

15.5.1 Microplastics in Soils and Vegetation

Plastics enter soils from a variety of sources, including landfill sites that contain 79 percent of the plastic produced worldwide (Geyer, Jambeck and Law, 2017), wastewater treatment plants, manufacturing activities, and through the use of plastics as a mulch material (Azeem et al., 2021). Small plastic fragments, or **microplastics**, with a size between 1 μm and 5 mm are small enough to travel deep into the rooting depth layers of soil, and even smaller **nanoplastics** (<1 μm) can enter plants through the stomata. Microfibers produced by the textile industry, as well as other sources from vehicle road tire abrasion and artificial turf surfaces, are sources of microplastics to the environment and soils (Guo et al., 2020). Here, the term **microplastics** is used for any plastic material less than 5 mm in size.

Plastics are so widely used not only because of their versatility and low production cost, but also because of their toughness and durability. Plastic **degradation**, the changing of its physical structure or chemical properties, occurs slowly, with UV radiation being mainly responsible for the initiation of degradation (Zhang et al., 2021). Plastics containing biodegradable materials can increase the degradation rate. Strips of standard plastic shopping bags placed underwater were still intact after 40 weeks of submersion, whereas biodegradable (compostable) plastic bags disappeared after 16–24 weeks of submersion (Brine and Thompson, 2010). Other studies have shown that the half-life for biodegradable plastics ranged from weeks to years depending on the type of plastic and marine conditions (Lott et al., 2021).

Most large plastic items such as bags, bottles, or other packaging materials are eventually buried in landfill waste sites. In agriculture and gardening, plastic sheets or material containing shredded plastic material are commonly used as mulch to conserve soil moisture and reduce weeds. These large plastic items can be progressively broken down into smaller fragments by physical or chemical processes. Plastics are also manufactured at the micro- or nanoscale for industrial, engineering, cosmetic, or pharmaceutical purposes. Soil drying cracks, root channels, earthworm paths, and the digging behavior of various insects and animals all provide potential pathways for microplastics to penetrate deeper into soils with rain, irrigation, or wind. Agricultural practices such as plowing, tilling, and harvesting can also relocate surface microplastic deposits deeper into the soil. Since the production of plastics is likely to increase (Geyer, Jambeck and Law, 2017), and the decomposition rate of microplastics is poorly understood, the accumulation of microplastics in soils is likely to increase (Rillig, 2012), with no known removal method.

Once in soil, microplastics can affect properties that influence water flow and retention in several ways. Abel de Souza Machado et al. (2018) added various quantities of microplastic contaminates such as polyester fibers, polyacrylic fibers, and polyethylene fragments to a loamy sand soil and measured the soil properties relevant to soil water capacity and flow: bulk density, water-holding capacity, hydraulic conductivity, soil aggregation, and microbial activity. Although clear relationships between the microplastic type and concentration and the soil properties were not always apparent, the most noticeable impacts were associated with the addition of polyester fibers. These fibers account for roughly 70 percent of the global plastic fiber production. As polyester fibers were added to the soil, the fraction of water stable soil aggregates decreased, and the bulk density decreased, resulting in an increase in the soil's water-holding capacity. Other studies found no significant changes, or even increases, in the soil's bulk density and other soil structural properties related to water-holding capacity and flow (see Wang et al., 2022). Given the large variation in plastic composition, size, and shape, coupled with the large variation in soil's physical and chemical properties, inconsistent results are not surprising. What is clear is that the addition of microplastics to soils alters the soil's physical, chemical, and microbiological properties, with either beneficial or detrimental consequences (Dulanja et al., 2022).

When soil properties are altered by microplastics, vegetation is also affected. Research in this area is in early stages, and given the large variation in effects of microplastics on soil properties, their effects on plants are likely even more complex. Studies have found that extremely small nanoplastics in the soil can enter plants through roots. Small (5–50 μm) microplastic beads tagged with fluorescent dye were found 5 months later in birch tree roots, and the authors suggest that birch trees could be used to remove microplastics from contaminated soils (Austen et al., 2022). The uptake of small microplastics via roots has also been observed in plants that humans and animals consume, such as carrot, wheat, and lettuce (Ullah et al., 2021). The nanoplastic pathway from soil to shoot is through the xylem tissue, so the transpiration rate likely plays an important role (Li et al., 2020; Azeem et al., 2021). The greater the transpiration rate, the greater the likelihood that these nanoplastics will travel from the soil into shoots and stems, then possibly into people and animals, with health impacts yet to be fully determined.

15.5.2 Microplastics in Rivers and Lakes

Freshwater rivers and lakes connect terrestrial surfaces to the oceans, so rivers should provide efficient transport of plastics deposited on land and entering soils into oceans with temporary storage in lakes. Indeed, microplastics have been found in many of the world's largest rivers. Along an 820-km stretch of the Rhine River between Basel, Switzerland, and Rotterdam, the

Netherlands, microplastics were found in every sample collected at 11 different locations, with an average 892,777 plastic particles per square kilometer and the highest concentrations located near densely populated areas (Mani et al., 2016). Near Chicago, Illinois, the 12-km-long human-made North Shore Channel receives water from Lake Michigan and joins the Chicago River with a wastewater treatment plant releasing effluent roughly 6 km along the channel's course. Microplastics were found in all net-based samples collected from the Channel, with higher microplastic concentrations found downstream of the wastewater treatment plant (17.93 per km^2) than upstream (1.94 per km^2), indicating that the effluent from the treatment plant was a plastic source (McCormick et al., 2014). Many of the microplastics were colonized by dense, diverse, bacterial biofilms, so the water transported not only the microplastics but also the bacterial colonies. The most abundant family of bacteria found, *Campylobacteraceae*, includes multiple taxa associated with human gastrointestinal infections (McCormick et al., 2014).

Microplastics have been found in lakes worldwide. The North American (Laurentian) Great Lakes contain roughly 20% of the world's accessible surface fresh water, used by roughly 48 million people. River tributaries from several urban and industrial regions drain into the Lakes. Of the 107 water samples collected from 29 tributaries, all contained plastics, with a maximum and median concentration of 32 and 1.9 particles per cubic meter (Baldwin, Corsi and Mason, 2016). Almost all (98%) were microplastics, with most (71%) in the form of fibers or lines, most of which will likely settle and accumulate in lakebed sediment and affect benthic organisms (Baldwin, Corsi and Mason, 2016). Although some of the microplastics entering the Lakes will eventually settle, some will remain buoyant. Of 21 offshore net tow samples on Lakes Superior, Huron, and Erie, all but one contained plastics with an average concentration equivalent to 43,157 per square kilometer (0.043 per m^2), with microplastic pellets and fragments representing 81% of all the plastics (Eriksen et al., 2013). Downstream, Lake Ontario water samples collected near the city of Toronto contained 0.8 microplastic particles per liter of water (800 per m^3), with stormwater runoff samples containing 15.4 particles per liter (15,400 per m^3), wastewater effluent containing 13.3 particles per liter (13,300 per m^3), and agricultural runoff containing 0.9 particles per liter (900 per m^3) (Helm, Athey and Rochman, 2020).

Similar results have been found in lakes in other locations, even in remote regions. For example, all 22 surface water samples collected from three deep southern subalpine lakes in the Lake District of Italy (Lakes Maggiore, Garda, and Iseo) that serve as essential water supplies for the surrounding densely populated region were all contaminated with microplastics. The mean plastic concentrations were 25,000 particles per square kilometer (0.25 per m^2), 39,000 particles per square kilometer (0.39 per m^2), and 40,000 particles per square kilometer (0.40 per m^2), for Lakes Garda, Maggiore, and Iseo, respectively (Sighicelli et al., 2018). The majority (74 percent) of the plastics were microplastics, and samples collected near major inflows had, by far, the greatest microplastic concentrations (Sighicelli et al., 2018). Even remote high-altitude lakes in the **Tibetan Plateau** region contained microplastics. Six of seven sample sites of lakeshore sediment from four lakes contained microplastics with concentrations ranging from 8 to 563 particles per m^2 (Zhang et al., 2016). River inflows were suggested to be the source of the microplastics to these remote high-altitude lakes (Zhang et al., 2016).

15.5.3 Microplastics in Oceans

The accumulation of plastic waste in oceans is a consequence of the production and disposal of plastics on land. The majority (80%) of all the plastic waste found in oceans comes from land and is transported to the oceans by rivers (Alimi, Hernandez and Tufenkji, 2018). As in the terrestrial environment, the amount of microplastics in the oceans is increasing. Far offshore in the Pacific Ocean, plastics, with the majority (60%) less dense than seawater, entered the ocean from the thousands of rivers located along the highly populated Pacific Ocean basin and eventually accumulated, concentrated, and remain trapped within the **North Pacific Subtropical Gyre** (Chapter 5). This accumulation of floating plastic debris, 46% consisting of fishing nets, now covers an area of 1.6 million square kilometers with at least 79 thousand tonnes of plastic, and is infamously referred to as the **Great Pacific Garbage Patch** (Lebreton et al., 2018). The mass concentration (mg of plastic per m^3 of water) of **pelagic** (at or near the ocean surface in offshore regions) microplastics is predicted to double between 2016 and 2030, and quadrupled by 2060 (Isobe et al., 2019) as the existing larger floating plastic materials eventually break down into smaller microplastics. Although microplastics currently comprise only 8% of the total mass of plastics in the Great Pacific Garbage Patch, they represent 94% by number of the estimated 1.8 trillion pieces of floating plastic (Lebreton et al., 2018).

The impacts of microplastics on the biotic marine environment are not well known. As with plastics in soils, initial studies report results that are highly variable depending on the location, size, shape, texture, and chemistry of the microplastics and the organism itself. For example, many marine organisms, ranging in size from microscopic plankton and zooplankton to massive whale sharks, are filter feeders susceptible to ingesting microplastics. Cole et al. (2013) exposed 15 zooplankton taxa representative of those found in the northeast Atlantic coastal system to minute polystyrene microplastic spheres ranging in size from 7.3 to 30.6 μm suspended in natural seawater. Using a bioimaging technique, Cole et al. (2013) found that 13 of the 15 zooplankton ingested the microplastics (Figure 15.6). Microplastics ingestion affected the zooplankton's ability to ingest algae.

As larger organisms consume smaller organisms (e.g., zooplankton) that have ingested microplastics, over time the microplastics can **bioaccumulate**. In addition, the decreased quality and quantity of zooplankton have detrimental impacts on the larger organisms that consume them. Along the Atlantic Ocean off the coast of the United States, for example, temperate northern star coral (*Astrangia poculata*) that ingested microplastics failed to consume brine shrimp eggs (Rotjan et al., 2019). The pelagic filter-feeding red crabs (*Pleuroncodes planipes*) in the Pacific Ocean, off the coast of California, United States, had a median of five microplastic particles in their gastrointestinal track (Choy et al., 2019). These red crabs are often fed upon by predators such as fish, squid, sea turtles, and sea birds, resulting in bioaccumulation in these species. In the Labrador Sea off the coast of Eastern Canada, 79 percent of the 70 sea birds *Fulmarus glacialis* sampled contained an average of 11.6 pieces of microplastics in their digestive tracts (Avery-Gomm et al., 2018). Overall, studies such as these indicate that the primary consequence of ingesting microplastics is the filling or alteration of the digestive track leading to compromised ingestion or assimilation of nutritious food (Hale et al., 2020).

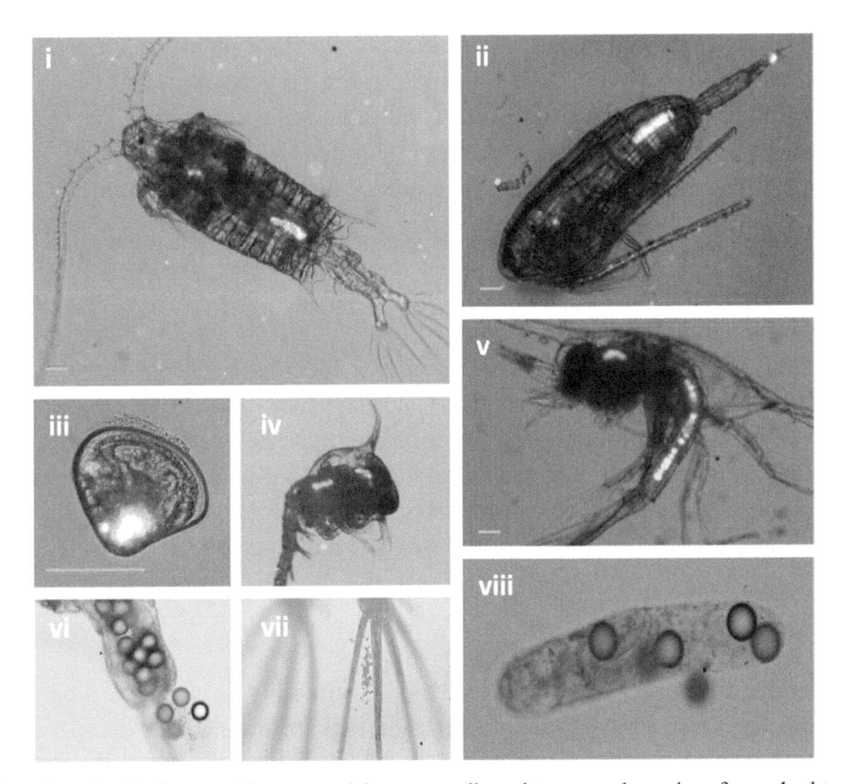

Figure 15.6 Microplastics ingested by, egested from, or adhered to several species of zooplankton organisms. See Cole et al. (2013) for species descriptions. Scale (gray lines) is 100 μm. Adapted with permission from Cole et al. (2013). © American Chemical Society.

15.5.4 Microplastics and Human Health

Since microplastics are now found in soil, water, air (Gasperi et al., 2018), plants and animals around the globe (Hamid et al., 2018), it is reasonable to wonder about the quantity of microplastics that humans ingest, and subsequently, the impacts on human health. To address the first question, Nor et al. (2021) coupled a "plastic model" that calculates the abundance of microplastics entering the human body with a "chemical model" that calculates the chemical transformation of the ingested microplastics. The estimated median total microplastic intake rates for adults were a shocking 883 particles per day, and 553 for children. Of the nine intake sources of microplastics considered (e.g., consumption of fish, shellfish, tap or bottled water, salt, beer, milk, and air), air was the highest median contributor by mass, and consumption of fish was the lowest. The abundance of microplastics in the gut and tissue compartments was 2 to 6 times higher in adults (after 70 years) than children (after 18 years).

What are the observed and potential impacts of microplastic ingestion? Although it is clear that humans are ingesting microplastics in food (46,000–52,000 particles per person per year, adult women and men, respectively), and inhaling microplastics in the air (48,000–62,000 particles per person per year, adult women and men, respectively) (Cox et al., 2019), little is known about the health impact(s). The likely health impacts are often based on the observed impacts to other organisms. Oxidative stress, inflammatory lesions, chronic inflammation, and the release of chemicals and any adsorbed pathogenic organisms are potential detrimental human health consequences since they have been observed in several other organisms (Prata et al., 2020). Ingesting seafood contaminated with microplastics exposes humans to the release of the chemicals or bacteria contained on the microplastics' surface, and the indigestible plastic itself will likely enhance inflammation and disrupt the gut microbiome (Smith et al., 2018). The inhalation of microplastics has the potential to introduce chemical toxins, and pathogens and parasites adhered to the microplastics' surfaces (Karbalaei et al., 2018). Respiratory illnesses such as asthma and other inflammatory lung diseases, especially for those with preexisting conditions, have been reported (Lu et al., 2022). Studies on the impacts of pollution exposure on human health require years of research, and hopefully the health effects can be understood while there is still time to reduce the use of and remove plastics from the environment.

15.6 SUMMARY

Water quality can be broadly defined as the suitability of water for a particular use. Since the use of water varies by organism, a characteristic of water that makes it unsuitable for use by one may not necessarily make it unsuitable for another. Therefore, the influence of a pollutant, the substance or condition that makes water unsuitable for an organism, can vary widely in terms of the characteristics of the pollutant. Thus, quantifying limits and forming policies on issues surrounding water quality can be complex and challenging.

Salts are prevalent on Earth's surface, and vegetation exposed to soil water containing excessive quantities of dissolved salts will have limited growth and productivity. A decrease in the water potential in the soil due to an increase in solute concentration to values less than the water potential in the root cells could dehydrate the plant. In arid and semi-arid regions where salts are often present in the soils, irrigation can result in further salt accumulation at the surface as high evaporation rates and capillary rise can move water to the surface to form salt crusts as the water evaporates. Such areas then often become unsuitable for crop growth. Sea level rise and the use of groundwater containing dissolved salts have introduced saline water to coastal agricultural regions and forests, with negative ecological and economic consequences.

Decreased biodiversity associated with an increased dependence on a few crops for food production, coupled with an increase in crop irrigation (and therefore sometimes an increase in soil salinity and reduced crop yield) and climate change, all in the context of an increasing human population, have resulted in the widespread and increased use of chemical water-soluble fertilizers. Consequently, runoff from both agricultural and urban landscapes rich in dissolved nitrogen and phosphate accumulates in rivers, ponds, lakes, and oceans. Aquatic algae, plankton, and bacteria thrive under the influence of the fertilizers, resulting in excessive growth creating harmful agal blooms. These blooms pose a serious risk to human and other species' health due to the production of toxins. Anthropogenic eutrophication, the overproduction of organic matter in aquatic systems stemming from human activities, results in a depletion of dissolved oxygen in the water as the aquatic material eventually decomposes. Hypoxic or anoxic (dead zone) conditions may then occur, resulting in mass die-off events. The socioeconomic costs of eutrophication in several locations were provided.

Since the 1950s when plastics first became popular as a low-cost, long-lasting, multi-purpose material, they have been found in nearly every environment (air, water, soil) on a global scale. As larger plastic material is eventually broken down into the size categories of microplastics (<5 mm) and nanoplastics (<1 μm), the ability to migrate through the environment, and be ingested by plants and organisms including humans, increases. Plastics commonly first enter the environment from discarded materials in terrestrial landfill sites. Once in the soil, plastics influence soil properties that in turn influence water movement. Nanoplastics have been found inside crops and trees, with transfer of plastics into the plants influenced by the transpiration rate. The health effects of consuming crops containing plastic fragments are not yet known.

Runoff from terrestrial surfaces, and effluent from wastewater treatment plants, provide pathways for plastics to accumulate in rivers and lakes. Nearly all samples collected in lakes and rivers worldwide, from the North American Great Lakes, the European Alps, and even the Tibetan Plateau, were contaminated with plastics. Rivers connect the land to the sea, and the plastics they contain eventually reach and accumulate in the oceans. Large ocean circulation patterns such as the North Pacific Subtropical Gyre have formed massive floating islands of plastic debris and fishing nets known as the Great Pacific Garbage Patch, covering an area of 1.6 million km^2. As these plastics are broken down to the smaller size by UV radiation and mechanical wave action, the various marine filter feeders ranging from microscopic plankton and zooplankton to whale sharks can ingest the fragments. Microplastics have been found in several small marine species (e.g., crabs), and larger marine species (e.g., sea turtles and birds), providing evidence of the bioaccumulation of plastics through the food chain.

Some studies have estimated the microplastic intake by humans through the consumption of sea food, vegetables, beverages, and breathing. The negative health impacts of microplastic ingestion have been documented in small organisms, so, following the same biological and chemical processes, it is reasonable to assume that health impacts on humans will be similar. Given the recency of the use and discovery of microplastics in the environment, long-term studies of the human health impacts are not yet available. With the continued use of nonbiodegradable plastics and the global distribution, remediation and removal strategies will be costly and difficult.

15.7 QUESTIONS

15.1 Traditionally, we often think of a pollutant as chemical substance or material that is detrimental to the health of an organism. Could water temperature alone be considered a pollutant? Provide an example of how water temperature could fulfill the definition of a pollutant for a specific aquatic organism.

15.2 How have climate change and the associated changes in the water cycle resulted in an increase in the extent of the soil salinity problem for agriculture?

15.3 How does increasing soil salinity affect the photosynthetic energy used by crops for growth, cell maintenance, and stress tolerance? Calculate the decreased yield for a crop of wheat growing in soil with an electrical conductivity of 10 dS m^{-1}. What level of soil salinity, as measured by the electrical conductivity of the soil water extracts, would result in a 50 percent decrease in wheat yield?

15.4 What are the factors that have resulted in the widespread and increased application of chemical fertilizers to agricultural lands? How do changes in the water cycle under the influence of a changing climate increase fertilizer application rates and therefore exacerbate issues associated with decreased water quality?

15.5 What is (are) the difference(s) between an algal bloom, a harmful algal bloom, and eutrophication? Provide an example of a sequence of events that could result in the occurrence of all three events at one location.

15.6 On August 2, 2014, roughly half a million residents of Toledo, Ohio, United States, were unable to access water in an event known as the Toledo Water Crisis. Summarize the reasons why this crisis occurred and propose some realistic practical strategies to reduce the likelihood of a similar future crisis.

15.7 As air temperatures increase, an increase in the frequency of extreme precipitation events is likely to also occur. What is the basis behind this prediction, and what are the implications for these changes in temperature and precipitation on the frequency and harmful agal blooms?

15.8 Small fragments of plastics, or microplastics, have been found inside the tissue of plant roots. Describe the role liquid water plays in the relocation of plastics from their source (typically in waste landfill sites) to the soil, the breakdown of the plastics to smaller sizes, and the transfer of the plastics from the soil to inside the roots. What are some possible implications of the transfer of microplastics into plant tissue for the plant or those that consume these plants?

15.9 All studies reviewed in this chapter found microplastics in nearly all of the samples collected from freshwater lakes worldwide. Generally speaking, what were the sources of plastic pollutants? What could possibly be done to remove plastics, and prevent more from entering these large bodies of fresh water that often serve as a critical supply of fresh water for millions of people?

15.10 Most of the plastics discarded on the terrestrial land surface will eventually reach the oceans before they decompose. Studies mentioned in this chapter have found that plastics bioaccumulate up through the food chain and into humans through the ingestion of contaminated seafood, and also through the air we breathe. How much is known about the impacts, observed or postulated, about the impact of ingestion or breathing microplastics on human health? In your opinion, and based on studies of other organisms mentioned in this chapter, is the ingestion of microplastics a concern that needs to be addressed with further study and research? Why or why not?

REFERENCES

Abel de Souza Machado, A., Lau, C. W., Till, J. et al. (2018) 'Impacts of microplastics on the soil biophysical environment', *Environmental Science and Technology*, 52, pp. 9656–9665. doi: 10.1021/acs.est.8b02212.

Alimi, O. S., Hernandez, L. M. and Tufenkji, N. (2018) 'Microplastics and nanoplastics in aquatic environments: aggregation, deposition, and enhanced contaminant transport', *Environmental Science and Technology*, 52, pp. 1704–1724. doi: 10.1021/acs.est.7b05559.

Austen, K., MacLean, J., Balanzategui, D. and Hölker, F. (2022) 'Microplastic inclusion in birch tree roots', *Science of the Total Environment*, 808, p. 152085. doi: 10.1016/j.scitotenv.2021.152085.

Avery-Gomm, S., Provencher, J. F., Liboiron, M., Poon, F. E. and Smith, P. A. (2018) 'Plastic pollution in the Labrador Sea: An assessment using the seabird northern fulmar *Fulmarus glacialis* as a biological monitoring species', *Marine Pollution Bulletin*, 127, pp. 817–822. doi: 10.1016/j.marpolbul.2017.10.001.

Azeem, I., Adeel, M., Ahmad, M. A. et al. (2021) 'Uptake and accumulation of nano/microplastics in plants – A critical review', *Nanomaterials*, 11, p. e2021WR030772. doi: 10.3390/nano11112935.

Baldwin, A., Corsi, S. and Mason, S. (2016) 'Plastic debris in 29 Great Lake tributaries—Relations to watershed attributes and hydrology', *Environmental Science and Technology*, 50, pp. 10377–10385.

Brine, T. O. and Thompson, R. C. (2010) 'Degradation of plastic carrier bags in the marine environment', *Marine Pollution Bulletin*, 60, pp. 2279–2283. doi: 10.1016/j.marpolbul.2010.08.005.

Choy, C. A., Robison, B. H., Gagne, T. O. et al. (2019) 'The vertical distribution and biological transport of marine microplastics across the epipelagic and mesopelagic water column', *Scientific Reports*, 9, p. 7843. doi: 10.1038/s41598-019-44117-2.

Cole, M., Lindeque, P., Fileman, E. et al. (2013) 'Microplastic ingestion by zooplankton', *Environmental Science and Technology*, 47, pp. 6646–6655. doi: 10.1021/es400663f.

Corwin, D. L. (2019) 'Climate change impacts on soil salinity in agricultural areas', *European Journal of Soil Science*, 72, pp. 842–862.

Cox, K. D., Covernton, G. A., Davies, H. L. et al. (2019) 'Human consumption of microplastics' (Correction), *Environmental Research Letters*, 53, pp. 7068–7074. doi: 10.1021/acs.est.9b01517.

Dodds, W. K., Bouska, W. W., Eitzmann, J. L. et al. (2009) 'Eutrophication of U.S. freshwaters – analysis of potential economic damages', *Environmental Science and Technology*, 43, pp. 12–19. doi: 10.1021/es801217q.

Dulanja Dissanayake, P., Kim, S., Sarkar, B. et al. (2022) 'Effects of microplastics on the terrestrial environment: A critical review', *Environmental Research*, 209, p. 112734. doi: 10.1016/j.envres.2022.112734.

Eriksen, M., Mason, S., Wilson, S. et al. (2013) 'Microplastic pollution in the surface waters of the Laurentian Great Lakes', *Marine Pollution Bulletin*, 77, pp. 177–182. doi: 10.1016/j.marpolbul.2013.10.007.

Fleming, L. E., Kirkpatrick, B., Backer, L. C. et al. (2011) 'Review of Florida red tide and human health effects', *Harmful Algae*, 10, pp. 224–233. doi: 10.1016/j.hal.2010.08.006.

Flewelling, L., Naar, J., Abbott, J. et al. (2005) 'Red tides and marine mammal mortalities', *Nature*, 435, pp. 755–756. doi: 10.1038/nature435755a.

Garcia-Hernandez, J. A., Brouwer, R. and Pinto, R. (2022) 'Estimating the total economic costs of nutrient emission reduction policies to halt eutrophication in the Great Lakes', *Water Resources Research*, 58(4). doi: 10.1029/2021WR030772.

Gasperi, J., Wright, S. L., Dris, R. et al. (2018) 'Microplastics in air: Are we breathing it in?', *Current Opinion in Environmental Science & Health*, 1, pp. 1–5. doi: 10.1016/j.coesh.2017.10.002.

Geyer, R., Jambeck, J. R. and Law, K. L. (2017) 'Production, use, and fate of all plastics ever made', *Science Advances*, 3, pp. 25–29. doi: 10.1126/sciadv.1700782.

Glibert, P. M., Allen, J. I., Bouwman, A. F. et al. (2010) 'Modeling of HABs and eutrophication: Status, advances, challenges', *Journal of Marine Systems*, 83, pp. 262–275. doi: 10.1016/j.jmarsys.2010.05.004.

Guo, J., Huang, X. P., Xiang, L. et al. (2020) 'Source, migration and toxicology of microplastics in soil', *Environment International*, 137, p. 105263. doi: 10.1016/j.envint.2019.105263.

Hale, R., Seeley, M. E., LaGuardia, M. J., Mai, L. and Zeng, E. Y. (2020) 'A global perspective on microplastics', *Journal of Geophysical Research Oceans*, 125, p. e2018JC014719. doi: 10.1029/2018JC014719.

Hamid, F. S., Bhatti, M. S., Anuar, N. et al. (2018) 'Worldwide distribution and abundance of microplastics – How dire is the situation?', *Waste Management & Research*, 36, pp. 873–897. doi: 10.1177/0734242X18785730.

Helm, P., Athey, S. and Rochman, C. M. (2020) 'Microplastics entering northwestern Lake Ontario are diverse and linked to urban sources', *Water Research*, 174. doi: 10.1016/j.watres.2020.115623.

Ho, J. C. and Michalak, A. M. (2015) 'Challenges in tracking harmful algal blooms: A synthesis of evidence from Lake Erie', *Journal of Great Lakes Research*, 41, pp. 317–325. doi: 10.1016/j.jglr.2015.01.001.

Hoagland, P. and Scatasta, S. (2006) 'The economic effects of harmful algal blooms', in Granéli, E. and Turner, J. T. (eds.) *Ecology of Harmful Algae*. Springer, pp. 391–402 doi: 10.1007/987-3-540-32210-8_30.

Howell, T. (2001) 'Enhancing water use efficiency in irrigated agriculture', *Journal of Agronomy*, 93, pp. 281–289.

International Development Research Centre (2010) *Facts & Figures on Food and Biodiversity*. Available at: https://idrc-crdi.ca/en/research-in-action/facts-figures-food-and-biodiversity.

Isobe, A., Iwasaki, S., Uchida, K. and Tokai, T. (2019) 'Abundance of non-conservative microplastics in the upper ocean from 1957 to 2066', *Nature Communications*, 10, pp. 1–3. doi: 10.1038/s41467-019-08316-9.

Karbalaei, S., Hanachi, P., Walker, T. R. and Cole, M. (2018) 'Occurrence, sources, human health impacts and mitigation of microplastic pollution', *Environmental Science and Pollution Research*, 25, pp. 36046–36063. doi: 10.1007/s11356-018-3508-7.

Le Moal, M., Gascuel-Odoux, C., Ménesguen, A. et al. (2019) 'Eutrophication: A new wine in an old bottle?', *Science of the Total Environment*, 651, pp. 1–11. doi: 10.1016/j.scitotenv.2018.09.139.

Lebreton, L., Slat, B., Ferrari, F. et al. (2018) 'Evidence that the Great Pacific Garbage Patch is rapidly accumulating plastic', *Scientific Reports*, 8, pp. 1–15. doi: 10.1038/s41598-018-22939-w.

Leff, B., Ramankutty, N. and Foley, J. A. (2004) 'Geographic distribution of major crops across the world', *Global and Planetary Change*, 18, pp. 1–27. doi: 10.1029/2003GB002108.

Li, L., Luo, Y., Li, R. et al. (2020) 'Effective uptake of submicrometre plastics by crop plants via a crack-entry mode', *Nature Sustainability*, 3, pp. 929–937. doi: 10.1038/s41893-020-0567-9.

Lott, C., Eich, A., Makarow, D. et al. (2021) 'Half-life of biodegradable plastics in the marine environment depends on material, habitat, and climate zone', *Frontiers in Marine Science*, 8, pp. 1–19. doi: 10.3389/fmars.2021.662074.

Lu, K., Zhan, D., Fang, Y. et al. (2022) 'Microplastics, potential threat to patients with lung diseases', *Frontiers in Toxicology*, 4, 958414. doi: 10.3389/ftox.2022.958414.

Maas, E.V. and Hoffman, G. J. (1977) 'Crop salt tolerance – current assessment', *Journal of the Irrigation and Drainage Division*, 103, pp. 115–134.

Mani, T., Hauk, A., Walter, U. and Burkhardt-Holm, P. (2016) 'Microplastics profile along the Rhine River', *Scientific Reports*, 5, p. 17988. doi: 10.1038/srep17988.

McCormick, A., Hollein, T. J., Mason, S. A., Schluep, J. and Kelly, J. J. (2014) 'Microplastic is an abundant and distinct microbial habitat in an urban river', *Environmental Science and Technology*, 48, pp. 11863–11871. doi: 10.1021/es503610r.

Moftakhari, H., AghaKouchak, A., Sanders, B. F., Allaire, M. and Matthew, R. A. (2018) 'What is nuisance flooding? Defining and monitoring an emerging challenge', *Water Resources Research*, 54, pp. 4218–4227. doi: 10.1029/2018WR022828.

Moftakhari, H. R., AghaKouchak, A., Sanders, B. F. et al. (2015) 'Increased nuisance flooding along the coasts of the United States due to sea level rise: Past and future', *Geophysical Research Letters*, 42, pp. 9846–9852. doi: 10.1002/2015GL066072.

Munns, R. and Gilliham, M. (2015) 'Salinity tolerance of crops – what is the cost?', *New Phytologist*, 208, pp. 668–673. doi: 10.111/nph.13519.

Nor, N. H. M., Kooi, M., Diepens, N. J. and Koelmans, A. A. (2021) 'Lifetime accumulation of microplastics in children and adults', *Environmental Research Letters*, 55, pp. 5084–5096. doi: 10.1021/acs.est.0c07384.

Panta, S., Flowers, T., Lane, P. et al. (2014) 'Halophyte agriculture: Success stories', *Environmental and Experimental Botany*, 107, pp. 71–83. doi: 10.1016/j.envexpbot.2014.05.006.

Portmann, F. T., Siebert, S. and Döll, P. (2010) 'MIRCA2000 – Global monthly irrigated and rainfed crop areas around the year 2000: A new high-resolution data set for agricultural and hydrological modeling', *Global Biogeochemical Cycles*, 24, pp. 1–24. doi: 10.1029/2008GB003435.

Prata, J. C., da Costa, J. P., Lopes, I. et al. (2020) 'Environmental exposure to microplastics: An overview on possible human health effects', *Science of the Total Environment*, 702, p. 134455. doi: 10.1016/j.scitotenv.2019.134455.

Pretty, J. N., Mason, C. F., Nedwell, D. B. et al. (2003) 'Environmental costs of freshwater eutrophication in England and Wales', *Environmental Science and Technology*, 2, pp. 201–208. doi: 10.1021/es020793k.

Pulido, O. M. (2016) 'Phycotoxins by harmful algal blooms (HABS) and human poisoning: An overview', *International Clinical Pathology Journal*, 2. doi: 10.15406/icpjl.2016.02.00062.

Qin, Y., Abatzoglou, J. T., Siebert, S. et al. (2020) 'Agricultural risks from changing snowmelt', *Nature Climate Change*, 10, pp. 459–465. doi: 10.1038/s41558-020-0746-8.

Rillig, M. (2012) 'Microplastics in terrestrial ecosystems and the soil?', *Environmental Science and Technology*, 42, pp. 6453–6454. doi: 10.1021/es302011r.

Ritchie, H. and Rose, M. (2021) *Clean Water and Sanitation*. Available at: https://ourworldindata.org/clean-water-sanitation (Accessed: October 7, 2022).

Ritchie, H., Rose, M. and Rosado, P. (2013) *Fertilizers*. Available at: https://ourworldindata.org/fertilizers (Accessed: October 7, 2022).

Rotjan, R. D., Sharp, K. H., Gauthier, A. E. et al. (2019) 'Patterns, dynamics and consequences of microplastic ingestion by the temperate coral, *Astrangia poculata*', *Proceedings of the Royal Society of London, Series B.*, 286, pp. 1–9. doi: 10.1098/rspb.2019.0726.

Schaible, G. D. and Aillery, M. P. (2012) *Water Conservation in Irrigated Agriculture: Trends and Challenges in the Face of Emerging Demands.* Economic Information Bulletin No. EIB-99. USDA Economic Research Service.

Schindler, D. W., Hecky, R. E. and McCullough, G. K. (2012) 'The rapid eutrophication of Lake Winnipeg: Greening under global change', *Journal of Great Lakes Research*, 38, pp. 6–13. doi: 10.1016/j.jglr.2012.04.003.

Seybold, E. C., Dwivedi, R., Musselman, K. N. et al. (2022) 'Winter runoff events pose an unquantified continental-scale risk of high wintertime nutrient export', *Environmental Research Letters*, 17, pp. 1–14. doi: 10.1088/1748-9326/ac8be5.

Shahid, S. A., Zaman, M. and Heng, L. (2018) 'Soil salinity: Historical perspectives and a world overview of the problem', in *Guideline for Salinity Assessment, Mitigation and Adaptation Using Nuclear and Related Techniques*. Springer. doi: 10.1007/978-3-319-96190-3_2.

Sighicelli, M., Pietrelli, L., Lecce, F. et al. (2018) 'Microplastic pollution in the surface waters of Italian Subalpine lakes', *Environmental Pollution*, 236, pp. 645–651. doi: 10.1016/j.envpol.2018.02.008.

Smith, M., Love, D. C., Rochman, C. M. and Neff, R. A. (2018) 'Microplastics in seafood and the implications for human health', *Current Environmental Health Reports. Current Environmental Health Reports*, 5, pp. 375–386. doi: 10.1007/s40572-018-0206-z.

Smith, R. B., Bass, B., Sawyer, D., Depew, D. and Watson, S. B. (2019) 'Estimating the economic costs of algal blooms in the Canadian Lake Erie Basin', *Harmful Algae*, 87, p. 101624. doi: 10.1016/j.hal.2019.101624.

Steidinger, K. A. (2009) 'Historical perspective on *Karenia brevis* red tide research in the Gulf of Mexico', *Harmful Algae*, 8, pp. 549–561. doi: 10.1016/j.hal.2008.11.009.

Ullah, R., Tsui, M. T.-K., Chen, H. et al. (2021) 'Microplastics interaction with terrestrial plants and their impacts on agriculture', *Journal of Environmental Quality*, 50, pp. 1024–1041. doi: 10.1002/jeq2.20264.

Ury, E., Yang, X., Wright, J. P. and Bernhardt, E. S. (2021) 'Rapid deforestation of a coastal landscape driven by sea-level rise and extreme events', *Ecological Applications*, 31, pp. 1–11. doi: 10.1002/eap.2339.

Wang, F., Wang, Q., Adams, C. A., Sun, Y. and Zhang, S. (2022) 'Effects of microplastics on soil properties: Current knowledge and future perspectives', *Journal of Hazardous Materials*, 424C. doi: 10.1016/j.jhazmat.2021.127531.

Wright, S. L. and Kelly, F. J. (2017) 'Plastic and human health: A micro issue?', *Environmental Science and Technology*, 51, pp. 6634–6647. doi: 10.1021/acs.est.7b00423.

Zhang, K., Hamidian, A. H., Tubić, A. et al. (2021) 'Understanding plastic degradation and microplastic formation in the environment: A review', *Environmental Pollution*, 274. doi: 10.1016/j.envpol.2021.116554.

Zhang, K., Su, J., Xiong, X. et al. (2016) 'Microplastic pollution of lakeshore sediments from remote lakes in', *Environmental Pollution*, 219, pp. 450–455. doi: 10.1016/j.envpol.2016.05.048.

Zörb, C., Geilfus, C. M. and Dietz, K. J. (2019) 'Salinity and crop yield', *Plant Biology*, 21, pp. 31–38.

16 Changes in Evaporation and Humidity

Key Learning Objectives

After reading this chapter, you will be able to:

1. Describe how an increase in air temperature affects the input of water vapor into the atmosphere.
2. Explain the measured and modeled trends in potential and actual evaporation rates, and the variables that influence these rates.
3. Discuss how the rate of transpiration will likely change through the response of vegetation to a warmer and drier atmosphere.
4. Provide examples of how ectotherms respond and adapt to changes in humidity.
5. Summarize the effects of heatwaves and the role that humidity plays in the ability of humans to tolerate excessive heat.

16.1 Introduction

Although the atmosphere contains the smallest volume of Earth's water, it could be argued that water vapor is the most important reservoir of water for Earth's climate and water cycle. As discussed in previous chapters, water vapor is an effective greenhouse gas, keeping Earth's mean annual surface temperature at close to +15 °C compared with −15 °C without. Water vapor completes the third phase of the water cycle, connecting surface liquid water to clouds and precipitation through the evaporation process. Energy is transferred as the energy used to evaporate water in one location and time is released at another location and time upon condensation. Water vapor and the clouds that form provide strong feedbacks in the climate system: an increase in water vapor increases surface temperature, promoting further evaporation. Water vapor thus increases, increasing surface temperature, and so on (a **positive feedback**). Cloud cover decreases surface temperature, decreasing evaporation and water vapor. This in turn decreases cloud cover, increasing surface temperature and promoting evaporation, increasing water vapor and cloud cover, and so on (a **negative feedback**).

Evaporation returns water to the atmosphere, so detecting and interpreting spatial and temporal trends in evaporation is important to understand how changes in humidity will impact not only the water balance but biota in terms of their ability to regulate temperature. In this chapter we will look at the connection between the increases in saturation and ambient vapor pressures with changes in temperature: this is important to understand how evaporation and humidity are changing in a warming world. We will review Penman's evaporation equation, which forms the basis for many studies examining spatial and temporal trends in the potential evaporation rate.

Vegetation can play a major role in reducing evaporation by reducing transpiration, and Monteith's modifications to Penman's equation are reviewed as this equation (and versions thereof) forms the basis for many of the calculations of the actual evaporation rate. Several studies show that the actual evaporation rate is increasing in some regions but decreasing in others where precipitation has been increasing. Changes in the evaporation rate, potential or actual, are strongly dependent on changes in the vapor pressure deficit, the difference between the saturation and ambient vapor pressure.

Most observations and model predictions show that the vapor pressure deficit is increasing as the rate of increase in the saturation vapor pressure with warming outpaces the increase in vapor pressure resulting from increased evaporation. Since transpiration is a major source of water vapor, the response of vegetation to increases in the vapor pressure deficit are discussed. Under indoor controlled conditions, many studies have shown that transpiration will indeed increase with a drier atmosphere. Under outdoor natural conditions, however, many studies show that transpiration will decrease with a drier atmosphere, owing to a decrease in stomatal conductance.

Ectotherms and endotherms respond to changes in humidity, and we will look at several examples of their responses to living in a warmer, drier world. Amphibians are especially sensitive to changes in humidity, and examples of their ability to tolerate and adapt to decreases in humidity are described. Arthropod populations have shown alarming declines in several regions, and we will discuss the roles of climate and habitat changes in these declines.

Heatwaves are responsible for thousands of human deaths annually. Classification of heatwaves and increases in their occurrence are reviewed, and the thermal response of humans to combined excessive temperature and humidity presented in the context of the wet-bulb temperature. A discussion of the ability of humans to tolerate excessive heat and humidity, in the context of climate change, heatwaves, and social factors, concludes the chapter.

16.2 Changes in Evaporation

Evaporation requires the availability of surface liquid water and sufficient energy to break the hydrogen bonds between water molecules at the water surface, both of which are affected by a warming atmosphere. As ice sheets, glaciers, and permafrost melt, there should be more surface liquid water available for evaporation. Higher air temperatures result in a slight decrease in the **latent heat of vaporization** (λ; J kg^{-1}) (Chapter 7, Eq. 7.4), and more importantly, the **saturation vapor pressure** (e_s; kPa) increases by roughly 7 percent for each °C increase in air temperature (Chapter 7, Eq. 7.2). Therefore, the **vapor pressure deficit** ($D = e_s - e_a$; kPa), the difference between the saturation and ambient vapor (e_a; kPa) pressures and the driving force behind evaporation (Chapter 7), increases exponentially as air temperature increases (above a wet surface, the vapor pressure deficit is equal to the vertical **vapor pressure gradient** since the surface vapor pressure is equal to e_s). As discussed in Chapter 7, wind plays an important role in the

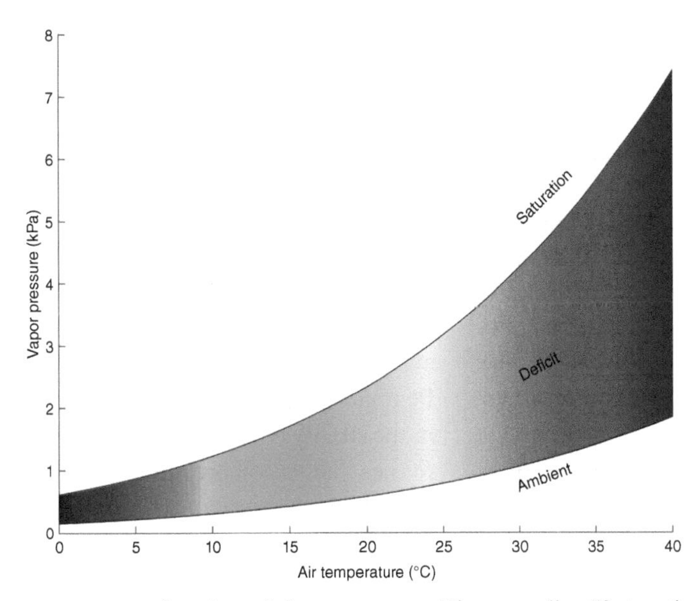

Figure 16.1 Vapor pressures as a function of air temperature. The upper line "Saturation" is saturation vapor pressure (e_s) calculated using Eq. (7.3) for each air temperature. The lower line "Ambient" is the ambient vapor pressure (e_a) calculated using Eq. (7.3) multiplied by a relative humidity of 25 percent (0.25). The colored region between the two lines is the difference between the saturation and ambient vapor pressures, the vapor pressure deficit (D). As air temperature increases, D increases rapidly (from blue to red) since the rate of increase of e_s exceeds e_a.

evaporation rate, as the replacement of humid with drier air can preserve high evaporation rates by maintaining a large vapor pressure deficit. In a warming world, high evaporation rates could continue, fueled by an increase in the temperature-dependent saturation vapor pressure increasing faster than the increase in ambient vapor pressure, if surface water is available (Figure 16.1).

Before looking at the evidence of any trends in the observed or modeled evaporation rates, let us review some terms, and add some new ones, for clarity. As discussed in Chapter 3, **evaporation** (E) is the change of state of water from a liquid to a gas, and the latent heat of vaporization (λ; J kg^{-1}) defines the energy required for this change of state. Evaporation is expressed as the volume of water (m^3) evaporated over a surface area (m^2) of water per elapsed time, equivalent to units of depth per time (e.g., mm per day). Equally valid is the expression of the evaporation rate in watts per square meter (W m^{-2}), as these units represent the rate of energy used to evaporate water (W = J s^{-1}) per unit surface area of water (m^2). This unit conversion can be attained by multiplying the evaporation rate (m s^{-1}) by the latent heat of vaporization (J kg^{-1}) and density of water (kg m^{-3}):

$$E\left(\frac{\mathrm{m}}{\mathrm{s}}\right) \times \lambda\left(\frac{\mathrm{J}}{\mathrm{kg}}\right) \times \rho\left(\frac{\mathrm{kg}}{\mathrm{m}^3}\right) = \lambda E\left(\frac{\mathrm{J}}{\mathrm{m}^2\mathrm{s}}\right) \mathrm{or} \left(\frac{\mathrm{W}}{\mathrm{m}^2}\right).$$

Different terms are used to describe evaporation. Evaporation occurs only at liquid water surfaces exposed to air, at any temperature. **Transpiration** refers to evaporation which occurs within leaf tissue of vascular plants, and is a major component of the terrestrial water cycle (Chapter 12). **Potential evaporation** (E_p) is the rate of evaporation that occurs with no limits on the supply of water, as for lakes or a pan filled and resupplied with water. Potential evaporation is also the maximum possible evaporation rate that could occur for the ambient conditions at a given location and time with unlimited water. The **actual evaporation** (E_a), as implied, refers to the actual observed evaporation rate at a given location and time. The actual evaporation rate should not exceed the potential evaporation rate.

Based on the nature of the chemical bonds within and between water molecules (Chapter 2) and our understanding of the processes controlling the evaporation rate (Chapter 7), the potential evaporation rate can be estimated using the observations of Dalton (1799) as quantified by the experiments performed by Penman (1948) under "well-watered" conditions. Under such conditions with no restrictions on the supply of water, the evaporation rate is simply a function of the dryness of the atmosphere (i.e., the vapor pressure gradient) and the rate at which the humidified air is replaced by dry air (i.e., the ventilation or vapor transfer term that is a function of the horizontal wind speed). This framework to calculate the actual evaporation rate (E_a) under well-watered conditions (hence $E_a \sim E_p$) is what Penman (1948) referred to as the **sink–strength formula** (see Chapter 7, Eq. 7.5) given in generic form as Eq. (16.1):

$$E_a = f(u) \times (e_0 - e_a) \tag{16.1}$$

where E_a equals $f(u)$ – a ventilation (vapor transfer) term that is a function of the horizontal wind velocity (i.e., the **sink term**) – multiplied by the difference between the vapor pressure at the evaporation surface (e_0; kPa) and the atmosphere above (e_a; kPa) (i.e., the **strength term**). The wind function is an empirical function that can vary with location, thus requiring site-specific calibration. Penman (1948) combined the energy available to evaporate water with the sink–strength equation to develop the well-known **Penman equation** (Eq. 16.2):

$$E_a = \frac{\Delta/\gamma \times R_a + f(u)(e_0 - e_a)}{\Delta/\gamma + 1} \tag{16.2}$$

where E_a is the actual evaporation rate (in units of mm per day in this equation), Δ is the slope of the vapor pressure (e; kPa) versus temperature (T; °C) relationship (kPa °C^{-1}) ($\Delta = (e_0 - e_a)/(T_0 - T_a)$, where subscript 0 is the evaporation surface and a is the atmosphere above), γ is the psychrometric constant (0.067 kPa °C^{-1} at an atmospheric pressure of 101.3 kPa), R_a is the net available energy (W m^{-2}), and $f(u)$ is the ventilation (vapor transfer) function multiplied by the vapor pressure gradient (mm per day). To calculate the potential evaporation rate using Eq. (16.2), the appropriate wind function would need to be used, as derived over a uniform open-water surface (e.g., a lake), or a well-watered vegetated surface (e.g., an irrigated crop), as was represented by the cylinders used in Penman's experiment (Chapter 7). The vapor pressure at the surface (e_0) could be represented by the saturation vapor pressure (e_s) calculated at the surface temperature (T_0). In this case, E_a then represents E_p using Eq. (16.2). Regardless, Eq. (16.2) shows that as the energy available to drive evaporation increases, so does the evaporation rate.

The available energy (R_a) and the wind function term ($f(u)$) are seldom known or measured at most locations, thus limiting the practical use of Eq. (16.2). To overcome this limitation, several

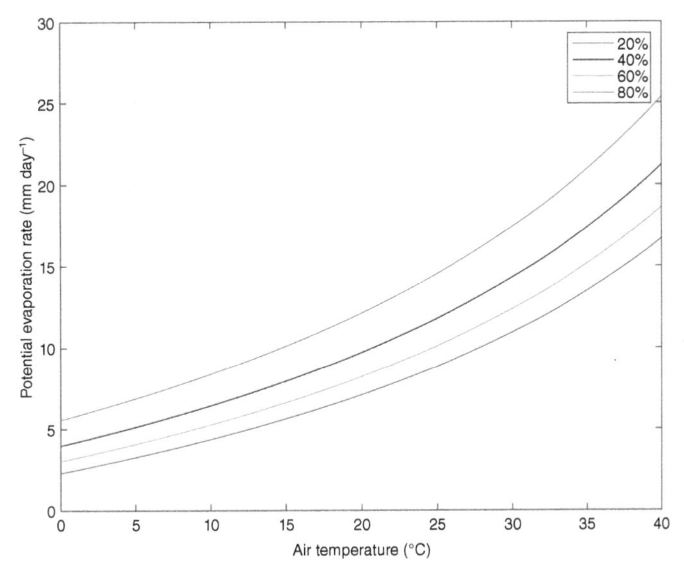

Figure 16.2 Effect of increasing the daily mean air temperature on the daily potential evaporation rates predicted using Eq. (16.4) for an open-water lake surface with an albedo of 0.05 at 40° latitude and an elevation of 2,000 m above sea level. Each plot represents the potential evaporation rate at a different relative humidity (%) shown in the legend. Dew-point temperatures were calculated using Eq. (7.7).

equations have been developed that replace R_a and $f(u)$ with the commonly measured daily mean air (T_a; °C) and daily mean dew-point temperatures (T_d; °C). Empirical equations based on routine measurements to estimate E_a across wide regions have been proposed. For example, for well-watered vegetation, E_a (mm day^{-1}) with an albedo of 0.25 can be estimated as (Eq. 16.3):

$$E_a = \frac{\left(500\,T_m \big/ (100 - A) \right) + 15(T_a - T_d)}{80 - T_a} \tag{16.3}$$

and for an open-water lake surface with an albedo of 0.05, it can be estimated as (Eq. 16.4) (Linacre, 1977):

$$E_a = \frac{\left(700\,T_m \big/ (100 - A) \right) + 15(T_a - T_d)}{80 - T_a} \tag{16.4}$$

where T_m (°C) is the sea-level equivalent of T_a (°C) estimated as $T_m = T_a + 0.006\,h$, h is the elevation in meters above sea level, and A is the latitude (in degrees). In both Eqs. (16.3) and (16.4), E_a should be close to E_p since in both cases the assumption is that there is no restriction on water supply. Equation (16.4) shows that, as air temperatures increase, resulting in a greater difference between the air and dew-point temperatures, the potential evaporation rate increases dramatically (Figure 16.2). Figure 16.2 also shows that Eq. (16.4) correctly predicts

Figure 16.3 A Class A evaporation pan, 25.4 cm deep and 120.7 cm diameter. Water levels are measured with a small float and data logger (right side), and refilling occurs once per day with a water level gauge and solenoid switch attached to a water line (left side). Photo credit: P. D. Blanken.

that as relative humidity decreases, E_p increases. Despite the empirical forms, simplifications, and no explicit representation of wind speed, Linacre (1977) and Anyadike (1987) found that evaporation calculated using Eq. (16.4) compared well with Class A evaporation pan measurements (see Figure 16.3), and also performed better than other estimates of E_p at several locations worldwide.

16.3 Trends in Potential Evaporation

Warming and drying of the atmosphere should increase the potential evaporation rate as predicted by equations shown above. Climate models can use these equations not only to predict, but also to estimate past trends in potential or actual evaporation in locations or times with no or few observations. In Britain, for example, there are no long-term observations of the actual evaporation rate (ironic given the pioneering evaporation studies that were performed there). Kay et al. (2013) reviewed historic trends and future projections for evaporation based on models using versions of Penman's equation to calculate E_p and E_a from 1961 through 2012 over England and Wales, and Scotland. As expected, with increasing temperature both E_p and E_a showed positive linear trends in both regions. In England and Wales, E_p and E_a increased at roughly 1.0 and 0.7 mm yr^{-1}, respectively, and in Scotland, 0.6 mm yr^{-1} for both E_p and E_a. Most model projections for Britain indicate increases in E_p depending on the equation(s) used to calculate E_p (Kay et al., 2013). Such results agree with climate model predictions of what should happen under an intensified water cycle in response to higher temperatures and increased precipitation (e.g., Huntington, 2006). Trends in evaporation based not on models but on the few direct evaporation observations available, however, disagree with the model calculations.

For decades, measurements of the daily decrease in water level from **evaporation pans**, small shallow water-filled pans, have been used to represent the local potential evaporation rate (Figure 16.3). Pan evaporation measurements (E_{pan}) have been incorrectly interpreted to represent E_p from nearby open-water bodies such as lakes or reservoirs as seen in the background of Figure (16.3). The rate of evaporation from the pan and nearby bodies of water (e.g., a lake) can differ for several reasons. The energy available to evaporate water, R_a, is seldom equal, owing to the large storage of heat in lakes with a greater depth and surface area than an evaporation pan. As a result, E_{pan} can overestimate lake E_p in spring when the pan's water warms more quickly than a lake, and underestimate in fall when the pan's water cools more quickly than a lake. Wind speed, air temperature, and humidity can also vary between the pan's location and across a lake, and therefore so will evaporation. Over time, vegetation growth near the pan can create conditions that are not representative of conditions over a lake. Site-specific corrections to E_{pan} can be developed based on independent evaporation measurements such as the eddy covariance (Chapter 7). A **pan coefficient** ($k_{pan} = E_a / E_{pan}$), the ratio of the independently measured actual evaporation (E_a) to the pan-measured evaporation (E_{pan}), can then be used to correct the pan evaporation estimates. The pan coefficient is typically equal to ~0.70 when averaged over long time periods (e.g., the warm season of May to September in the Northern Hemisphere midlatitudes) but often has significant short-term (monthly) variation due to differences in heat storage between the small volume of water in the pan and the large volume in a lake or reservoir.

Some studies that examined decadal-long trends found that E_{pan} is decreasing, not increasing as is predicted by Penman's equation (Eq. 16.2) in a warmer, drier world. Peterson, Golubev and Groisman (1995) noted "evaporation is losing its strength" based on the observation that the warm-season (May–September) pan evaporation decreased by 97 mm over a 45-year period (21.6 mm per decade) in the western United States, with significant downward trends in the eastern United States, Europe, Asia, and Siberia. Chattopadhyay and Hulme (1997) reported that despite an increase in air temperature, pan evaporation decreased in almost all parts of India between 1961 and 1992. Similar findings of large decreases in E_{pan} were reported across China (−29.3 mm per decade; Liu et al., 2004), Australia (−0.5 to −4.0 mm per year; see references within Stephens et al., 2018), and many other locations worldwide (Fu, Charles and Yu, 2009). Apparently there was an **evaporation paradox** between the expected increase in E_p and evaporation-pan decrease in E_p (Brutsaert and Parlange, 1998).

To understand this evaporation paradox, issues surrounding the quality and accuracy of the pan evaporation measurements need examination, as well as any plausible explanations for why E_p could decrease. As mentioned, uncertainties and errors in E_{pan} can result from the pan's location, installation, maintenance and observation practices, length of the seasonal record (warm season only), number of years of observations, and other factors (Fu, Charles and Yu, 2009). Therefore, E_{pan} may not represent the actual potential evaporation. If E_{pan} measurements are accurate, then the meteorological variables that affect potential evaporation, such as the available energy, the horizontal wind speed, and the vapor pressure deficit, need to be examined for any long-term trends.

Could meteorological trends explain the evaporation paradox? The surface must have energy available for evaporation, and much of this energy is provided by the incident surface solar radiation. A sufficient increase in cloud cover could decrease the available energy, thus decreasing E_{pan}. To determine the decrease in available energy required to explain the E_{pan} trends, Roderick and Farquhar (2002) used pan evaporation data from the Northern Hemisphere in locations where the surface global (direct and diffuse radiation) solar radiation decreased by 2–4 percent per decade (between 1960 and 1990) owing to increased cloud cover and/or aerosol pollution. They calculated that this decrease in solar radiation is equivalent to a decrease in the annual pan evaporation of roughly 90 to 155 mm year^{-1}, agreeing with the 110 mm year^{-1} decrease observed at seven locations. Roderick and Farquhar (2002) concluded that the decrease in surface solar irradiance due to clouds and pollution (**global dimming**; Wild, 2009) explained the evaporation paradox. This explanation, however, has been questioned because of inconsistencies. In some locations where E_{pan} has decreased, the actual evaporation has not, and some locations show no positive correlation between solar radiation and E_{pan} (Fu, Charles and Yu, 2009).

Dry air supplied by the horizontal wind to remove the air humidified by evaporation (the ventilation term) is required to maintain evaporation. To determine the relative roles of energy available to drive evaporation and the ventilation to remove humidified air, Roderick et al. (2007) developed a model to calculate pan evaporation based not only on solar radiation, but also temperature, humidity, and wind speed. This "preferred approach" to calculating pan evaporation ($E_{pan;}$ kg m^{-2} s^{-1}) was the **PenPan model** with the radiative ($E_{R,pan}$) and aerodynamic ($E_{A,pan}$) components explicitly represented (Eq. 16.5) (Rotstayn, Roderick and Farguhar, 2006):

$$E_{pan} = E_{R,pan} + E_{A,pan} = \left(\frac{\Delta}{\Delta + a\gamma}\frac{R_n}{\lambda}\right) + \left(\frac{a\gamma}{\Delta + a\gamma} f(u) D\right) \tag{16.5}$$

where Δ is the slope of the saturation vapor pressure gradient versus temperature gradient relationship (kPa °C^{-1}) evaluated at air temperature, R_n is the net radiation at pan (W m^{-2}), λ is the latent heat of vaporization (J kg^{-1}), a is the ratio of the effective surface area for heat and vapor transfer (a can vary from 2.1 to 2.5), γ is the psychrometric constant (0.067 kPa °C^{-1} at an atmospheric pressure of 101.3 kPa), D is the vapor pressure deficit (kPa), and $f(u)$ is the horizontal wind speed (u; m s^{-1})-dependent vapor transfer function expressed in the form $f(u) = 1.39\times10^{-8}(1 + 1.35u)$ (Rotstayn, Roderick and Farguhar, 2006). Roderick et al. (2007) used the PenPan model to attribute trends in E_{pan} based on radiative ($E_{R,pan}$) or aerodynamic ($E_{A,pan}$) contributions at 41 sites distributed across Australia for the period 1975–2004. Decreasing solar radiation played a minor role, and changes in temperature and humidity were generally too small to affect the pan evaporation rates. Decreases in pan evaporation were mostly due to decreases in wind speed.

Could decreases in surface horizontal wind speed explain the evaporation paradox? McVicar et al. (2012) reviewed 148 studies from across the globe with long-term measurements and found that wind speed was decreasing at an average rate of 0.014 m s^{-1} per year. This **stilling effect** of a reduced wind speed reducing E_{pan} was often the cause of the largest decline of evaporative demand. Stephens et al. (2018) later confirmed that in southern and western Australia the decreasing trends in E_{pan} between the 1970s and mid-2000s were mainly driven by decreasing wind speed. Starting in the early 1990s, however, the trends in E_{pan} reversed owing to increasing air temperatures resulting in greater vapor pressure deficits (Stephens et al., 2018). This E_{pan} trend reversal points to the significance of increased atmospheric evaporative demand as air temperatures and vapor pressure deficits increase.

A closer look at the recent trends in pan evaporation measurements in several regions now shows that the rate has indeed been increasing. Increasing air temperatures should result in an increase in potential evaporation due to an increase in the vapor pressure deficit. Pan evaporation should therefore be increasing, not decreasing, if the increase in the vapor pressure deficit acting to increase evaporation offsets any decreases in evaporation from decreases in solar radiation and wind speed. Indeed, 24 of 37 evaporation pan sites in Australia shifted the trend from decreasing to increasing E_{pan} beginning in the mid-1990s, primarily owing to increased vapor pressure deficits driven by increased air temperature, not reduced moisture (Stephens et al., 2018). In China, where decreasing trends in E_{pan} were reported during the 1960–2000 period, recent studies now report a significant upward trend between 1988 and 2017 primarily driven by increased air temperature (Shen et al., 2022). In the Canadian Prairies, trends in E_{pan} from pan evaporation measurements supplemented with calculations of the potential evaporation rate revealed that northern prairie regions showed increasing trends in potential evaporation, especially in April, whereas southern prairie regions showed decreasing trends, especially in the summer months (Burn and Hesch, 2007). The results were mixed but explainable; decreases in wind speed explained the decreasing evaporation trend and increases in the vapor pressure deficit explained the increasing evaporation trend.

The evaporation paradox is not a paradox at all. Clearly all four meteorological variables – available energy (radiation), horizontal wind speed, air temperature, and humidity – need to be included in any calculation or prediction of evaporation, or interpretation of measured or modeled potential evaporation trends. Locations that have reported decreases in the potential evaporation rate show decreases in solar radiation and/or wind speed, and locations that have reported increases in the potential evaporation rate show increases in air temperature and the vapor pressure deficit. For locations that have shifted from a decreasing to an increasing trend, the implication is that vapor pressure deficits have increased as a result of warming, but is there any evidence of increases in vapor pressure deficits based on trends of the actual evaporation rate?

16.4 Trends in Actual Evaporation

Vegetation can reduce the potential evaporation rate by restricting the amount of water available for evaporation through the control of the stomata guard cells (Chapters 11 and 12). With a limited water supply imposed by vegetation or surface drying, evaporation proceeds at the **actual evaporation rate**, which is the usual case over any terrestrial surface. Vegetation can also restrict the availability of water for evaporation at the ground surface by reducing the solar radiation through shading, reducing wind speed, reducing the vapor pressure deficit by lowering air temperature and increasing humidity, and by adding organic material that acts as a mulch. As shown in Chapter 7, transpiration represents the majority of water vapor over terrestrial surfaces, and much effort has been directed towards measuring and understanding this active, biological control of the water cycle. As shown in Chapter 11, there is a significant decrease in water potential at the liquid–air interface within the leaf as humidity decreases. When this demand for water vapor is sufficient to decrease the xylem water potential to a point that places the water column in the xylem at risk of cavitation, the stomata will close to reduce the water tension inside the xylem tissue.

In recognition of the active role of vegetation in reducing the potential evaporation rate, Monteith modified Penman's equation (Eq. 16.2) to develop the **Penman–Monteith equation** derived in Chapter 12 (Eq. 16.6):

$$\lambda E = \frac{\Delta \times H + \rho c_p \left\{ (e_s - e_a) \big/ r_a \right\}}{\Delta + \gamma \left(1 + r_s \big/ r_a \right)} \tag{16.6}$$

where the actual evaporation rate is expressed here as the equivalent latent heat flux (λE; W m^{-2}), H is the surface available energy $R_n - G_0$ (W m^{-2}), the net radiation minus any soil heat flux, r_a (s m^{-1}) is the **aerodynamic resistance** to vapor transfer (analogous to the wind-based vapor transfer function), and r_s (s m^{-1}) is the **surface resistance** to water vapor transfer. As discussed in Chapter 12, the r_s term represents both the stomatal resistance from the entire plant canopy and any resistance to water vapor transfer imposed by the soil. Equation (16.6) provides both the means to calculate r_s when λE and the other variables have been measured or to predict λE based on measurements or estimates of r_a and r_s. With the Penman–Monteith equation, the role of vegetation in limiting the actual evaporation to rates less than the potential evaporation rate is explicitly recognized through the r_s term, or its reciprocal, the **canopy conductance** g_s (m s^{-1}).

The Penman–Monteith equation has been used in many simulations and calculations of the actual evaporation rate. In fact, the **Food and Agricultural Organization of the United Nations (FAO)** recommends that the Penman–Monteith equation be used as the standard method to compute the **reference crop evaporation** (Allen et al., 1998), commonly referred to as "FAO 56 PM", as a common basis to compare climate change simulations (Allen et al., 2006). Many well-known climate projection models, such as the United Kingdom's Climate Projection Weather Generator, and Meteorological Office Rainfall and Evaporation Calculation System, use versions of the Penman–Monteith equation based on FAO recommendations (Kay et al., 2013).

Observations and simulations of changes in the actual evaporation rate have overall shown that rates are increasing. A report published by the **Intergovernmental Panel on Climate Change (IPCC)** stated:

> It is virtually certain that evaporation will increase over the oceans and very likely that evapotranspiration will increase over land with regional exceptions in drying areas (Douville et al., 2021).

Long-term actual evaporation measurements are sparse compared with evaporation-pan potential evaporation measurements. Estimates of E_a, however, have been made using a water balance approach in river basins where precipitation (P), river discharge and all other water losses and storage terms (R) are measured (i.e., $E_a = P - R$). According to this water balance approach over the 3,000,000-km^2 Mississippi River Basin in the United States, E_a (from all sources) was increasing at a rate of 0.95 mm yr^{-1} during the period 1947–1997 (Milly and Dunne, 2001). Widening the spatial scale to six large river basins in the United States, Walter et al. (2004) calculated an increase in E_a over the period 1950–2000 at a rate of 1.04 mm yr^{-1}. These large basin water balance studies point to an increase in precipitation as the primary cause of increased evaporation. Brutsaert and Parlange (1998) noted that such increases in the actual evaporation rates complement the pan evaporation studies that report a decreasing trend in the potential evaporation rate, since an increase in humidity in response to the increased regional actual evaporation would act to reduce pan evaporation.

In some locations, studies have found that the actual evaporation rate is decreasing. Using a Penman–Monteith approach over the agricultural Platte River Basin in Central Nebraska in the United States from 1893 to 2008, Irmak et al. (2012) found that precipitation was increasing by 0.87 mm yr^{-1}. The calculated reference evaporation rate, however, was decreasing by 0.36 mm yr^{-1} owing to a decrease in the available energy, stemming from increased cloud cover associated with the increased precipitation. Such studies highlight the importance of changes in both precipitation and evaporation in the regional water balance.

16.5 Trends in Vapor Pressure Deficit

If actual evaporation is increasing, then water vapor should also be increasing as more water vapor is added to the atmosphere. With increasing air temperatures, the saturation vapor pressure deficit will increase. Since a primary control on evaporation is the difference between saturation and actual vapor pressures, knowing how atmospheric humidity is changing is key to predicting changes in evaporation in a warming world. Although there are some differences between measurements and models, Allan et al. (2022) concluded with confidence that the atmosphere's water vapor content is increasing with climate warming at a rate of 4–5% near the surface and 10–15% at the 300-hPa pressure height for each 1 °C increase in the global mean surface temperature. With warming, the saturation vapor pressure increases 7% with each 1 °C increase in temperature (Eq. 7.2). Whether the difference between the saturation vapor pressure (e_s) and the ambient vapor pressure (e_a) (the vapor pressure deficit, D) increases or not depends on whether the increase in water vapor content from increases in evaporation can continue to keep pace with the exponential increase in the saturation vapor pressure. This depends not only on the rate of temperature increase (which affects e_s) but also on the availability of surface water for evaporation (which affects e_a). In a warming world, as the surface water continues to evaporate, eventually surface drying will occur, and the vapor pressure deficit will then increase dramatically.

Studies have reported mixed results on trends in surface vapor pressure deficits over time and across locations. In the United States between 1960 and 2012, D increased across the western United States owing to increasing temperature and dry conditions, and in other regions D decreased owing to increases in the ambient vapor pressure (Seagar et al., 2015). Between 1979 and 2013 in the United States, Ficklin and Novick (2017) found D increased in all seasons across most regions at a rate of 0.007 kPa per year, driven by an increase in e_s and a decrease in e_a. Their model predicted a 51% increase in the summertime D between 2065 and 2099, resulting in a 9–51% reduction in stomatal conductance. In China between 1994 and 2017, the increase in e_s due to rising temperatures and decreased e_a due to drying surface conditions resulted in a rapid increase in D (Li et al., 2021). Even in the humid tropical South American Amazon region, D increased between 1987 and 2016 (Barkhordarian et al., 2019). Overall, in most regions of the world where the water supply cannot keep up with the evaporative demand, D has been increasing (Figure 16.4).

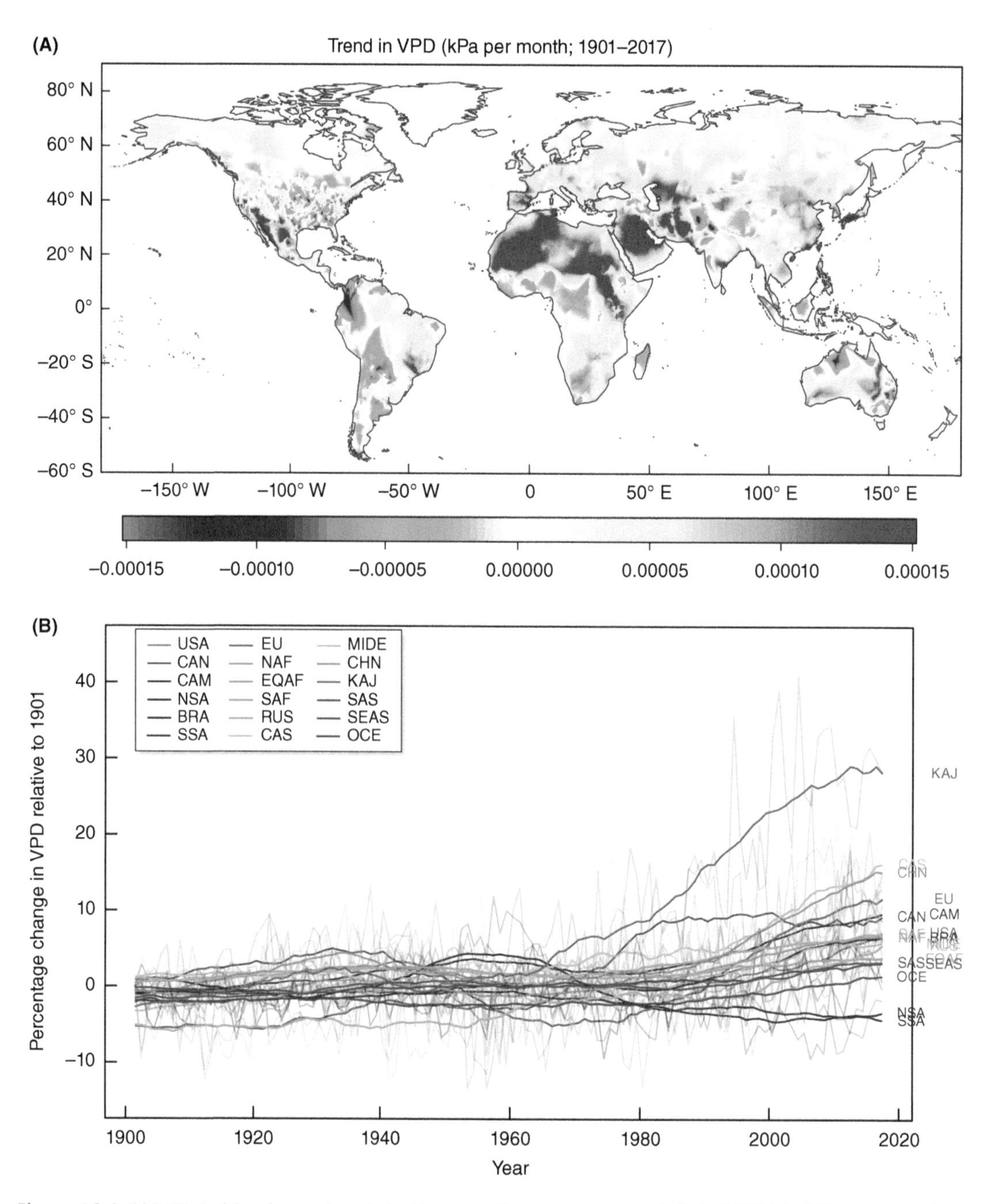

Figure 16.4 (A) Global land-area trends in the annual vapor pressure deficit (VPD) in kPa per month over the period 1901–2017. (B) Annual percent change in VPD relative to 1901 averaged for the regions shown (KAJ is Korea and Japan). Bold lines are 10-year running means. See Grossiord et al. (2020) for details. Reproduced with permission from Grossiord et al. (2020).

16.6 Trends in Transpiration

Increasing vapor pressure deficits could increase transpiration stemming from the increased atmospheric demand for water vapor (*D*) if there is no water stress. To help determine whether this was true, López, Way and Sadok (2021) performed a quantitative literature review of studies that examined the temperature-independent *D* acclimation (long-term) response for 112 species spanning from 1970 to 2018. In those studies that compared a large number of species grown under either low or high *D* conditions, the leaf and whole-plant transpiration rates increased when grown under the high *D* condition. The stomatal conductance decreased, as did the rate of photosynthesis, for species grown under high *D* conditions. Overall, 84 percent of the species exhibited a significant response to *D*. It is important to note that most of the species were examined under controlled to semi-controlled conditions, with most plants grown indoors under well-watered conditions.

In typical outdoor conditions where the availability of water for plant growth is at times limited, plants can respond by decreasing their **stomatal conductance** (Chapters 11 and 12), thus reducing transpiration. As discussed in Chapter 11, the purpose of decreasing stomatal conductance under periods of high *D* is to reduce the water tension inside the xylem tissue, thereby lowering the risk of cavitation. Therefore, the stomatal response to *D* is a response to both the water supply to the roots and water demand by the atmosphere.

What is the general response to increasing *D* on the actual evaporation rate from terrestrial vegetated ecosystems, an increase or decrease? An increase in *D* caused by an increase in air temperature has resulted in a 1.5 percent per decade (1985–2018) increase in lake evaporation worldwide (Zhao, Li and Gao, 2022). In contrast, transpiration will probably not continue to increase. Novick et al. (2016) showed that evaporation was limited more by stomatal conductance than by soil moisture in many vegetated ecosystems. In a literature review on plant response to increasing *D*, Grossiord et al. (2020) stated that the abundance of evidence suggests that transpiration will increase in most species until a threshold *D* is reached. When reached, the decrease in stomatal conductance will then reduce transpiration, reduce photosynthesis (and therefore reduce growth productivity), and therefore reduce carbon dioxide uptake. The magnitude of this threshold *D* will vary depending on the species' sensitivity to *D* and therefore with numerous physiological and climate factors, a result highlighted by Massmann, Gentine and Lin (2019).

The response of many of the world's forests to a drying atmosphere has been a reduction in stomatal conductance with implications that extend beyond a reduction in transpiration. From 122 diverse landscape locations across the western United States, the tree-ring width of **Douglas fir** (*Pseudotsuga menziesii*) showed a significant decrease as *D* increased across all regions, with decreased growth projected to continue with continued warming and drying, impacting both water and carbon cycling in these forested ecosystems (Restaino, Peterson and Littell, 2016). In the Mediterranean region, where *D* is one of the most crucial drivers of drought-induced dieback and tree mortality, with corresponding impacts on water and carbon cycling, increasing *D* led to a reduction in tree and shrub growth (Castellaneta et al., 2022). From the central Amazon rainforests (Antezana-Vera and Marenco, 2019) to the North American boreal aspen forests (Hogg et al., 1997) increasing *D* has had negative effects on tree growth.

16.7 The Effects of Changes in Humidity on Ectotherms: Amphibians

The ability of ectotherms to regulate water loss and temperature lies in their skin's resistance to vapor transfer (Chapter 13). Those with a large skin resistance such as tortoises, desert lizard species, and many insect species inhabit extremely hot and dry regions. In contrast, those ectotherms that inhabit cooler, humid regions have a small skin resistance to water vapor loss. Amphibians that rely on the availability of water to complete their life cycle, such as many frog and salamander species, are especially vulnerable to a drying atmosphere.

Frogs, especially sensitive to humidity since moist skin is required for gas exchange, have been the subject of many desiccation studies. The activity pattern of wood frogs (*Rana sylvatica*) in northern Minnesota, USA, as they emerged from winter hibernation to commence springtime breeding in upland ponds, decreased significantly as the vapor pressure deficit increased (Bellis, 1962). There was no correlation between frog activity and either air or soil temperature. Although other field studies have shown the necessity of access to surface water for reproductive success, this early study showed that atmospheric humidity also plays an important role. Other amphibian species such as salamanders showed a similar response of decreased activity as D increases. In North Carolina, USA, the response of two lungless salamander species, one living in low elevations (*Plethodon teyahalee*) and one living in higher elevations (*Plethodon metcalfi*), to D were examined (Riddell and Sears, 2015). For both species, high vapor pressure deficits reduced the probability of finding surface-active salamanders. In an attempt to generalize the response of amphibians to changing air temperatures and humidity, Greenberg and Palen (2021) examined three ecologically distinct amphibian species found in the North American Pacific Northwest region: the wet-adapted **coastal tailed frogs** (*Ascaphus truei*), the dry-adapted **spadefoot toads** (*Spea intermontana*), and the **Pacific chorus frogs** (*Pseudacris regilla*) that have habitats with a wide range of temperature and moisture. They found that hydration played a larger role than temperature in all three species regardless of habitat. Performance (jumping) for all three species declined abruptly after a 20–30% decrease in mass from dehydration. In terms of activity, dehydration alone restricted activity by up to an estimated 60% for all three species, regardless of their habitation in distinct climatological niches.

Whereas some amphibian species have little tolerance or short-term adaptability to drier conditions, other species can adapt much more effectively. Species adapted to warm and often dry environments tend to have a significant survival advantage with changes in climate and land use. Frishkoff, Hadly and Daily (2015) found that tropical amphibian and reptile species adapted to warmer climate zones were found more frequently in warmer, recently deforested areas than other species. Any increases in temperature associated with climate change and/or land use change will therefore favor those species that are already adapted to warm conditions. The ability of these warm-adapted species to resist dehydration as humidity decreases will also be a factor. Another example of an ectothermic species found across a wide range of temperature and humidity environments is lizards. Weaver et al. (2022) found rapid and significant acclimatory changes in cutaneous evaporative water loss of the **western fence lizard** (*Sceloporus occidentalis*) in response to changes in ambient humidity.

These and other studies suggest that in the ectothermic community there are both winners and losers in a warmer, drier world. Amphibian species will suffer dire consequences, whereas reptiles that are quickly adaptable and already exposed to warm, dry conditions may indeed expand their activity and habitat range.

16.8 The Effects of Changes in Humidity on Ectotherms: Insects

Arthropods, including the subphyla insects, account for the majority of species on Earth, with many species probably still undiscovered, and they too are susceptible to changes in climate and land cover. Given their large surface area relative to small body size, insects are especially sensitive to changes in humidity and air temperature that influence thermoregulation and therefore activities ranging from development to reproduction and habitat selection. Many insect species prefer humid habitats such as wetlands and forests, and the loss of these habitats has been cited as a significant cause of insect species decline. The decline of insect communities worldwide has reached alarming levels, with impacts in nearly all terrestrial and aquatic ecosystems at all levels of the food chain. The global extent of insect decline was summarized by Forister, Pelton and Black (2019). Populations of native bees (28% of species threatened), butterflies (19% of species at risk of extinction), and tiger beetles (33% of species threatened or endangered) across much of North America have decreased to the extent that several species are now considered threatened or at risk of extinction. The declines are even higher in Europe. In Germany, the total flying insect biomass decreased by more than 70% between 1989 and 2016 (Hallmann et al., 2017). Pesticides and pollution (including light pollution) have played major roles in the decline of many insect species, but so have changes in climate and microclimate associated with changes in land use and land cover.

With habitat loss come changes in microclimate, as decreasing areas of wetlands and deforestation tend to further increase temperature and decrease humidity alongside increases imposed by climate change. The impact of these two factors on insect species decline – habitat loss versus climate change – can be difficult to separate given the vast geographic distribution of insects and their ability to adapt to change. Regardless, attempts have been made. Based on monitoring studies of at least 10 insect species spanning at least 10 years, Halsch et al. (2021) concluded that climate change impacts are considerable when compared with those resulting from changes in land use. Butterfly species found across an elevational gradient in Northern California in the United States suffered population declines at high elevations in locations where there were no immediate effects of habitat loss. Some insect species will likely benefit in a warmer and drier climate. As ectotherms, an increase in temperature, as long as humidity does not decrease too much, should be beneficial to them. Tropical regions are already warm, so further increases in temperature may offer no additional benefit and decreases in humidity associated with warming may do more harm than good. In temperate regions, shorter and warmer winters could result in earlier and faster development and emergence, which could then increase the number of generations per year. The number of generations or offspring produced per year is referred to as **voltinism**; a **univoltine** species produces one generation per year, and a **multivoltine** species produces two or more per year. In North America with warmer winters, the **spruce beetle** (*Dendroctonus rufipennis*) can complete one generation in one year instead of two (or more) years, and the **mountain pine beetle** (*Dendroctonus ponderosae*) has now been able to complete two generations per year instead of one (Forrest, 2016). In boreal and alpine regions, higher temperatures and the melting of ice and snow should benefit insect populations. Overall, the response of insects to climate change is complex. Added to this complexity is that many insect species can adapt and evolve to environmental changes such as changes in humidity quickly, owing to their high reproductive rate.

The sheer number of insect species, their habitat range, and their existence on Earth for nearly 500 million years says volumes about their success in adapting to changing environmental conditions. The capability of an organism to change its behavior, morphology, physiology, and/or even gene expression is known as **phenotypic plasticity** (Lewis and Pfenning, 2021). There is strong evidence for the importance of plasticity in insects to heat tolerance, as shown by Noer et al. (2022). Over several days, seven insect species were collected in temperate Melbourne, Australia, under ambient conditions of air temperature ranging between 10 °C and 35 °C and relative humidity ranging from 25% to 100%. Five insect species were collected in tropical Cape Tribulation, Australia, under ambient conditions of a nearly constant air temperature of ~20 °C and a relative humidity of 75% to 100%. Insects were then tested for heat tolerance by placing individuals in glass vials submerged in water baths at various temperatures. The insects' adaptability, or plasticity, to changing temperature was demonstrated by all insects regardless of where they were collected. Insect species collected from the temperate location, however, were more heat-tolerant than those from the tropical location. This study showed that species adapted to tropical regions with a stable but high temperature and relative humidity were less adaptable to any further heat (and dehydration) stress than their temperate counterparts (i.e., the tropical species were at their thermal limit). Perhaps more importantly, this study showed a very fast (i.e., days) and species-specific adaptive response.

16.9 The Effects of Changes in Humidity on Endotherms: Humans

In sharp contrast to insects, endotherms such as humans are much less adaptable and less tolerant of changes in temperature and humidity. Heat and humidity directly affect the ability of humans to regulate body temperature (Chapter 14), and excessive heat results in thousands of deaths annually. The definition of a **heatwave**, a prolonged period (e.g., several days) of dangerously high air temperatures, varies widely and lacks a universal definition. And the temperature used to define dangerously high air temperatures and issue public warnings, a **heat index**, also varies but is almost always based on when the air temperature exceeds some threshold. The threshold can be absolute (i.e., exceeding a specific temperature) or relative (i.e., temperature exceeds a value defined as a percentile based on long-term climate data). Both the absolute air temperature used, and the percentile used for the exceedance air temperature, can vary by location or reporting agency. Which statistic is used for the air temperature (daily average, maximum, or minimum) varies; and how the temperature was measured, and the number of consecutive days the chosen air temperature statistic is reached, if any, can also vary among heat index calculations. Additional meteorological variables, such as radiation, wind speed, and especially humidity, are sometimes incorporated in a heat index as these can affect human temperature regulation (Chapter 14). How heatwaves and heat indices are defined can influence the interpretation of heat-related death rates and other health measures, and thus has implications for operational heat warning systems (Kent et al., 2014) and for interpreting spatial and temporal trends and patterns (Smith, Zaitchik and Gohlke, 2013). Despite variations in heatwave definitions, high air temperature is the main criterion, and the heatwave intensity has a greater impact on mortality than the duration (Xu et al., 2016).

Heatwaves are increasing in frequency, duration, and/or intensity in many regions, especially Europe. In 2003, exceptional heat enveloped much of Europe, breaking air temperature records in June and again in August. In August 2003, maximum daily air temperatures were 7.5 °C to 12.5 °C above average, reaching or exceeding 40 °C in many locations (47 °C in Portugal) (García-Herrera et al., 2010). As a consequence, there were an estimated 70,000 deaths from exposure to heat (Robine et al., 2008). Europe has experienced heatwaves every summer from 2018 through 2022. In 2022, record-high temperatures as high as 20 °C above average were recorded late into the fall (October–November) across much of Europe. European summer heatwave events that would have been expected twice per century in the early 2000s are now expected to occur twice per decade (Christidis, Jones and Stott, 2015). This rapid increase in the frequency and magnitude of heatwaves in Europe compared with other midlatitude locations may be due to changes in the behavior of the upper-atmosphere jet stream in the region (Rousi et al., 2022).

In North America, heatwaves have also been increasing. In late June through mid-July 2021, the western North America heatwave affected southwestern Canada and the northwestern United States. This event was one of the most extreme global events ever recorded, with the town of Lytton, British Columbia, setting a Canadian record-high air temperature of 49.6 °C on June 29, 2021 (Thompson et al., 2022). The next day, the town was destroyed by wildfire with two fatalities, and more than 1,400 fatalities over the affected region (Lin, Ruping and Vitart, 2022). Across the United States in 50 large cities between 1961 and 2010, the frequency, duration, intensity, and length of the heatwave season have all increased significantly, with the average number of heatwaves increasing by 0.6 per decade (Habeeb, Vargo and Stone, 2015). Interestingly, minimum nighttime air temperatures, not daytime maximum temperatures, were used to classify heatwaves, since nighttime air temperature accounts for humidity (without solar radiation at night, high nighttime air temperatures are a result of increased incident longwave radiation emitted from higher water vapor concentrations). Habeeb, Vargo and Stone (2015) noted also that persistent exposure to heat through the night has a greater negative physiological impact on human health than daytime heat. Humidity should be included as a metric to quantify a heatwave and its impact on human health.

Excessive air temperatures clearly have a negative impact on humans' and other endotherms' ability to regulate body temperature, but how do changes in humidity that often accompany increases in air temperature affect thermal regulation? Based on a review of existing studies, Asseng et al. (2021) found that the preferable (i.e., comfortable so as to not induce thermal stress) ambient temperature ranges for humans, livestock, and several agricultural crops were all similar: 17 °C to 24 °C. Stress temperature thresholds, above which physiological stress, bodily damage and even death can occur, varied with humidity; as relative humidity increased, the stress temperature threshold decreased for humans, livestock, and crops. For example, in humans, lethal heat events occur either with extended exposure to air temperatures above

50 °C and low humidity, or extended exposure to air temperature above 32 °C and high humidity. Sherwood and Huber (2010) also found that humidity had a significant effect on the stress temperature threshold for humans, stating that a **wet-bulb temperature** (T_w) of 35 °C should not be exceeded for extended periods of time. The T_w is the temperature measured by a standard thermometer's bulb covered by a wetted cloth wick (as used in a psychrometer to measure humidity; Chapter 7). This temperature measurement includes both the air temperature and humidity since the rate of evaporation from the wet bulb varies with temperature and humidity with the evaporation rate influencing the wet-bulb temperature. Sherwood and Huber (2010) stated that any object cannot lose heat to an environment if T_w in the air exceeds the object's skin temperature. Since a human's skin temperature is roughly 35 °C, this temperature defines the 35 °C stress threshold T_w.

Although the wet-bulb temperature can be directly measured, air temperature (the dry-bulb temperature) and relative humidity are common meteorological measurements and forecast outputs. Stull (2011) developed an equation to calculate T_w from air temperature and relative humidity measurements. The variation in T_w with changes in air temperature and relative humidity, and when dangerously high combinations of air temperature and relative humidity are reached, is shown in Figure 16.5. The wet-bulb temperature increases as both air temperature and relative humidity increase, but the increase in T_w is rapid at high temperatures and relative humidities.

Changes in humidity will influence the ability of humans to tolerate excessive heat. For example, T_w reaches the critical temperature with a dry-bulb temperature T_a of 40 °C and relative humidity of 75%. At 45 °C, the critical T_w is reached with a relative humidity of 50%. As vapor pressure deficits are generally increasing with warming as the increase in the saturation vapor pressure outpaces the increase in vapor pressure, the relative humidity will also decrease. As shown in Figure 16.5, a decrease in relative humidity should increase the maximum tolerable air temperature for humans by 1 or 2 °C, thus providing some relief from warming if relative humidity decreases. At 75 locations across Canada, Vincent, Wijngaarden and Hopkinson (2007) reported a national average 0.6% decrease in relative humidity between 1953 and 2005. Over the same period, air temperatures increased nationally by 1.2 °C. Based on Figure 16.5 (i.e., the slope of the line of the relative humidity versus air temperature plot when $T_w = 35$ °C) a 0.6% decrease in relative humidity would *decrease* the air temperature at which the critical T_w is reached by only roughly 0.14 °C, which is not sufficient to offset the observed increase in air temperature. Globally, a negative trend in relative humidity has been reported in many regions, especially in boreal regions in the summer due to a decrease in terrestrial evaporation (e.g., transpiration) (Vicente-Serrano et al., 2018).

What are the consequences of people suffering from dangerously high combinations of air temperatures and humidity? It is likely that with increasing air temperatures the atmosphere's demand for water vapor will increase and relative humidity will decrease. Despite the decrease in the wet-bulb temperature from decreasing relative humidity, the thermal benefit is small, and there are several factors that make people susceptible to the effects of coupled high temperatures and humidity. First, 80 percent of the world's population lives within 100 m of sea level where relative humidity and nighttime air temperatures tend to be high. Second, sea levels are rising, exposing a greater proportion of the population to humid conditions. Third, there continues to be a shift in the global population from rural to urban regions, and urban regions tend to have

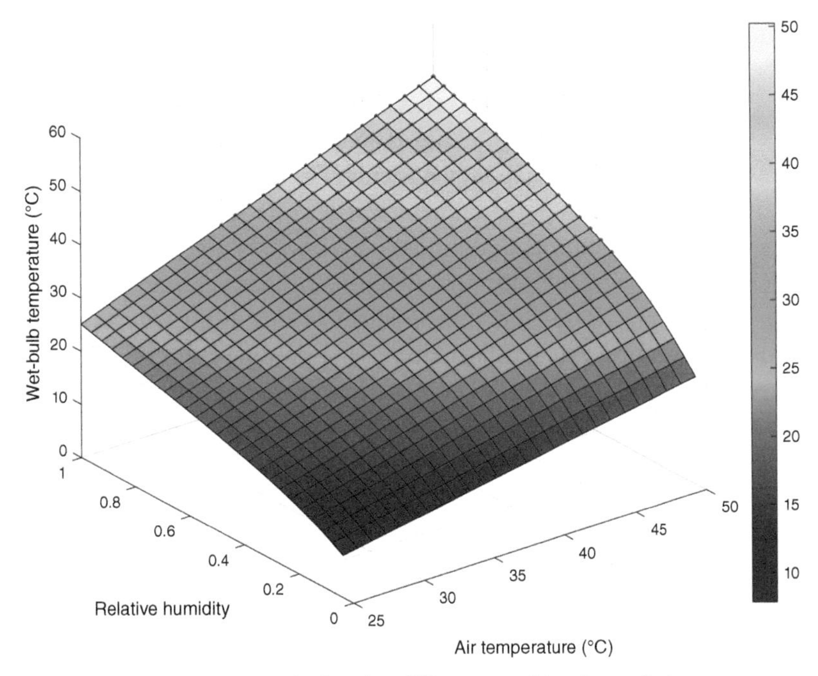

Figure 16.5 The wet-bulb temperature calculated at different combinations of air temperature and relative humidity. Red dots indicate critical wet-bulb temperature of 35 °C or higher, considered dangerous to humans. The wet-bulb temperature was calculated using the equation provided in Stull (2011).

high air temperature (especially at night) and humidity due to air pollution and other factors associated with urban heat islands. These factors create conditions where millions of people are vulnerable to heat stress.

Increasing heat and humidity have an unequal effect on populations and regions. Many studies, as summarized by Romanello et al. (2021), found that the young (younger than 1 year) and old (older than 65 years) were much more susceptible to the effects of heatwaves. Also, countries with a low HDI score (United Nation's Human Development Index) suffered greater losses than those countries with a higher HDI. In 2020, 79 percent of the 295 billion hours of potential work lost due to extreme heat exposure were in the agricultural sector of low HDI countries. The impacts of heatwaves are not experienced uniformly across the human population.

16.10 SUMMARY

Compared with the volumes of water stored in liquid or solid forms, a small volume is stored as vapor, yet this reservoir plays a significant role in the Earth's energy and water cycles. Humidity influences air temperature, cloud cover, precipitation, and thermal regulation for ectotherms and endotherms. With an unrestricted water supply, evaporation can occur at the potential evaporation rate: the upper limit of evaporation, restricted by the amount of energy available to change state and the vapor pressure deficit that is influenced by the wind-driven supply of dry air. The actual evaporation rate is controlled by these same factors, and a surface resistance to water vapor transfer that represents restrictions on the availability of water for evaporation. Therefore, the actual evaporation rate cannot exceed the potential evaporation rate.

Both evaporation rates have shown trends over time at several locations worldwide. Historic and future trends in the potential evaporation rate have been calculated using various versions of Penman's equation, and in most regions, past and predicted trends are positive with evaporation increasing at rates of roughly 1 mm per year, as expected with a warming and drying atmosphere. Based on measurements made using evaporation pans, however, the potential evaporation rate has been decreasing in some regions. This discrepancy between the expected increase in the potential evaporation rate and the observed decrease led to a series of publications and discussions on this evaporation paradox. Various errors associated with pan evaporation measurements, and possible meteorological reasons, could explain the paradox, such as decreases in surface solar radiation (global dimming), decreases in surface wind speed (the stilling effect), and changes in humidity. These factors explain both the decrease and the subsequent (upon updating the pan evaporation records) shift to an increasing trend in the potential evaporation rate at several locations, thereby reconciling theory with observations.

The key variable that controls the actual evaporation rate is the vapor pressure deficit, the difference between the temperature-dependent saturation vapor pressure and the evaporation-dependent ambient vapor pressure. Several studies reported that at locations where increases in precipitation have kept surfaces moist, the actual evaporation rate has been increasing at rates of the order of 1 mm per year. In drier regions, the rate has been decreasing, but given the contribution of evaporation from the vast oceans, the atmosphere's water vapor content has been increasing with climate warming. Concurrently, the saturation vapor pressure has been increasing at a much faster rate (7% per °C) than surface ambient vapor pressure (4–5% per °C), so the saturation deficit has been increasing with warming. Rapid increases in the saturation deficit should increase evaporation, yet over terrestrial surfaces, vegetation has been found to restrict evaporation (transpiration) through its biological control via stomatal pores. Vegetation's ability to conserve water by reducing transpiration in response to drier conditions will likely further increase saturation deficits and exacerbate the positive feedback between warming and drying.

Humidity affects the ability of both ectotherms and endotherms to thermoregulate. Ectotherms, such as several frog and lizard species, are especially sensitive to decreasing humidity associated with higher temperatures. Movement and activity levels for frogs, lizards, and salamanders have been shown to decrease as vapor pressure deficits increase. Those species already adapted to warm conditions are much more resistant to decreases in humidity than those species adapted to cooler, more humid environments. Insect species, generally adaptable to dry environments, are also susceptible to changes in humidity associated with warming. Alarming declines in insect populations have been reported in several regions, and several factors including habitat loss, pollution, and pesticide use are responsible. Several studies have shown that some insect species can adapt to and benefit from a warmer and drier environment, while others cannot and will not benefit.

We, *Homo sapiens* endotherms, cannot tolerate excessive heat or cold, and humidity strongly affects our ability to maintain a constant temperature. Periods of excessive heat, or heatwaves, affect and kill thousands annually. A precise definition of a heatwave, however, is lacking despite increases in the impacts of excessive heat. Some of this lack of clarity in the definition, and therefore ability to provide forecasts and public warnings of heatwaves, arises since humidity in addition to air temperature should be considered as part of the definition. Humidity and air temperature can be combined in metrics such as the wet-bulb temperature, and the exceedance threshold temperature of 35 °C for human safety has been proposed. Observed increases in air temperature when coupled with changes in relative humidity are exposing a greater number of people to dangerously high wet-bulb temperatures. The young and old, and countries with a low Human Development Index, are disproportionately vulnerable to periods of excessive heat and humidity.

16.11 QUESTIONS

16.1 With water available at the surface to evaporate, the evaporation rate could accelerate as air temperature increases. Explain why this may occur. How may an increase in the ambient vapor pressure resulting from increased evaporation further increase air temperature, creating a positive feedback loop between evaporation and temperature?

16.2 Imagine a small lake surrounded by a forest. Describe the variables to consider when estimating the evaporation rates over the lake and forest. Are there conditions when the actual evaporation rate from the lake could be less than the potential evaporation rate? Are there conditions when the actual evaporation rate from the forest could equal the potential evaporation rate?

16.3 For the location you are at right now, estimate the actual evaporation rate for well-watered vegetation and open-water lake evaporation using Eqs. (16.3) and (16.4), respectively. Use the nearest meteorological station, or your own observations, for the required data. Useful equations can be found in Chapter 7, such as Eqs. (7.3) and (7.7). Explain your results, and whether you feel the calculations are realistic.

16.4 Evaporation pans have traditionally been used to measure the potential evaporation rate. What are some factors to consider when interpreting pan evaporation measurements that may result in errors and uncertainties in these measurements? If pan evaporation measurements are truly not representative of the actual potential evaporation rate, then why have so many studies used such measurements to report on regional long-term evaporation trends?

16.5 What is the "evaporation paradox"? Can it be explained by changes in the physical meteorological variables that are known to control the evaporation rate? If so, explain how changes in these variables can explain this apparent paradox.

16.6 What modifications did Monteith make to Penman's equation to make calculations and predications of the evaporation rate applicable to vegetated surfaces? Why was such a modification necessary?

16.7 Several studies show that the vapor pressure deficit in many regions is increasing despite increases in the ambient vapor pressure. An increased vapor pressure deficit should increase transpiration, but this has not been the case under natural, *in-situ* conditions. Explain why transpiration could decrease, not increase, despite an increase in the vapor pressure deficit.

16.8 How will ectothermic amphibian species such as frogs tolerate or adapt to a drier, warmer atmosphere? What changes in behavior have been observed in ectothermic amphibian species as conditions become drier and warmer?

16.9 Changes in land use and land cover, in addition to other factors such as climate change and pesticide use, have been mentioned as factors causing significant insect population declines. How could changes in land use and land cover mimic changes in climate, thus affecting insect populations? Could microclimates be created to help increase insect populations? If so, how?

16.10 The frequency, duration, and magnitude of heatwaves has been increasing in several regions worldwide. Based on your understanding of thermoregulation, what do you think is the most meaningful yet practical definition of a heatwave to reduce the number of heat-related deaths? How does humidity play a role in a human's ability to tolerate heat? How will human susceptibility to heatwaves change as climate, land use and land cover, and socioeconomic conditions are all changing rapidly and simultaneously?

REFERENCES

Allan, R. P., Willett, K. M., John, V. O. and Trent, T. (2022) 'Global changes in water vapor 1979–2020', *Journal of Geophysical Research Atmospheres*, 127, p. e2022JD036728. doi: 10.1029/2022JD036728.

Allen, R. G., Pereira, L. S., Raes, D. and Smith, M. (1998) *Crop Evapotranspiration – Guidelines for Computing Crop Water Requirements.* FAO Irrigation and Drainage Paper 56. FAO. doi: ISBN 92-5-104219-5.

Allen, R. G., Pruitt, W. O., Wright, J. L. et al. (2006) 'A recommendation on standardized surface resistance for hourly calculation of reference ET_0 by the FAO56 Penman-Monteith method', *Agricultural Water Management*, 81, pp. 1–22. doi: 10.1016/j.agwat.2005.03.007.

Antezana-Vera, S. A. and Marenco, R. A. (2019) 'Intra-annual tree growth responds to micrometeorological variability in the central Amazon', *Biogeosciences and Forestry*, 14, pp. 242–249. doi: 10.3832/ifor3532-014.

Anyadike, R. N. C. (1987) 'The Linacre evaporation formula tested and compared to others in various climates over West Africa', *Agricultural and Forest Meteorology*, 39, pp. 111–119.

Asseng, S., Spankuch, D., Hernandez-Ochoa, I. M. and Laporta, J. (2021) 'The upper temperature thresholds of life', *The Lancet Planetary Health*, 5, pp. e378–e385. doi: 10.1016/S2542-5196(21)00079-6. Open Access article under CC BY-NC-ND 4.0 license.

Barkhordarian, A., Saatchi, S. S., Behrangi, A., Loikith, P. C. and Mechoso, C. R. (2019) 'A recent systematic increase in vapor pressure deficit over tropical South America', *Science Reports*, 9, p. 15331. doi: 10.1038/s41598-019-51857-8.

Bellis, E. D. (1962) 'The influence of humidity on wood frog activity', *The American Midland Naturalist*, 68, pp. 139–148.

Brutsaert, W. and Parlange, M. B. (1998) 'Hydrologic cycle explains the evaporation paradox', *Nature*, 396, p. 30. doi: 10.1038/23845.

Burn, D. H. and Hesch, N. M. (2007) 'Trends in evaporation for the Canadian Prairies', *Journal of Hydrology*, 336, pp. 61–73. doi: 10.1016/j.jhydrol.2006.12.011.

Castellaneta, M., Rita, A., Camarero, J. J., Colangelo, M. and Ripullone, F. (2022) 'Declines in canopy greenness and tree growth are caused by combined climate extremes during drought-induced dieback', *Science of the Total Environment*, 813, p. 152666. doi: 10.1016/j.scitotenv.2021.152666.

Chattopadhyay, N. and Hulme, M. (1997) 'Evaporation and potential evapotranspiration in India under conditions of recent and future climate change', *Agricultural and Forest Meteorology*, 87, pp. 55–73.

Christidis, N., Jones, G. S. and Stott, P. A. (2015) 'Dramatically increasing chance of extremely hot summers since the 2003 European heatwave', *Nature Climate Change*, 5, pp. 3–7. doi: 10.1038/NCLIMATE2468.

Douville, H., Rhagavan, K., Renwick, J. et al. (2021) 'Water cycle changes', in Masson-Delmotte, V. et al. (eds.) *Climate Change 2021: The Physical Science Basis. Contribution of Working Group I to the Sixth Assessment Report of the Intergovernmental Panel on Climate Change*. Cambridge University Press, pp. 1055–1210. doi: 10.1017/9781009157896.010.

Ficklin, D. L. and Novick, K. A. (2017) 'Historic and projected changes in vapor pressure deficit suggest a continental-scale drying of the United States atmosphere', *Journal of Geophysical Research Atmospheres*, 122, pp. 2061–2079. doi: 10.1002/2016JD025855.

Forister, M. L., Pelton, E. M. and Black, S. H. (2019) 'Declines in insect abundance and diversity: We know enough to act now', *Conservation Science and Practice*, 1, p. e80. doi: 10.1111/csp2.80.

Forrest, J. R. K. (2016) 'Complex responses of insect phenology to climate change', *Current Opinion in Insect Science*, 17, pp. 49–54. doi: 10.1016/j.cois.2016.07.002.

Frishkoff, L. O., Hadly, E. A. and Daily, G. C. (2015) 'Thermal niche predicts tolerance to habitat conversion in tropical amphibians and reptiles', *Global Change Biology*, 21, pp. 3901–3916. doi: 10.111/gcb.13016.

Fu, G., Charles, S. P. and Yu, J. (2009) 'A critical overview of pan evaporation trends over the last 50 years', *Climate Change*, 97, pp. 193–214. doi: 10.1007/s10584-009-9579-1.

García-Herrera, R., Díaz, J., Trigo, R. M., Luterbacher, J. and Fischer, E. M. (2010) 'A review of the European summer heat wave of 2003', *Critical Reviews in Environmental Science and Technology*, 40, pp. 267–306. doi: 10.1080/10643380802238137.

Greenberg, D. A. and Palen, W. J. (2021) 'Hydrothermal physiology and climate vulnerability in amphibians', *Proceedings of the Royal Society of London, Series B*, 288. doi: 0.1098/rspb.2020.2273.

Grossiord, C., Buckley, T. N., Cernusak, L. A. et al. (2020) 'Plant responses to rising vapor pressure deficit', *New Phytologist*, 226, pp. 1550–1566. doi: 10.1111/nph.16485.

Habeeb, D., Vargo, J. and Stone, B. (2015) 'Rising heat wave trends in large US cities', *Natural Hazards*, 76, pp. 1651–1665. doi: 10.1007/s11069-014-1563-z.

Hallmann, C. A., Sorg, M., Jongejans, E. et al. (2017) 'More than 75 percent decline over 27 years in total flying insect biomass in protected areas', *PLoS ONE*, 12, e0185809. doi: 10.1371/journal.pone.0185809.

Halsch, C. A., Shapiro, A. M., Fordyce, J. A. et al. (2021) 'Insects and recent climate change', *Proceedings of the National Academy of Sciences USA*, 118, pp. 1–9. doi: 10.1073/pnas.2002543117.

Hogg, E. H., Black, T. A., den Hartog, G. et al. (1997) 'A comparison of sap flow and eddy fluxes of water vapor from a boreal deciduous forest', *Journal of Geophysical Research Atmospheres*, 102, pp. 28929–28937. doi: 10.1029/96JD03881.

Huntington, T. G. (2006) 'Evidence for intensification of the global water cycle: Review and synthesis', *Journal of Hydrology*, 319, pp. 83–95. doi: 10.1016/j.jhydrol.2005.07.003.

Irmak, S., Kabenge, I., Skaggs, K. E. and Mutiibwa, D. (2012) 'Trend and magnitude of changes in climate variables and reference evapotranspiration over 116-yr period in the Platte River Basin, central Nebraska – USA', *Journal of Hydrology*, 420–421, pp. 228–244. doi: 10.1016/j.jhydrol.2011.12.006.

Kay, A. L., Bell, V. A., Blyth, E. M. et al. (2013) 'A hydrological perspective on evaporation: historical trends and future projections in Britain', *Journal of Water and Climate Change*, 4, pp. 193–208. doi: 10.2166/wcc.2013.014.

Kent, S. T., McClure, L. A., Zaitchik, B. F., Smith, T. T. and Gohlke, J. M. (2014) 'Heat waves and health outcomes in Alabama (USA): The importance of heat wave definition', *Environmental Health Perspectives*, 122, pp. 151–158.

Lewis, N. and Pfenning, D. W. (2021) 'Phenotypic plasticity', in Futuma, D. J. (ed.) *Evolutionary Biology*. Oxford University Press. doi: 10.1093/obo/9780199941728-0093.

Li, M., Yao, J., Guan, J. and Zheng, J. (2021) 'Observed changes in vapor pressure deficit suggest a systematic drying of the atmosphere in Xinjiang of China', *Atmospheric Research*, 248, p. 105199. doi: 10.1016/j.atmosres.2020.105199.

Lin, H., Ruping, M. and Vitart, F. (2022) 'The 2021 western North American heatwave and its subseasonal predictions', *Geophysical Research Letters*, 49, p. e2021GL097036. doi: 10.1029/2021GL097036.

Linacre, E. T. (1977) 'A simple formula for estimating evaporation rates in various climates, using temperature data alone', *Agricultural Meteorology*, 18, pp. 409–424.

Liu, B., Xu, M., Henderson, M. and Gong, W. (2004) 'A spatial analysis of pan evaporation trends in China, 1955–2000', *Journal of Geophysical Research*, 109, p. D15102. doi: 10.1029/2004JD004511.

López, J., Way, D. A. and Sadok, W. (2021) 'Systemic effects of rising atmospheric vapor pressure deficit on plant physiology and productivity', *Global Biogeochemical Cycles*, 27, pp. 1704–1720. doi: 10.1111/gcb.15548.

Massmann, A., Gentine, P. and Lin, C. (2019) 'When does vapor pressure deficit drive or reduce evapotranspiration?', *Journal of Advances in Modeling Earth Systems*, 11, pp. 3305–3320. doi: 10.1029/2019MS001790.

McVicar, T. R., Roderick, M. L., Donohue, R. J. et al. (2012) 'Global review and synthesis of trends in observed terrestrial near-surface wind speeds: Implications for evaporation', *Journal of Hydrology*, 416–417, pp. 182–205. doi: 10.1016/j.jhydrol.2011.10.024.

Milly, P. C. D. and Dunne, K. A. (2001) 'Trends in evaporation and surface cooling in the Mississippi River basin', *Geophysical Research Letters*, 28, pp. 1219–1222.

Noer, N. K., Oersted, M., Schiffer, M. et al. (2022) 'Into the wild – A field study on the evolutionary and ecological importance of thermal plasticity in ectotherms across temperate and tropical regions', *Philosophical Transactions of the Royal Society of London, Series B*, 377, p. 20210004. doi: 10.1098/rstb.2021.0004.

Novick, K. A., Ficklin, D. L., Stoy, P. C. et al. (2016) 'The increasing importance of atmospheric demand for ecosystem water and carbon fluxes', *Nature Climate Change*, 6, pp. 1023–1027. doi: 10.1038/nclimate3114.

Penman, H. L. (1948) 'Natural evaporation from open water, bare soil and grass', *Proceedings of the Royal Society of London, Series A*, 193, pp. 120–145.

Peterson, T. C., Golubev, V. S. and Groisman, P. Y. (1995) 'Evaporation losing its strength', *Nature*, 377, pp. 687–688.

Restaino, C. M., Peterson, D. L. and Littell, J. (2016) 'Increased water deficit decreases Douglas fir growth throughout western US forests', *Proceedings of the National Academy of Sciences USA*, 113, pp. 9557–9562. doi: 10.1073/pnas.1602384113.

Riddell, E. A. and Sears, M. W. (2015) 'Geographic variation of resistance to water loss within two species of lungless salamanders – Implications for activity', *Ecosphere*, 6, pp. 1–16. doi: 10.1890/ES14-00360.1.

Robine, J., Cheung, S. L. K., Le Roy, S. et al. (2008) 'Death toll exceeded 70,000 in Europe during the summer of 2003', *Comptes Rendus Biologies*, 331, pp. 171–178. doi: 10.1016/j.crvi.2007.12.001.

Roderick, M. L. and Farquhar, G. D. (2002) 'The cause of decreased pan evaporation over the past 50 years', *Science*, 298, pp. 1410–1412. doi: 10.1126/science.1075390-a.

Roderick, M. L., Rotstayn, L. D., Farquhar, G. D. and Hobbins, M. T. (2007) 'On the attribution of changing pan evaporation', *Geophysical Research Letters*, 34, pp. 1–6. doi: 10.1029/2007GL031166.

Romanello, M., McGushin, A., Di Napoli, C. et al. (2021) 'Review: The 2021 report of the *Lancet* Countdown on health and climate change: code red for a healthy future', 398, pp. 1619–1662. doi: 10.1016/S0140-6736(21)01787-6.

Rotstayn, L. D., Roderick, M. L. and Farquhar, G. D. (2006) 'A simple pan-evaporation model for analysis of climate simulations: Evaluation over Australia', *Geophysical Research Letters*, 33, p. L17715. doi: 10.1029/2006GL027114.

Rousi, E., Kornhuber, K., Beobide-Arsuaga, G., Luo, F. and Coumou, D. (2022) 'Accelerated western European heatwave trends linked to more-persistent double jets over Eurasia', *Nature Communications*, 13, p. 3851. doi: 10.1038/s41467-022-31432-y.

Seagar, R., Hooks, A., Park Williams, A. et al. (2015) 'Climatology, variability, and trends in the U.S. vapor pressure deficit, an important fire-related meteorological quantity', *Journal of Applied Meteorology and Climatology*, 54, pp. 1121–1141. doi: 10.1175/JAMC-D-14-0321.1.

Shen, J., Yang, H., Li, S. et al. (2022) 'Revisiting the pan evaporation trend in China during 1988–2017', *Journal of Geophysical Research Atmospheres*, 127, p. e2022JD036489. doi: 10.1029/2022JD036489.

Sherwood, S. C. and Huber, M. (2010) 'An adaptability limit to climate change due to heat stress', *Proceedings of the National Academy of Sciences USA*, 107, pp. 9552–9555. doi: 10.1073/pnas.0913352107.

Smith, T. T., Zaitchik, B. F. and Gohlke, J. M. (2013) 'Heat waves in the United States: Definitions, patterns and trends', *Climatic Change*, 118, pp. 811–825. doi: 10.1007/s10584-012-0659-2.

Stephens, C. M., McVicar, T. R., Johnson, F. M. and Marshall, L. A. (2018) 'Revisiting pan evaporation trends in Australia a decade on', *Geophysical Research Letters*, 45, pp. 11164–11172. doi: 10.1029/2018GL079332.

Stull, R. (2011) 'Wet-bulb temperature from relative humidity and air temperature', *Journal of Applied Meteorology and Climatology*, 50, pp. 2267–2269. doi: 10.1175/JAMC-D-11-0143.1.

Thompson, V., Kennedy-Asser, A. T., Vosper, E. et al. (2022) 'The 2021 western North America heat wave among the most extreme events ever recorded globally', *Science Advances*, 8, pp. 1–11. doi: 10.1126/sciadv.abm6860.

Vicente-Serrano, S. M., Nieto, R., Gimeno, L. et al. (2018) 'Recent changes of relative humidity: regional connections with land and ocean processes', *Earth System Dynamics*, 9, pp. 915–937. doi: 10.5194/esd-9-915-2018.

Vincent, L. A., Wijngaarden, W. A. and Hopkinson, R. (2007) 'Trends in relative humidity in Canada from 1953–2003', *Journal of Climate*, 20, pp. 5100–5113. doi: https://doi.org/10.1175/JCLI4293.1.

Walter, M., Wilks, D. S., Parlange, J.-Y. and Schneider, R. L. (2004) 'Increasing evapotranspiration from the conterminous United States', *Journal of Hydrometeorology*, 5, pp. 405–408. doi: 10.1175/1525-7541(2004)005<0405:IEFTCU>2.0.CO;2.

Weaver, A., Edwards, H., McIntyre, T. et al. (2022) 'Cutaneous evaporative water loss in lizards is variable across body regions and plastic in response to humidity', *Herpetologica*, 78, pp. 169–183. doi: 10.1655/Herpetologica-D-21-00030.1.

Wild, M. (2009) 'Global dimming and brightening: A review', *Journal of Geophysical Research*, 114, pp. 1–31. doi: 10.1029/2008JD011470.

Xu, Z., FitzGerald, G., Guo, Y., Jalaludin, B. and Tong, S. (2016) 'Impact of heatwave on mortality under different heatwave definitions: A systematic review and meta-analysis', *Environment International*, 89–90, pp. 193–203. doi: 10.1016/j.envint.2016.02.007.

Zhao, G., Li, Y. and Gao, H. (2022) 'Evaporative water loss of 1.42 million global lakes', *Nature Communications*, 13, p. 3686. doi: 10.1038/s41467-022-31125-6.

Index